Host-Parasite Cellular and Molecular Interactions in Protozoal Infections

NATO ASI Series

Advanced Science Institutes Series

A series presenting the results of activities sponsored by the NATO Science Committee, which aims at the dissemination of advanced scientific and technological knowledge, with a view to strengthening links between scientific communities.

The Series is published by an international board of publishers in conjunction with the NATO Scientific Affairs Division

A Life Sciences **B Physics**	Plenum Publishing Corporation London and New York
C Mathematical and Physical Sciences	D. Reidel Publishing Company Dordrecht, Boston, Lancaster and Tokyo
D Behavioural and Social Sciences **E Applied Sciences**	Martinus Nijhoff Publishers Boston, The Hague, Dordrecht and Lancaster
F Computer and Systems Sciences **G Ecological Sciences** **H Cell Biology**	Springer-Verlag Berlin Heidelberg New York London Paris Tokyo

Series H: Cell Biology Vol. 11

Host-Parasite Cellular and Molecular Interactions in Protozoal Infections

Edited by

K.-P. Chang

Department of Microbiology and Immunology
UHS/Chicago Medical School
3333 Green Bay Road
North Chicago, Illinois 60064, USA

David Snary

Department of Molecular Biology
Wellcome Research Laboratories
Langley Court
Beckenham, Kent BR3 3BS, England

Springer-Verlag
Berlin Heidelberg New York London Paris Tokyo
Published in cooperation with NATO Scientific Affairs Divison

Proceedings of the NATO Advanced Research Workshop on Host-Parasite Cellular and Molecular Interactions in Protozoal Infections held at Hotel Villa del Mare, Acquafredda di Maratea, Italy, from September 27 – Oktober 1, 1986

ISBN-13:978-3-642-72842-6 e-ISBN-13:978-3-642-72840-2
DOI: 10.1007/978-3-642-72840-2

Library of Congress Cataloging in Publication Data. NATO Advanced Research Workshop on Host-Parasite Cellular and Molecular Interactions in Protozoal Infections (1986 : Acquafredda di Maratea, Italy) Host-parasite cellular and molecular interactions in protozoal infections. (NATO ASI series. Series H, Cell biology ; vol. 11) "Proceedings of the NATO Advanced Research Workshop on Host-Parasite Cellular and Molecular Interactions in Protozoal Infections held at Hotel Villa del Mare, Acquafredda di Maratea, Italy, from September 27–October 1, 1986"–T.p. verso. "Published in cooperation with NATO Scientific Affairs Division." Includes index. 1. Protozoa, Pathogenic–Congresses. 2. Host-parasite relationships–Congresses. I. Chang, K.P. II. Snary, David, 1945- . III. North Atlantic Treaty Organization. Scientific Affairs Division. IV. Title. V. Series. [DNLM: 1. Host-Parasite Relations–congresses. 2. Protozoan Infections–parasitology–congresses. WC 700 N279h 1986] QR251.N37 1986 616.9'36 87-20647
ISBN-13:978-3-642-72842-6 (U.S.)

Softcover reprint of the hardcover 1st edition 1987

2131/3140-543210

PREFACE

Tropical diseases such as leishmaniasis, malaria, trypanosomiasis, toxoplasmosis and amebiasis continue to plague the world, resulting in considerable morbidity and mortality, especially in the third world countries. These diseases are caused by a group of protozoa which have, over the years, undergone evolutionary adaptation to live often intracellularly in a parasitic way of life. So well-adapted have they become that they recognize the right hosts or cells to parasitize, yet at the same time they escape recognition and destruction by the host immune system. The mechanisms of such recognition and the escape of recognition are governed largely by host-parasite surface membrane interactions at the cellular and molecular level. Unique molecules produced by unusual pathways of these parasites have also been discovered and found to play important roles in their survival in the host. Understanding these mechanisms and pathways is essential not only to formulate a rational strategy for chemo- and immuno-prophylaxis and -therapy but also to unravel the mystery of biological evolution in symbiosis and parasitism.

In the advent of our knowledge on the molecular biology and biochemistry of parasite membrane and other molecules, it is opportune to examine and discuss their possible roles in host-parasite recognition and interaction in a comparative approach. To highlight the recent advances of this area in various host-parasite systems, a NATO advanced Research Workshop was held from September 27 to October 1, 1986 at Hotel Villa del Mare, Acquafredda di Maratea, Italy. Some sixty participants of the workshop are from 16 countries, mainly the NATO Alliance, and include scientists and graduate students in their respective fields from research laboratories of universities and governments as well as representatives from the World Health Organization and industry. This volume consists mostly of papers presented and discussed in the Workshop. It is hoped that central themes or concepts will emerge from diversity of the findings reported from various systems, thereby providing useful leads for future investigations to understand the fundamental mechanisms of parasitism in modern perspectives.

The Workshop and this post-workshop volume would not have been possible without the financial support from the NATO Scientific Affairs Committee and from the US National Science Foundation (for some of the US participants). Special thanks are due to: Drs. Aikawa, M. R. Hollingdale, D. Snary and S. Tzartos for valuable suggestions in serving as members of the Organizing Committee; session chairpersons and discussants, and all participants for their enthusiastic participation and encouragement; and the School of Graduate Studies, the Business Office and the Department of Microbiology and Immunology, UHS/The Chicago Medical School for their cooperation and support.

K.-P. CHANG
North Chicago, IL

D. SNARY
Buckenham, Kent

LIST OF CONTRIBUTORS

Aikawa, M.
Institute of Pathology
Case Western Reserve University
Cleveland, Ohio, 44106, USA

Allen, A.K.
Charing Cross and Westminster
Hospital Medical School
London, UK

Alunda, J.M.
Department of Zoology
University of Glasgow
Glasgow G12 8QQ
Scotland, UK

Andrutis, A.T.
Laboratory of Parasitic Diseases
National Institutes of Allergy and
Infectious Diseases
National Institutes of Health
Bethesda, MD 20892, USA

Barry, J.D.
Department of Genetics
University of Glasgow
Glasgow G12 8QQ, UK

Baudhuin, P.
International Institute of Cellular
and Molecular Pathology
Laboratoire de Chimie Physiologique
Universite Catholique de Louvain
BO-1200 Brussels, Belgium

Beals, T.P.
Department of Medical Microbiology
Stanford University School of
Medicine
Stanford, CA 94306, USA

Behrens, K.
Institute of General Microbiology
University of Bern
3012 Bern, Switzerland

Blackwell, J.M.
Department of Tropical Hygiene
London School of Hygiene and
Tropical Medicine
Keppel Street
London WC1E 7HT, UK

Boothroyd, J.C.
Department of Medical Microbiology
Stanford University School of Medicine
Stanford, CA 94306, USA

Bordier, C.
Institut de Biochimie
Universite de Lausanne
CH-1066 Epalinges, Switzerland

Bouvier, J.
Institut de Biochimie
Universite de Lausanne
CH-1066 Epalinges, Switzerland

Bracha, R.
Department of Biophysics and Unit for
Molecular Biology of Parasitic
Diseases
Weizmann Institute of Science
Rehovot 76100, Israel

Braun, R.
Institute of General Microbiology
University of Bern
3012 Bern, Switzerland

Brun, R.
Swiss Tropical Institute
Basel, Switzerland

Bulow, R.
Max-Planck-Institut fur Biologie
Corrensstrasse 38, D7400
Tubingen, F.R. Germany

Cardoso de Almeida, M.L.
The Molteno Institute
MRC Biochemical Parasitology Unit
University of Cambridge
Cambridge CB2 3EE, UK

Chang, K.-P.
Department of Microbiology/Immunology
UHS/The Chicago Medical School
North Chicago, IL 60064, USA

Chaudhuri, G.
Department of Microbiology/Immunology
UHS/The Chicago Medical School
North Chicago, IL 60064, USA

Chavez, B.
Center for Research and Advanced Studies
National Polytechnical Institute
Apartado Postal 14-740
07000 Mexico, D.F., Mexico

Chayen, A.
Department of Biophysics and Unit for Molecular Biology of Parasitic Diseases
Weizmann Institute of Science
Rehovot 76100, Israel

Cohen, L.A.
Laboratory of Chemistry
National Institute of Arthritis, Diabetes and Digestive and Kidney Diseases
National Institutes of Health
Bethesda, MD 20892, USA

Coombs, G.H.
Department of Zoology
University of Glasgow
Glasgow G12 8QQ
Scotland, UK

Coppens, I.
International Institute of Cellular and Molecular Pathology
Laboratory de Chimie Physioloque
Universite Catholique de Louvain
B-1200 Brussels, Belgium

Czichos, J.
Max-Planck-Institut fur Biologie
Corrensstrasse 38, D7400
Tubingen, F.R. Germany

Daniel, W.
Malaria Section
Laboratory of Parasitic Diseases
National Institutes of Allergy and Infectious Diseases
National Institutes of Health
Bethesda, MD 20892, USA

Darcy, F.
Centre d'Immunologie et de Biologie Parasitaire
Unite Mixte INSERM 167 CNRS 624
Institut Pasteur
1 Rue du Prof. Calmette
59019 Lille Cedex, France

Davey, L.S.
Department of Molecular Biology
Wellcome Biotech
Beckenham, Kent BR3 3BS, UK

Diamond, L.S.
Laboratory of Parasitic Diseases
National Institute of Allergy and Infectious Diseases
National Institutes of Health
Bethesda, MD 20892, USA

Dubremetz, J.F.
Unite de Biologie et Biochimie Parasitaires et Fongiques
INSERM U 42, 369, Rue Jules Guesde
59650 Villeneuve d'Ascq, France

Dwek, R.A.
Oxford Oligosaccharide Group
Department of Biochemistry
University of Oxford
South Parks Road
Oxford, UK

Dwyer, D. M.
Cell Biology and Immunology Section
Laboratory of Parasitic Diseases
National Institute of Allergy and Infectious Diseases
National Institutes of Health
Bethesda, MD 20892, USA

Ehlers, B.
Max-Planck-Institut fur Biologie
Corrensstrasse 38, D7400
Tubingen, F.R. Germany

Etges, R.
Institut de Biochimie
Universite de Lausanne
CH-1066 Epalinges, Switzerland

Fairlamb, A.H.
Laboratory of Medical Biochemistry
Rockefeller University
1230 York Avenue
New York, NY 10021, USA

Ferguson, M.A.J.
Department of Biochemistry
University of Oxford, UK

Fortier, B.
Unite de Biologie et Biochimie
Parasitaires et Fongiques
INSERM U 42, 369, Rue Jules Guesde
59650 Villeneuve d'Ascq, France

Geuze, H.J.
Department of Cell Biology
Medical Faculty
State University of Utrecht
Utrecht, Netherlands

Gonzalez-Robles, A.
Center for Research and
Advanced Studies
National Polytechnical Institute
Apartado Postal 14-740
07000 Mexico, D.F., Mexico

Guhl, F.
Department of Biological Sciences
University of the Andes
Bogota, Colombia

Goniou, M.G.
Hellenic Pasteur Institute
127 Vassilissis Sofias Avenue
Athens 11521, Greece

Hadziantoniou, M.G.
Hellenic Pasteur Institute
127 Vassilissis Sofias Avenue
Athens 11521, Greece

Hart, D.T.
International Institute for Cellular
and Molecular Pathology
B-1200 Brussels, Belgium

Hassan, H.F.
Department of Zoology
University of Glasgow
Glasgow G12 8QQ
Scotland, UK

Heidrich, H.-G.
Max-Planck-Institut fur Biochemie
D-8033 Martinsried bei Munchen
F.R. Germany

Henderson, G.B.
Laboratory of Medical Biochemistry
Rockefeller University
1230 York Avenue
New York, NY 10021, USA

Hillman, Y.
Department of Molecular Biology
Wellcome Biotech
Beckenham, Kent BR3 3BS, UK

Hodson, C.
National Blood Transfusion Service
Lancaster Centre Quernmore Road
Lancaster, LA1 3JP, UK

Holder, A.A.
Department of Molecular Biology
Wellcome Biotechnology
Beckenham, Kent BR3 3BS, UK

Hollingdale, M.R.
Biomedical Research Institute
Rockville, MD 20852, USA

Homans, S.W.
Oxford Oligosaccharide Group
Department of Biochemistry
University of Oxford
South Parks Road
Oxford, UK

Howard, R.J.
Laboratory of Parasitic Diseases
National Institute of Allergy &
Infectious Diseases
National Institutes of Health
Bethesda, MD 20892, USA

Hudson, L.
Department of Immunology
St. George's Hospital Medical School
London SW17 0RE, UK

Imboden, M.
Institute of General Microbiology
University of Bern
3012 Bern, Switzerland

Ingram, G.A.
Department of Biological Sciences
University of Salford
Salford M5 4WT, UK

Jaffe, C.L.
Weizmann Institute of Science
Rehovot, Israel

Jenni, L.
Swiss Tropical Institute
Basel, Switzerland

Johnson, C.L.
Department of Biochemistry
University of Kentucky Medical Center
Lexington, KY 40536, USA

Kahl, L.
Wellcome Research Laboratories
Beckenham, Kent BR3 3BS, UK

Kharazmi, A.
State Serum Institute
Department of Clinical Microbiology
Rigshospitalet, DK-2200
Copenhagen N, Denmark

Kilejian, A.
The Public Health Research Institute
New York, NY 10016, USA

Killick-Kendrick, R.
Medical Research Council External
Scientific Staff
Department of Pure and Applied Biology
Imperial College at Silwood Park
Ascot, Berks SL5 7PY, UK

King, D.L.
Department of Biochemistry
University of Kentucky Medical Center
Lexington, KY 40536, USA

Kink, J.A.
Department of Microbiology/Immunology
UHS/The Chicago Medical School
North Chicago, IL 60064, USA

Kirk, K.L.
Laboratory of Chemistry
National Institute of Arthritis,
Diabetes and Digestive Kidney Diseases
National Institutes of Health
Bethesda, MD 20892, USA

Kung, V.
Institute of General Microbiology
University of Bern
3012 Bern, Switzerland

Kweider, M.
Centre d'Immunologie et de Biologie
Parasitaire
Unite Mixte INSERM 167 CNRS 624
Institut Pasteur
1 Rue du Prof. Calmette
59019 Lille Cedex, France

Langer, P.
Department of Microbiology
and Immunology
Laramie, WY 82071, USA

Lemesre, J.L.
Centre d'Immunologie et de Biologie
Parasitaire
Unite Mixte INSERM 167 CNRS 624
Institut Pasteur
1 Rue du Prof. Calmette
59019 Lille Cedex, France

Lockwood, B.C.
Department of Biological Science
University of Stirling
Stirling FK9 4LA
Scotland, UK

Lockyer, M.J.
Department of Molecular Biology
Wellcome Biotech
Beckenham, Kent BR3 3BS, UK

Lohman, K.
Department of Epidemiology and Public
Health
Yale University School of Medicine
New Haven, CT 06510, USA

Lottspeich, F.
Max-Planck-Institut fur Biochemie
D-8033 Martinsried bei Munchen
F.R. Germany

Lyon, J.A.
Department of Immunology
Walter Reed Army Institute of Research
Washington, D.C. 20012, USA

Mallinson, D.J.
Department of Zoology
University of Glasgow
Glasgow G12 8QQ
Scotland, UK

Martinez-Palomo, A.
Center for Research
and Advanced Studies
National Polytechnical Institute
Apartado Postal 14-740
07000 Mexico, D.F., Mexico

McMahon-Pratt, D.
Department of Epidemiology & Public Health
Yale University School of Medicine
New Haven, CT 06510, USA

McMaster, R.
Department of Medical Genetics
University of British Columbia
Vancouver V6T 1W5, British Columbia
Canada

Meis, J.F.G.M.
Department of Medical Parasitology
Catholic University of Nijmegen
Nijmegen, Netherlands

Meuwissen, J.H.E.T.
Department of Medical Parasitology
Catholic University of Nijmegen
Nijmegen, Netherlands

Meza, I.
Center for Research
and Advanced Studies
National Polytechnical Institute
Apartado Postal 14-740
07000 Mexico, D.F., Mexico

Miettinen-Baumann, A.
Max-Planck-Institut fur Biochemie
D-8033 Martinsried bei Munchen
F.R. Germany

Miles, M.A.
London School of Hygiene and
Tropical Medicine
London, UK

Mirelman, D.
Department of Biophysics and Unit
for Molecular Biology of Parasitic
Diseases
Weizmann Institute of Science
Rehovot 76100, Israel

Molyneux, D.H.
Department of Biological Sciences
University of Salford
Salford M5 4WT, UK

Munoz, M.L.
Center for Research and Advanced
Studies
National Polytechnical Institute
Apartado Postal 14-740
07000 Mexico, D.F., Mexico

Newbold, C.I.
Tropical Medicine Unit
Nuffield Department of Clinical
Medicine
John Radcliffe Hospital
Oxford OX3 9DU, UK

Ng, W.
Department of Medicine
The University of British Columbia
and Vancouver General Hospital
Vancouver, Br. Columbia, Canada

Nicholls, S.C.
Department of Molecular Biology
Wellcome Biotech
Beckenham, Kent BR3 3BS, UK

North, M.J.
Department of Biological Science
University of Stirling
Stirling FK9 4LA
Scotland, UK

Olafson, R.W.
Department of Biochemistry and
Microbiology
University of Victoria
Victoria, British Columbia, Canada

Opperdoes, F.R.
International Institute of Cellular and
Molecular Pathology
Universite Catholique de Louvain
B-1200 Brussels, Belgium

Orlandi, P.A.
Department of Biochemistry
University of Kentucky Medical Center
Lexington, KY 40536, USA

Overath, P.
Max-Planck-Institut fur Biologie
Corrensstrasse 38, D7400
Tubingen, F.R. Germany

Pan, A.
Department of Epidemiology & Public Health
Yale University School of Medicine
New Haven, CT 06510, USA

Pasvol, G.
Tropical Medicine Unit
Nuffield Department of Clinical Medicine
John Radcliffe Hospital
Oxford OX3 9DU, UK

Pereira, M.E.A.
New England Medical Center Hospitals
Department of Geographic Medicine and Infectious Diseases
750 Washington Street
P. O. Box 041
Boston, MA 02111, USA

Pfefferkorn, E.R.
Department of Microbiology
Dartmouth Medical School
Hanover, NH 03756, USA

Posthuma, G.
Department of Cell Biology
Medical Faculty
State University of Utrecht
Utrecht, Netherlands

Prioli, R.P.
New England Medical Center Hospitals
Department of Geographic Medicine and Infectious Diseases
750 Washington Street
P. O. Box 041
Boston, MA 02111, USA

Rademacher, T.W.
Oxford Oligosaccharide Group
Department of Biochemistry
University of Oxford
South Parks Road
Oxford, UK

Reiner, N.E.
Department of Medicine
Department of Microbiology
The University of British Columbia and Vancouver General Hospital
Vancouver, Br. Columbia, Canada

Rijntjes, P.J.M.
Department of Gastroenterology
Catholoic University of Nijmegen
Nijmegen, Netherlands

Rivas, L.
Department of Epidemiology & Public Health
Yale University School of Medicine
New Haven, CT 06510, USA

Rizvi, F.S.
Centre d'Immunologie et de Biologie Parasitaire
Unite Mixte INSERM 167 CNRS 624
Institut Pasteur
1 Rue du Prof. Calmette
59019 Lille Cedex, France

Rojkind, M.
Center for Research and Advanced Studies
National Polytechnical Institute
Apartado Postal 14-740
07000 Mexico, D.F., Mexico

Rosales-Encina, J.L.
Center for Research and Advanced Studies
National Polytechnical Institute
Apartado Postal 14-740
07000 Mexico, D.F., Mexico

Rosenberg, I.
New England Medical Center Hospital
Department of Geographic Medicine and Infectious Diseases
750 Washington Street
P. O. Box 041
Boston, MA 02111, USA

Sadak, A.
Unite de Biologie et Biochimie Parasitaires et Fongiques
INSERM U 42, 369, Rue Jules Guesde
59650 Villeneuve d'Ascq, France

Santoro, F.
Centre d'Immunologie et de Biologie Parasitaire
Unite Mixte INSERM 167 CNRS 624
Institut Pasteur
1 Rue du Prof. Calmette
59019 Lille Cedex, France

Schneider, A.
Institute of General Microbiology
University of Bern
3012 Bern, Switzerland

Schwarz, R.T.
Laboratoire de Chimie Biologique
Universite des Sciences et Techniques de Lille
59655 Villeneuve D'Ascq, Cedex
France

Seebeck, T.
Institute of General Microbiology
University of Bern
3012 Bern, Switzerland

Sharma, Y.D.
The Public Health Research Institute
New York, NY 10016, USA

Sheppard, H.W.
Naval Biosciences Laboratory
Naval Supply Center
Oakland, CA 94625, USA

Sher, A.
National Institute of Allergy and Infectious Diseases
National Institutes of Health
Bethesda, MD 20892, USA

Sibley, L.D.
Department of Zoology/Physiology
Louisiana State University
Baton Rouge, LA 70803, USA

Sinden, R.E.
Molecular and Cellular Parasitology Research Group
Department of Pure and Applied Biology
Imperial College
London SW7 2AZ, UK

Snary, D.
Department of Molecular Biology
Wellcome Biotechnology
Langley Court
Beckenham, Kent BR3 3BS, UK

Soteriadou, K.Ph.
Hellenic Pasteur Institute
127 Vassilissis Sofias Avenue
Athens 11521
Greece

Strych, W.
Max-Planck-Institut fur Biochemie
D-8033 Martinsried bei Munchen
F.R. Germany

Taghy, Z.
Unite de Biologie et Biochimie Parasitaires et Fongiques
INSERM U 42, 369, Rue Jules Guesde
59650 Villeneuve d'Ascq, France

Tanner, M.J.A.
Department of Biochemistry
University of Bristol Medical School
University Walk
Bristol BS8 1TD, UK

Taylor, D.W.
Department of Biology
Georgetown University
Georgetown, D.C. 20057, USA

Tetley, L.
Department of Zoology
University of Glasgow
Glasgow G12 8QQ
Scotland, UK

Trager, W.
Department of Parasitology
The Rockefeller University
New York, NY 10021, USA

Turco, S.J.
Department of Biochemistry
University of Kentucky Medical Center
Lexington, KY 40536, USA

Turner, C.M.R.
Department of Zoology
University of Glasgow
Glasgow G12 8QQ
Scotland, UK

Turner, M.J.
The Molteno Institute
MRC Biochemical Parasitology Unit
University of Cambridge
Cambridge CB2 3GG, UK

Tzartos, S.J.
Hellenic Pasteur Institute
127 Vassilissis Sofias Avenue
Athens 11521, Greece

Tzinia, A.K.
Hellenic Pasteur Institute
127 Vassilissis Sofias Avenue
Athens 11521, Greece

Uni, S.
Institute of Pathology
Case Western Reserve University
Cleveland, Ohio 44106, USA

Verhave, J.P.
Department of Medical Parasitology
Catholic University of Nijmegen
Nijmegen, Netherlands

Vickerman, K.
Department of Zoology
University of Glasgow
Glasgow G12 8QQ, UK

Wallbanks, K.R.
Department of Biological Sciences
University of Salford
Salford M5 4WT, UK

Wallis, A.E.
Department of Medical Genetics
University of British Columbia
Vancouver V6T 1W5
British Columbia, Canada

Ward, J.
The Molteno Institute
MRC Biochemical Parasitology
Unit
University of Cambridge
Cambridge CB2 3EE, UK

Wiser, M.F.
Max-Planck-Institut fur
Zellbiologie
Rosenhof, D-6802
Ladenburg bei Heidelberg, F.R. Germany

Wozencraft, A.O.
Department of Tropical Hygiene
London School of Hygiene and Tropical
Medicine
Keppel Street
London WC1E 7HT, UK

Wright, B.L.
Department of Biochemistry
University of Kentucky Medical Center
Lexington, KY 40536, USA

Yang, Y.F.
The Public Health Research Institute
New York, NY 10016, USA

Yap, S.H.
Department of Gastroenterology
Catholic University of Nijmegen
Nijmegen, Netherlands

Zilberstein, D.
Department of Biology
Technion-Israel Institute of Technology
Haifa 32000, Israel

TABLE OF CONTENTS

Preface V
List of Contributors...VII
Table of Contents XV

Trypanosoma

Pattern of variant surface glycoprotein coating in nascent metacyclic *Trypanosoma brucei* in the salivary glands of *Glossina morsitans*
K. Vickerman, L. Tetley, C. M. R. Turner and J. D. Barry 1

Evolution of antigenic variation in African trypanosomes
J. C. Boothroyd and T. P. Beals 9

Structural studies on the glycophospholipid membrane anchor of *Trypanosoma brucei* variant surface glycoprotein
M. A. J. Ferguson, R. A. Dwek, S. W. Homans and T. W. Rademacher 19

Metabolism of trypanothione and glutathionylspermidine in trypanosomatids
A. H. Fairlamb and G. B. Henderson 29

Differentiation of *Trypanosoma brucei* from bloodstream to procyclic trypomastigotes
P. Overath, R. Bulow, J. Czichos and B. Ehlers 41

Digestive enzymes, receptor-mediated endocytosis and their role in the nutrition of the bloodstream-form trypanosome
F. R. Opperdoes, I. Coppens and P. Baudhuin 51

The Cytoskeleton of *Trypanosoma brucei*: Genes and proteins
R. Braun, T. Seebeck, K. Behrens, M. Imboden, A. Schneider and V. Kung 67

Genetic re-assortment of African trypanosomes in *Glossina*
L. Jenni and R. Brun 73

Cell surface glycoproteins of *Trypanosoma cruzi*
D. Snary, M. A. J. Ferguson, A. K. Allen, M. A. Miles and A. Sher 79

A specific inhibitor of *Trypanosoma cruzi* neuraminidase found in human plasma
R. P. Prioli, I. Rosenberg and M. E. A. Pereira 89

Definition of *Trypanosoma cruzi* antigens by antibodies and T cells
L. Hudson and F. Guhl 97

Leishmania

Molecular pathogenicity of *Leishmania* in intracellular Parasitism of macrophages: Association of virulent phenotype with increased expression of a major membrane glycoprotein and microsomal N-acetylglucosaminyltrans-ferase
K.-P. Chang and J. A. Kink 107

Infectivity of *Leishmania* promastigotes is dependent on the expression of a 65,000 dalton surface antigen
F. Santoro, M. Kweider, J.L. Lemesre, F. Darcy and F.S. Rizvi...117

Characterization of developmentally regulated molecules of Leishmania
D. McMahon-Pratt, C.L. Jaffe, L. Kahl, P. Langer
K. Lohman, A. Pan and L. Rivas.................................123

Leishmania antigens probed by monoclonal antibodies to leishmania membrane and to acetylcholine receptor
K. Ph. Soteriadou, A. K. Tzinia, M. G. Hadziantoniou, and S. J. Tzartos 137

Structural studies on surface glycoproteins of *Leishmania* promastigotes: Isolation, amino acid composition and amino terminal sequence studies
R. W. Olafson, A. E. Wallis and R. McMaster 149

Membrane attachment of the major surface protein of *Leishmania* promastigotes
C. Bordier, R. Etges, J. Bouvier, J. Ward, M. J. Turner and M. L. Cardoso de Almeida 157

The promastigote surface protease of *Leishmania*
R. Etges, J. Bouvier and C. Bordier 165

Acid Protease activity of *Leishmania* gp63 and its possible role in virulence
G. Chaudhuri and K. -P. Chang 169

The roles of surface membrane enzymes and transporters in the survival of Leishmania
D. M. Dwyer.................c..............................175

The plasma membrane H^+-ATPase of *Leishmania donovani* promastigotes
D. Zilberstein, H. W. Sheppard and D. M. Dwyer 183

Phosphomonoesterases and proteinases of leishmanias
G. H. Coombs, H. F. Hassan, B. C. Lockwood, D. J. Mallinson, M. J. North, J. M. Alunda and L. Tetley 189

The structure, localization, and function of the lipophosphoglycan of *Leishmania donovani*
S. J. Turco, C. L. Johnson, D. L. King, P. A. Orlandi, and B. L. Wright 197

Leishmania host-parasite interactions: The development of chemotherapeutic targets and specific drug delivery systems
I. Lipoprotein-mediated antileishmanial chemotherapy
D. T. Hart 203

Recognition processes involved in binding and ingestion of *Leishmania donovani* by host macrophages
A. O. Wozencraft and J. M. Blackwell 213

Interaction of *Leishmania* and membrane-stabilizing drugs with human phagocytic cells
A. Kharazmi 223

Suppression of macrophage accessory functions by *Leishmania donovani*
N. E. Reiner, W. Ng and W. R. McMaster 229

Plasmodium & Babesia

Extracellular development of erythrocytic malaria parasites
W. Trager 235

The relative roles of N- and O-linked carbohydrate in the invasion of human red cells by merozoites of *Plasmodium falciparum*
G. Pasvol, C. Hodson, M. J. A. Tanner and C. I. Newbold 245

Antigens of *Plasmodium falciparum* merozoites: Identification, isolation and functional characterization
H. -G. Heidrich, W. Strych, F. Lottspeich and A. Miettinen-Baumann 255

Studies on the *Plasmodium falciparum* 195,000 molecular mass merozoite surface antigen precursor
A. A. Holder, Y. Hillman, S. C. Nicholls and L. S. Davey 265

Aspects of posttranslational modification of a major *Plasmodium falciparum* merozoite surface antigen
R. T. Schwarz, M. J. Lockyer and A. A. Holder 275

Export of *Plasmodium falciparum* proteins to the host erythrocyte membrane: Special problems of protein trafficking and topogenesis
R. J. Howard, S. Uni, J. A. Lyon, D. W. Taylor, W. Daniel and M. Aikawa 281

Inhibition of knob formation on *Plamodium falciparum*-infected erythrocytes by 2-fluoro-L-histidine
M. Aikawa, S. Uni, A. T. Andrutis, K. L. Kirk, L. A. Cohen and R. J. Howard 297

The knob protein gene of *Plasmodium falciparum*
A. Kilejian, Y. D. Sharma and Y. F. Yang 307

Phosphoproteins associated with the host erythrocyte membrane during *Plasmodium chabaudi* infection
M. F. Wiser 315

Malarial sporozoite - hepatocyte interactions mediating invasion and exoerythrocytic development
M. R. Hollingdale 321

Immuno-electron microscopic observations on *Plasmodium falciparum* sporozoites using ultracryomicrotomy
G. Posthuma, J. F. G. M. Meis, P. J. M. Rijntjes, J. P. Verhave, M. R. Hollingdale, J. H. E. T. Meuwissen, H. J. Geuze and S. H. Yap 329

Structural homology of membrane proteins of *Babesia rodhaini*
D. Snary...335

Toxoplasma & Entamoeba

The mechanism by which interferon gamma blocks the growth of *Toxoplasma gondii* in cultured fibroblasts
E. R. Pfefferkorn 345

Modification of host cell phagocytic compartments by intracellular *Toxoplasma gondii*
L. D. Sibley 355

Characterization of a 42 kilodalton rhoptry antigen of *Toxoplasma gondii*
J. F. Dubremetz, A. Sadak, Z. Taghy and B. Fortier 365

Entamoeba histolytica: Activation and release of membrane dense bodies
A. Martinez-Palomo, I. Meza, B. Châvez, J. L. Rosales-Encina, M. L. Munõz, A. Gonzalez-Robles and M. Rojkind 371

Entamoeba histolytica: Effect of bacterial associates on isoenzyme patterns and virulence
D. Mirelman, R. Bracha, A. Chayen and L. S. Diamond 377

Insect vector

Trypanosomatid-vector interfaces - *In vitro* studies on parasite substrate interactions
D. H. Molyneux, K. R. Wallbanks and G. A. Ingram 387

The microecology of *Leishmania* in the gut and proboscis of the sandfly
R. Killick-Kendrick 397

The cellular and molecular interactions of malaria species with their mosquito vectors
R. E. Sinden 407

Concluding Remarks

W. Trager, E. R. Pfefferkorn and R. T. Schwarz..............416

Index..419

PATTERN OF VARIANT SURFACE GLYCOPROTEIN COATING IN NASCENT METACYCLIC TRYPANOSOMA BRUCEI IN THE SALIVARY GLANDS OF GLOSSINA MORSITANS

K. Vickerman[1], L. Tetley[1] C.M.R. Turner[1] and J.D. Barry[2],

Departments of Zoology[1] and Genetics [2], University of Glasgow G12 8QQ, U.K.

INTRODUCTION

The African trypanosomes are protected against their mammalian hosts' non-specific and specific defence mechanisms by a glycoprotein coat. The antigenic nature of the glycoprotein can be changed by a gene-switching mechanism enabling the parasite population to avoid destruction by host antibodies, and a succession of variable antigen types (VATs) appears in the blood of an infected animal as the parasitaemia fluctuates. New VATs arise regardless of the host mounting an immune response. These can be detected by immunofluorescence or immune lysis reactions as "heterotypes" (minority populations) in a "homotype" population (i.e. one composed predominantly of trypanosomes belonging to the same VAT). In infections of Trypanosoma brucei, homotype VATs appear in a semi-predictable sequence, but what selects out a particular heterotype to become the next homotype is not clear, and whether the genes coding for the various variant-surface glycoproteins (VSGs) are activated in a particular sequence is not known (reviewed Boothroyd 1985, Barry 1986).

On ingestion by the tsetse fly, the VSG coat of T. brucei is lost and is missing from subsequent stages of development of the trypanosome in the vector, but is reacquired at the mammal-infective metacyclic stage in the fly's salivary glands, so that the metacyclic is preadapted to life in its mammalian host. The metacyclics of a trypanosome clone do not all belong to the same VAT, but express a characteristic range of VATs - the M-VATs - representing only a small proportion (10 to 20) of the entire VAT repertoire (1000-2000) of the clone as expressed in the mammalian host. As the discharged metacyclics are heterogenous with respect to VAT, the question arises as to whether this heterogeneity is present from the time that the naked parasites acquire their VSG coat or whether sequential antigenic change occurs in the gland once the non-dividing metacyclics have put on the same VSG coat. The answer to this question is relevant to the wider question of whether antigenic variation at the level of the individual trypanosome consists of an ordered serial activation of VSG genes, or whether it is a stochastic process, with each gene having a discrete probability for activation.

This paper sumarises evidence from immunogold labelling of M-VATs in conjunction with scanning and transmission electron microscopy of infected salivary glands that (1) VSG coating of metacyclic T. brucei occurs while the parasites are attached to the microvillar border of the fly's salivary glands; (2) M-VAT heterogeneity is present ab initio suggesting that VSG gene activation is a stochastic process; (3) VSG gene activation occurs at the time of, or just after, the

NATO ASI Series, Vol. H11
Host-Parasite Cellular and Molecular Interactions
in Protozoal Infections. Edited by K.-P. Chang and D. Snary

last trypanosome division; (4) some metacyclics may express more than one VSG on their surface, suggesting that transcription may occur simultaneously from at least 2 expression sites in the genome of the individual trypanosome.

MATERIALS AND METHODS

Electron microscopy and the elucidation of stages in *T. brucei* metacyclic development in the salivary glands of *Glossina morsitans morsitans* were undertaken using trypanosomes of the ANTAR 1 and ETAR 1 serodemes as described by Tetley and Vickerman (1985). For the study of M-VATs *in situ*, IgG monoclonal antibodies against GUTats 7.1 and 7.2 were produced as described by Crowe *et al.* (1984) and complexed to 25 nm and 5 nm colloidal gold particles as described by De Mey (1983). Specificity of the immunogold (IG) probe was checked with homologous and heterologous bloodstream VATS. Infected salivary glands were slit along their length to expose the attached trypanosomes for fixation in glutaraldehyde. After rinsing in 10 mM Tris-HC1 1% BSA (pH 8.0) the gland preparation was exposed to the IG probe for 1 h at 20°C. Subsequent processing for TEM and SEM was as described by Tetley and Vickerman (1985) except that for SEM specimens were carbon coated (20 nm) instead of gold coated (to avoid obscuring adsorbed IG particles) and the microscope operated at 25 kv.

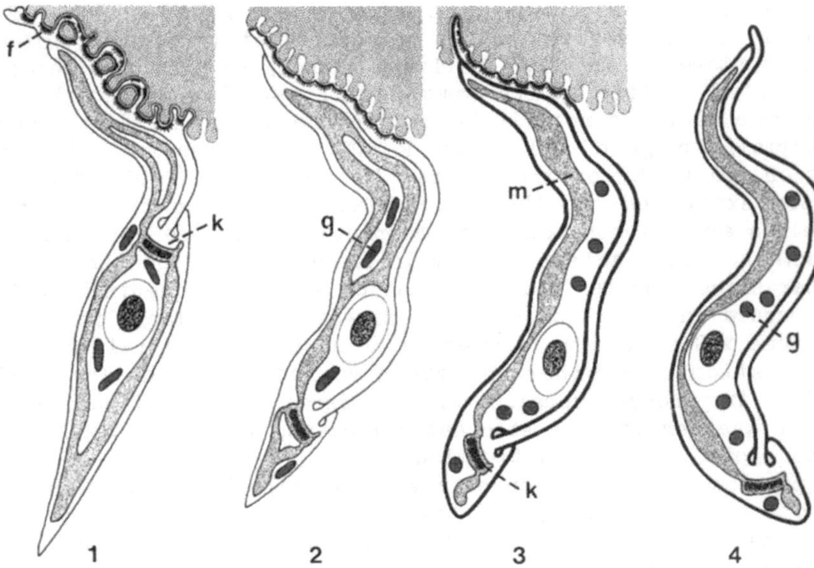

Fig. 1. Diagram of stages in development of *Trypanosoma brucei* in tsetse salivary gland. (1) Epimastigote* (2) premetacyclic* (3) nascent metacyclic+ (4) mature metacyclic+. Note changes in flagellar (f) attachment to host microvilli; kinetoplast (k)-mitochondrion (m); glycosomes (g); shape of posterior extremity. *Multiplicative stages; +coated stages.

RESULTS AND DISCUSSION

Developmental Stages in the Fly Salivary Glands

The principal stages of development of cloned T. brucei in the fly's salivary gland as revealed by electron microscopy (Tetley and Vickerman 1985), are shown in fig. 1. The uncoated multiplicative epimastigotes (with prenuclear kinetoplast) are attached to the gland epithelium by elaborate flagellar outgrowths ("flagellipodia") which insinuate between the microvilli and form punctate hemidesmosome-like attachment plaques where the microvilli indent them. Flagellipodia become reduced as the epimastigotes transform to uncoated trypomastigotes (with postnuclear kinetoplast) which remain attached and capable of binary fission. Attachment plaques persist, despite the disappearance of flagellipodia, when the uncoated trypomastigotes (premetacyclics) acquire the surface coat and cease dividing to become nascent metacyclics (fig. 2). The fact that coat acquisition occurs in the attached trypanosome has allowed us to look at the VAT of coating metacyclics using VAT-specific IG probes.

Unlike T. congolense and T. vivax, T. brucei appears to require attachment to living cells rather than an inert substratum for differentiation to the metacyclic stage in the life cycle to occur. Freeze fracture replicas of the host parasite attachment plaques show membrane particle aggregates on the host membrane but not on the opposed parasite's membrane (fig. 3), suggesting possible molecular modification of the host membrane by the parasite.

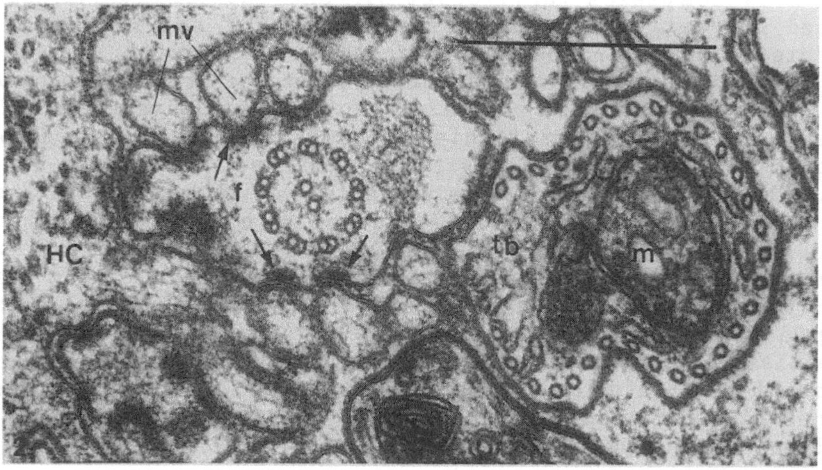

Fig. 2. Transmission electron micrograph (TEMG) of TS anterior end of coating trypanosome (nascent metacyclic) showing attachment plaques (arrowed) on flagellar membrane where host cell (HC) microvilli (mv) impinge. tb - trypanosome body; other abbreviations as in fig. 1. Scale bar 0.5 µm.

Fig. 3. TEMG of freeze-fracture replica of nascent metacyclic trypanosome. Note difference in density of intramembranous particles (IMPs, integral proteins) between flagellum and body PF; indented (attachment plaque) regions on flagellum membrane (large arrow); array of IMPs (arrowhead) on host microvillus PF; other abbreviations as in fig. 1. Scale bar 0.5 µm.

VAT Heterogeneity of Nascent Metacyclics

Both IG complexes (anti-GUTat 7.1 and anti-GUTat 7.2) bound to coated nascent metacyclics as shown by SEM and TEM of labelled infected glands, indicating that VAT heterogeneity is present *ab initio*. Homologous VAT nascent metacyclics were readily recognised in sections by their almost continuous covering of IG particles (fig. 4). Some uncoated trypomastigotes (premetacyclics) also showed IG adsorption, especially along the flagellum (fig. 4). We interpret such labelled flagellates as being in the act of coating, with surface VSG not having reached sufficient density to form a stainable coat. It is possible that the flagellum acquires its VSG coat before the body; the metacyclic's flagellar membrane is qualitatively different from the body membrane (fig. 3) and differential coating of body and flagellum is well known in other flagellates. IG was also observed in transmission micrographs attached to empty liposome-like structures associated with the gland microvilli (fig. 4). Whether this represents non-specific binding or binding to shed trypanosome membrane has not yet been ascertained.

VSG Gene Activation and the Expression Site

To determine the point in differentiation of the metacyclic when VSG gene transcription is initiated the pattern of distribution of nascent metacyclics of a particular VAT (GUTat 7.2) as shown by SEM of IG-labelled trypanosomes (fig 6) was studied; if VSG transcription is initiated in one of the dividing stages of trypanosome development, then clustering of labelled nascent metacyclics is to be expected. A montage of a 3600 um^2 area of an infected salivary gland was mapped onto a grid of rectangles, each rectangle corresponding to 338 um^2 and

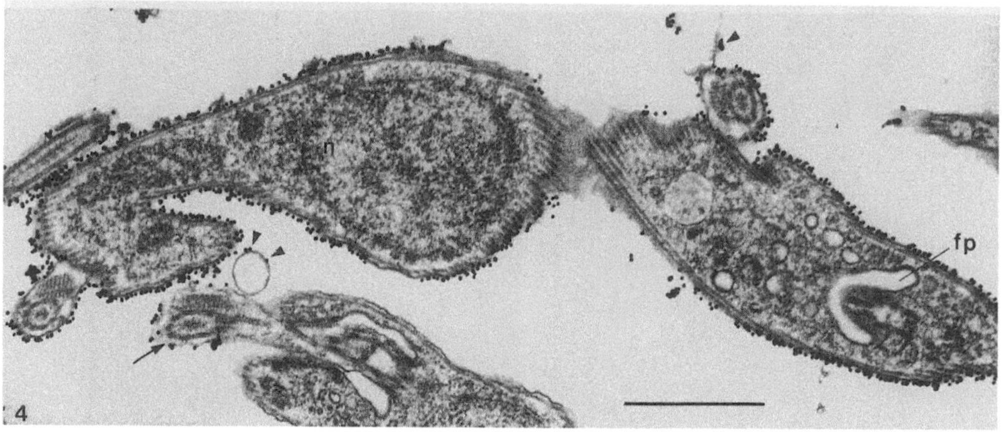

Fig. 4. TEMG of IG-labelled GUTat 7.2 nascent metacyclic. Note uncoated trypanosome alongside with IG particles on flagellum only (arrowed), also IG particles adhering to liposome-like structures (arrowheads). n - nucleus; p - flagellar pocket. Scale bar 1 µm.

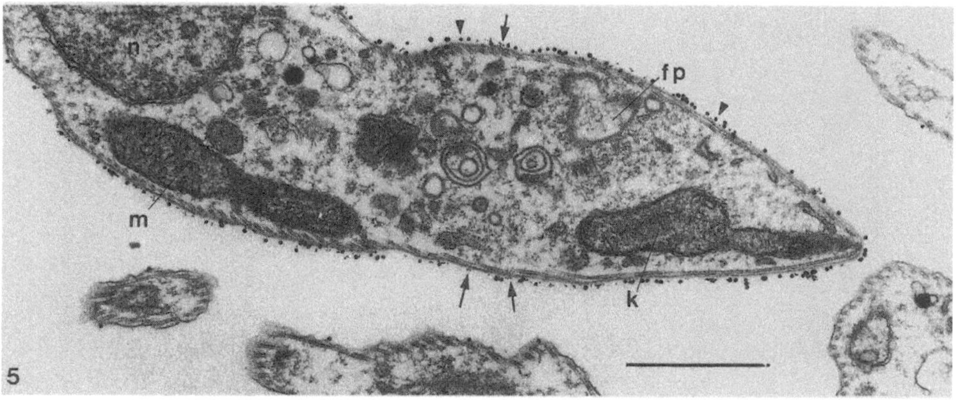

Fig. 5. TEMG of coating trypomastigote doubly labelled with 25 nm anti-GUTat 7.1 IG (arrowheads) and 5 nm anti-GUTat 7.2 IG (arrows), suggesting presence of mixed VSG coat; unlabelled trypanosomes lie alongside. Abbreviations as in figs 1 and 5. Scale bar 1 µm.

the frequency distribution of labelled trypanosomes determined. The corresponding theoretical frequency distribution for randomly distributed labelled trypanosomes was calculated from a Poisson series and the "goodness of fit" of theoretical and observed distributions tested by Chi-squared analysis. The observed distribution of nascent metacyclics was in agreement with the Poisson series, indicating randomness, i.e. VSG gene activation occurs at or after the last division of the differentiating trypanosome (data to be presented elsewhere).

Fig. 6. Survey scanning electron micrograph of trypanosome carpet on salivary gland epithelium after IG labelling for GUTat 7.2. Note IG labelled blunt-ended nascent metacyclics (large arrows) and drawn out posterior ends of unlabelled epimastigotes. Scale bar 10 µm.

The proportion of the different M-VATs in the metacyclic population is more or less constant (Hajduk et al. 1981) regardless of the bloodstream VAT ingested by the fly. This and the observation that VSG heterogeneity is present ab initio suggests that VSG gene activation is a stochastic rather than a deterministic event.

The fact that M-VAT heterogeneity occurs in nascent metacyclics does not rule out the possibility that M-VAT switching occurs in the lumen of the gland. Using different sized immunogold particles for the two IG probes, double labelling of some nascent metacyclics was noted while others in the same section showed only single labelling (fig. 5). Double labelling has been observed in immunofluorescence studies of individual bloodstream T.b. rhodesiense in the act of switching VATs (Esser and Schoenbechler, 1985) and as a stable character in an in vitro cultured line of T. equiperdum (Baltz et al., 1986). Studies on bloodstream trypanosomes suggest that VSG gene switching is linked to trypanosome division and mature metacyclic trypanosomes do not divide; it is unlikely therefore that metacyclic double labelling reflects M-VAT switching, though switching of one M-VAT to another occurs in the early bloodstream infection (Barry et al., 1985). But just as in vitro the absence of antibody-mediated selection permits the survival of double expressors, so its absence in the fly salivary gland will allow the same.

ACKNOWLEDGEMENTS

We thank Dr A.M. Jordan, Tsetse Research Laboratory, Bristol, for supplying tsetse, Cathy Cameron, Helen Hendry and Paul Tindall for skilled technical assistance and the U.K. Medical Research Council for financial support.

REFERENCES

Baltz T, Giraud C, Baltz D, Roth C, Raibaud A, Eisen H (1986) Stable expression of two variant-surface glycoproteins by cloned Trypanosoma equiperdum. Nature 317: 602-604.

Barry JD, (1986) The molecular biology of African trypanosomes. Trop Dis Bull 83: R1-R25

Barry JD, Crowe JS, Vickerman K (1985) Neutralisation of individual variable antigen types in metacyclic populations of Trypanosoma brucei does not prevent their subsequent expression in mice. Parasitology 90: 79-88.

Boothroyd JC (1985) Antigenic variation in African trypanosomes. Ann Rev Microbiol 39: 475-502.

Crowe JS, Lamont AG, Barry JD, Vickerman K (1984) Cytotoxicity of monoclonal antibodies to Trypanosoma brucei. Trans R Soc Trop Med Hyg 78: 508-513.

De Mey J, (1983) Colloidal gold probes in immunocytochemistry. In Immunocytochemistry: practical applications in Pathology and Biology (eds JM Polack, S Van Noorden). Wright; Bristol.

Esser KM, Schoenbechler MJ (1985) Expression of two variant surface glycoproteins on individual African trypanosomes during antigen switching. Science 229: 190-193.

Hajduk SL, Cameron CR, Barry JD, Vickerman K (1981) Antigenic variation in cyclically-transmitted Trypanosoma brucei. Parasitology 83: 595-607

Tetley L, Vickerman K (1985) Differentiation in Trypanosoma brucei: host-parasite cell junctions and their persistance during acquisition of the variable antigen coat. J Cell Sci 74: 1-19

EVOLUTION OF ANTIGENIC VARIATION IN AFRICAN TRYPANOSOMES

John C. Boothroyd and Thomas P. Beals.

Department of Medical Microbiology,
Stanford University School of Medicine,
Stanford, CA 94306, U.S.A.

INTRODUCTION

The african trypanosomes are responsible for several important diseases of humans and animals. One of the ways in which these parasitic protozoa successfully compete with the immune system of their mammalian hosts is through a phenomenon known as antigenic variation (reviewed in Donelson and Rice-Ficht, 1985; Pays, 1985; Boothroyd, 1985). This describes their ability to alter the antigenic composition of the surface coat which completely envelops the parasites while in the bloodstream of the host. On any one trypanosome, at any one time, this coat is essentially homogeneous, being comprised of a single glycoprotein species, known as the variant surface glycoprotein or VSG (Cross, 1975; Vickerman and Luckins, 1969).

The african trypanosomes are unique amongst the Kinetoplastida in possessing this phenomenon. It is known that the antigenic repertoire is stored as several hundred genes, each one coding for a distinct VSG (Capbern et al., 1977; Van der Ploeg et al., 1982). The extent of homology between these different VSG genes varies from essentially none to a clear and extensive relationship. In the latter case, the genes can be shown to exist as distinct VSG gene families (Pays et al., 1981; Allison et al., 1984). It seems likely that these gene families represent relatively recent stages in the evolution of the repertoire as a whole. As such, we have chosen to examine one VSG gene family in complete detail as a way of understanding how the enormous repertoire of VSG genes has evolved. In this paper we will specifically address the linkage of the members of the gene family being studied, the size of the region which has been duplicated in the generation of the different members and the extent of the homology between each member.

The particular gene family which we have chosen for study is that identified as homologous to VSG117. This gene family was chosen because of the extensive information on the prototypic VSG117 (it has been completely sequenced at both the protein (Allen et al., 1982; Holder, 1983) and nucleic acid level (Boothroyd et al., 1982)) and because it is one of the larger, more discrete families so far described (Pays et al., 1983).

NATO ASI Series, Vol. H11
Host-Parasite Cellular and Molecular Interactions
in Protozoal Infections. Edited by K.-P. Chang and D. Snary

RESULTS

Characterization of the VSG117 Gene Family.

Figure 1 shows a restriction map for the prototypic VSG117 gene (Bernards et al., 1981). This is the so-called Basic Copy (BC) of the gene which codes for VSG117 and which must be duplicated and transposed as a prerequisite to activation (Hoeijmakers et al., 1980). Several important regions are indicated. These are:

1. the 5'- and 3'-limits of the region duplicated and transposed as part of the activation process.
2. the upstream repeat sequences (URS) thought to be involved in this transposition process.
3. the co-transposed region which has no known coding function.
4. the 3'-splice site separating the upstream intron and the downstream protein-coding exon (there are no introns within the protein-coding region).
5. the initiation codon for translation (ATG).
6. the site coresponding to where the signal peptide is removed from the precursor polypeptide as part of its processing en route to the surface of the cell.

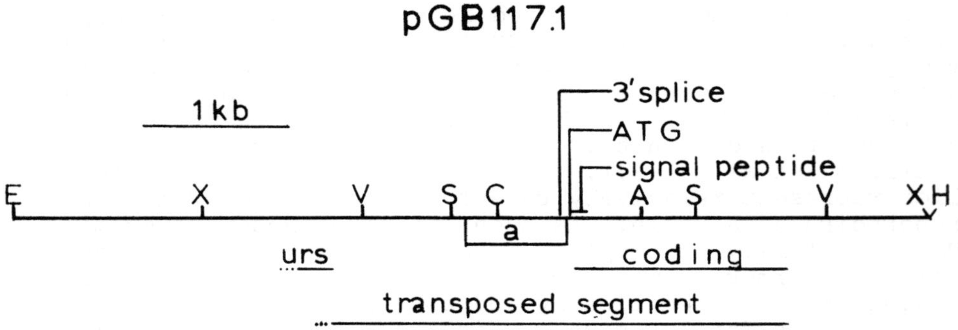

Figure 1. Map of the plasmid (pGB117.1 (Bernards et al., 1981)) containing the cloned VSG117 BC gene. The major sites involved in activation, transcription and translation of the gene are shown (see text). Restriction sites are abbreviated as A, AvaI; C, ClaI; E, EcoRI; H, HindIII; S, SalI; V, EcoRV; X, XbaI. The boxed region ('a') represents the probe used in generating the autoradiogram of Figure 2.

Figure 2 (lane G) presents a southern blot of disgested genomic DNA from variant MITat 1.5 expressing VSG118. This variant was derived from the same original cloned line as MITat 1.4 (expressing VSG117) and is thus "isogenic" with that variant. The hybridization probe used in this experiment is probe 'a' (Fig. 1) and derives from the co-transposed region of the VSG117 gene. This probe is used for defining the family because, as described below, it is derived from a region which has undergone the least variation and is thus well conserved between all family members. From this figure, it can be seen that the family apparently consists of a total of nine members. These are defined by the size of the HindIII fragment containing them. Digestion with other enzymes and similar hybridization studies also indicate that this is the total size of the family; that is, each band represents a discrete member (data not shown). Analysis of other variants derived from this line give the same pattern indicating that the family is stable even after multiple passages of the parasites in the laboratory over a period of years (data not shown).

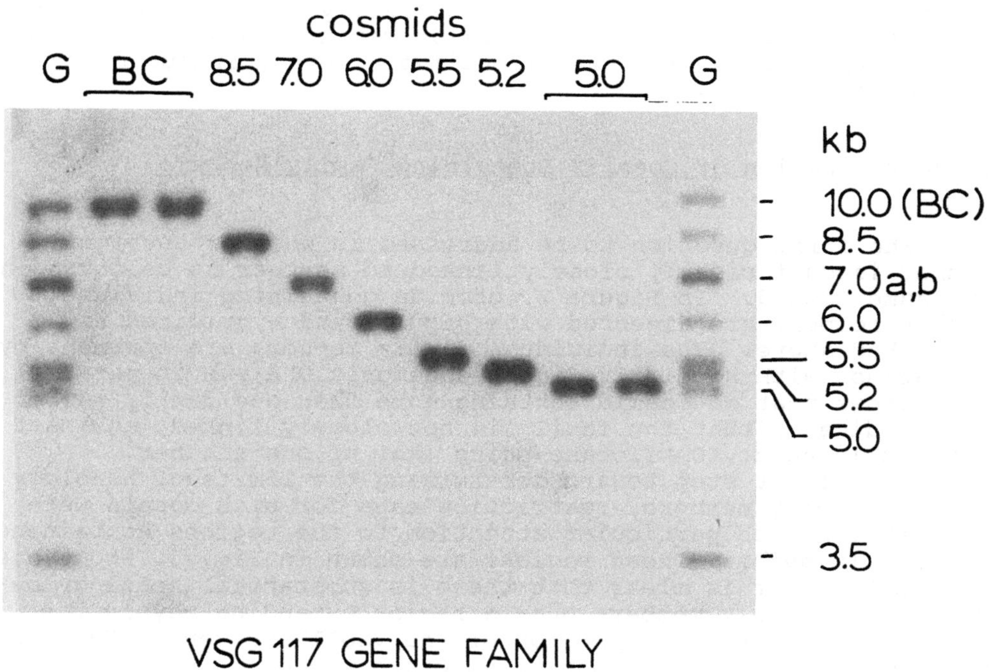

Figure 2. Homology between the VSG117 Family Members. Genomic DNA was digested with HindIII and loaded in the outside wells of an agarose gel ('G'). Each of nine cosmids containing seven of the VSG117 gene family members was also digested with HindIII and run in the indicated lanes. For two of the family members (the Basic Copy (BC) and the '5.0' gene), two cosmids containing these genes have been identified and these are run in parallel lanes. Two of the family members are not represented in the cosmids (one of those giving a 7.0 kb band and the smallest at 3.5 kb). The amount of cosmid DNA used was adjusted to give about the same molar amount as the genomic DNA. The gel was bidirectionally blotted to nitrocellulose paper and and one such blot hybridized with the probe ('a') shown in Fig. 1.

Molecular Cloning of Individual Family Members in Cosmids.

To study the family members in detail, a cosmid bank was generated using the vector c2xB (Bates and Swift, 1983) and genomic DNA from Trypanosoma brucei MITat1.5 (expressing VSG 118), partially digested with Sau3A and size fractionated by sucrose gradient centrifugation. This library was screened with a radio-labelled fragment from a recombinant clone of the VSG117 BC (probe 'a', Fig. 1). Although this fragment is not within the protein-coding region, it was known to effiently and specifically detect the complete family on Southern blots (Fig. 2). From this library a total of about 50 positive recombinants were identified and isolated for further study. These were further screened by digestion with HindIII and Southern blotting as this enzyme resolves the family into discrete fragments (Fig. 2). In this way, cosmids containing all but one of the family members have been identifed. Seven, including the BC are shown in Fig. 2.

Characterization of Cosmids Containing Family Members.

The first question to be addressed is whether any family member is sufficiently closely linked to another to be found on the same cosmid. In Figure 2, cosmids containing individual family members are digested with HindIII and hybridized as described above. The individual family members are apparent by comparison with similarly digested genomic DNA run in parallel. This shows that no cosmid contains more than one family member. This indicates that the family is not closely linked, at least not within about 40 kilobase-pairs (kb) of one another.

As a first step toward determining the limits of homology between family members, restriction maps for each cosmid were constructed with particular attention to the regions containing the family member. These regions are shown in Fig. 3. From this information, it is clear that there is substantial homology between most family members over a region extending beyond the ends of the transposed portion of the VSG117 BC gene. This is particularly evident at positions where there are conserved clusters of sites. It is interesting to note, however, that the cluster upstream of the transposed region (comprised of sites for XbaI, ClaI and EcoRV) is not present in the corresponding region of the VSG117 gene. From these data, it would appear that the family members are generally homologous over a region extending at least 2 kb upstream and 1 kb downstream of the transposed portion of the VSG117 gene. There are clear exceptions to this, however. For example, several family members do not have the downstream cluster (XbaI and HindIII), whereas the VSG117 BC as well as the '8.5' kb family member lack the upstream cluster described above. These results indicate that the family has not arisen by transposition of the unit which is known to be functional as a transposable segment in the duplication/transposition which leads to activation of the VSG117 gene. Further work is needed to determine more precisely the limits of homology between the family members to see if there is a discrete unit which has been duplicated several times in the evolution of this family. It is worth noting in this regard that this gene family is different from those often found in other systems in being not tandemly arrayed as a closely linked group.

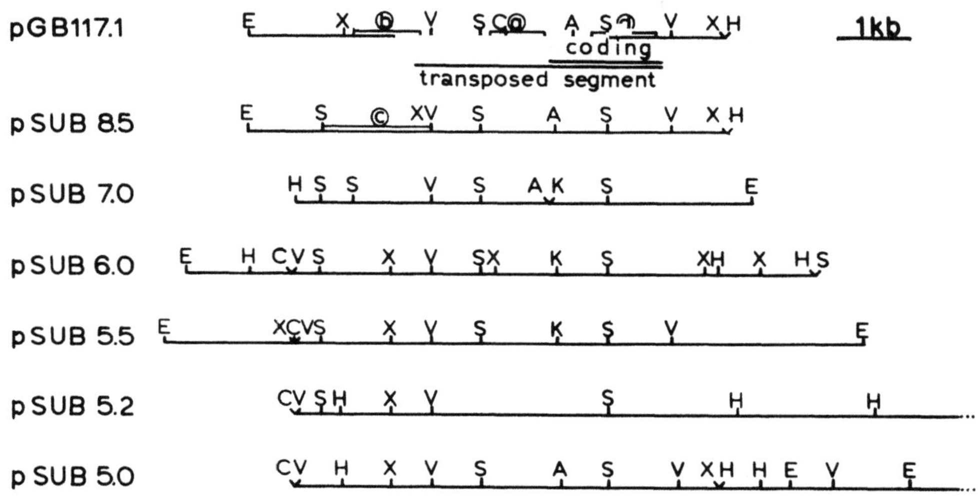

Figure 3. Maps of seven VSG117 gene family members. Fragments from cosmids containing family members as shown in Fig. 2 were subcloned in plasmids and mapped. The fragments used for each subcloning varied according to the location of convenient restriction sites. Each subclone is indicated by the size of the genomic HindIII fragment containing the family member (see Fig. 2). pGB117.1 contains the Basic Copy gene. Restriction sites are abbreviated as in Fig. 2. with the addition of K, KpnI. Regions used as probes in the hybridization studies of Fig. 4 are indicated by boxes and circled letters above.

<u>Extent of Homology Between Family Members Within the Transposed Segment.</u>

As a rapid means of determining whether the family members all have the complete gene defined by the VSG117 transposed segment, Southern blots of each cosmid were probed with a selection of different probes (Fig. 4). The regions represented by each probe are shown in Fig. 3.

Fig. 4a shows the same autoradiogram as that shown in Fig. 2. This uses a conserved region within the cotransposed portion as probe and hybridizes to all family members as already discussed. In Fig. 4b, a fragment containing the URS elements of the VSG117 gene is used as probe. Only the two cosmids containing the VSG117 BC gene are efficiently detected with this probe. This shows that, as predicted by the restriction mapping, only the VSG117 BC has these elements in detectable

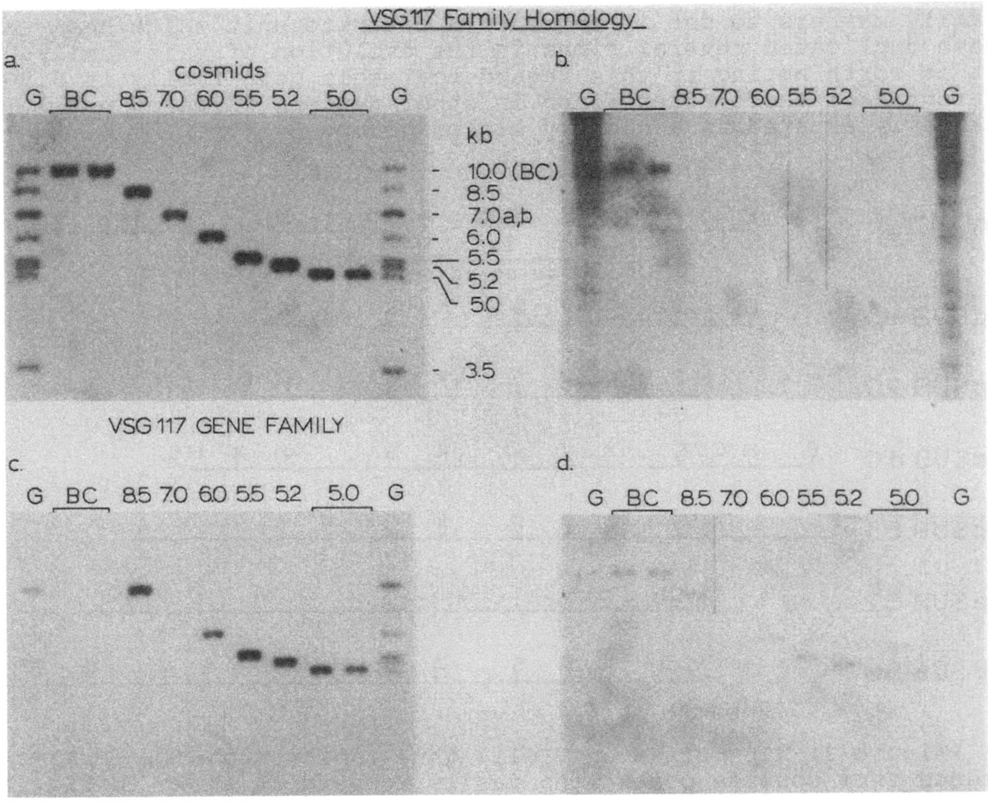

Figure 4. VSG117 Family Homology. The gel already described for Fig. 2 was bidirectionally blotted and each nitrocellulose sheet probed twice. The probe used for each hybridization ('a' through 'd') is indicated in Fig. 3. The smeared hybridization seen in the genomic lane (G) of part b is due to the fact that the URS elements of the VSG117 BC gene present in this probe are highly repeated throughout the genome of trypanosomes.

amounts (previously identified as four URS elements of about 81 bp each (Campbell et al., 1984)). This portion of the figure further shows that, with one exception, there is no detectable homology with these repeats anywhere else within the approximately 40 kb of genomic DNA present in each cosmid. The one exception is a light signal of hybridization in the cosmid containing member '5.5' in a band not closely adjacent to the family member itself. The exact position of this homology within the cosmid has not yet been identified.

To determine whether there is some different homology between the family members (other than the VSG117 BC, itself), a fragment from family member 8.5 from the region corresponding to the URS portion of the 117BC was used as a probe. The results are shown in Fig. 4c. This shows that there is substantial homology between this region and most of the family members, 7.0a and the BC being notable and clear exceptions.

This is again consistent with the results predicted from the restriction mapping of Fig. 3.

Finally, to determine whether the homology extends through the entire coding region of the VSG117 BC, a probe representing about 60% from the C-terminal coding region was used in hybridization studies. The results, shown in Fig. 4d, demonstrate that with the exception of member 7.0a, the entire family is homologous throughout this region. It is worth noting that member 7.0a is also exceptional in being the only member which lacks the region represented by probe b (Fig. 4b). The possibility that this family member has been segmentally gene converted is currently being investigated.

DISCUSSION

We have shown that there are eight closely related genes defined by their homology to the VSG117 Basic Copy gene. These genes are not closely linked, no two being found on the same cosmid. The homology betwen family members extends beyond the region which is known to be the functional unit in activation of the VSG117 BC.

Within the family, two members are particulary notable: the VSG117 BC gene, itself, in having a smaller region of overall homology with the other family members and in possessing the URS elements; and family member 7.0a in lacking homology at an upstream and downstream position otherwise observed in all the other family members (except the VSG117 BC).

Clearly one goal of this work is to ascribe function to structure. In this regard, there may be an important correlation between the fact that the VSG117 BC is the only member of the family which has been demonstrated to be functional in the process of antigenic variation. As this is also the only member of the family which has the URS elements which are known to be involved in its activation (directly or indirectly), it seems reasonable to suppose that these URS elements are essential for at least one mode of activation of VSG genes. It has previously been shown, however, that in a different trypanosome stock, VSG genes lacking the URS elements can be activated by duplication/transposition clearly indicating that alternative pathways exist which do not require these URS elements (Murphy et al., 1984; Pays et al., 1983). Furthermore, others have recently shown that portions of genes may be used in the construction of chimeric VSG genes late in the course of an infection (Longacre and Eisen, 1986; Roth et al., 1986; Pays et al., 1983). The frequency and timing of activation of the family members is currently being investigated.

The results presented here also address the question of homology on the smaller scale, i.e., the percentage homology between the family members. Sequencing within the regions shown in Fig. 1 is now in progress for the cloned family members. Although not yet complete, these results indicate that there is much greater homology within the cotransposed region than within that coding for the protein. This is consistent with the notion that there is a strong selective pressure for variation within the coding region, whereas no such selection exists for the cotransposed region. The intriguing possibility that there is some sort of directed mutagenesis cannot be excluded, however. Complete sequencing of these regions will help to address this point. Of particular interest will be whether there is some sort of conserved framework structure required for these VSGs to function in forming a surface coat and/or whether particular regions of the predicted polypeptides are hypervariable as might be predicted if only limited portions of the VSG molecule is exposed for binding of antibody on the surface of the parasite.

ACKNOWLEDGEMENTS

We thank Makio Iwashima for help in characterizing one of the subcloned family member genes. This work was supported in part by grants from National Institutes of Health (AI21025) and the MacArthur Foundation. T.P.B. was supported in part by a Cell and Molecular Biology Training Grant from the N.I.H. (GM07276) and a Training Grant from Johnson and Johnson Co. J.C.B. is a Burroughs Wellcome Scholar in Molecular Parasitology.

REFERENCES

Allen G, Gurnett LP, Cross GAM (1982) The complete amino acid sequence of a variant surface glycoprotein (VSG 117) from Trypanosoma brucei. J. Molec. Biol. 157: 527-546

Allison J, Rothwell V, Newport G, Agabian N, Stuart K (1984) The IsTat 1.3 VSG multigene family in Trypanosoma brucei: retention of the expression linked copy through multiple antigenic switches. Nucl Acids Res 12: 9051-9066

Bates PF, Swift RA (1983) Double cos site vectors: simplified cosmid cloning vectors. Gene 26: 137-146

Bernards A, Van der Ploeg LHT, Frasch ACC, Borst P, Boothroyd JC, Coleman S, Cross GAM (1981) Activation of trypanosome surface glycoprotein genes involves a duplication-transposition leading to an altered 3' end. Cell 27: 497-505

Boothroyd JC (1985) Antigenic variation in african trypanosomes. Ann Rev Micro 39: 475-502

Boothroyd JC, Paynter CA, Coleman SL, Cross GAM (1982) Complete nucleotide sequence of complementary DNA coding for a variant surface glycoprotein from Trypanosoma brucei. J Mol Biol 157: 547-556

Campbell DA, vanBree M, Boothroyd JC (1984) The 5'-limit of transposition and upstream barren region of a trypanosome VSG gene: Tandem 76 base-pair repeats flanking (TAA)90. Nucl Acids Res 12: 2759-2774

Capbern A, Biroud C, Baltz T, Maltern P (1977) Trypanosoma equiperdum: etude des variations antigeniques au cours de la trypanosome experimentale du lapin. Exp. Parasitol. 42: 6-13

Cross GAM (1975) Identification, purification and properties of clone-specific glycoprotein antigens constituting the surface coat of Trypanosoma brucei. Parasitology 71: 393-417

Donelson JE, Rice-Ficht AC (1985) Molecular biology of trypanosome antigenic variation. Microbiol. Rev. 49: 107-125

Hoeijmakers JHJ, Frasch ACC, Bernards A, Borst P, Cross GAM (1980) Novel expression-linked copies of the genes for variant surface antigens in trypanosomes. Nature 284: 78-80

Holder AA (1983) Characterisation of the cross-reacting carbohydrate groups on two variant surface glycoproteins of Trypanosoma brucei. Mol Biochem Parasit 7: 331-338

Longacre S, Eisen H (1986) Expression of whole and hybrid genes in Trypanosoma equiperdum antigenic variation. EMBO J 5: 1057-1063

Murphy WJ, Brentano ST, Rice-Ficht AC, Dorfman DM, Donelson JE (1984) DNA rearrangements of the variable surface antigen genes of trypanosomes. J. Protozool. 31: 65-73

Pays E (1985) Selective telomere activation and the control of antigen gene expression in trypanosomes. In Ann. Immunol. Inst. Pasteur, Paris

Pays E, Dekerck P, Van Assel S, Babiker E, Le Ray D, Van Meirvenne N, Steinert M (1983) Comparative analysis of a Trypanosoma brucei gambiense antigen family and its potential use in epidemiology of sleeping sickness. Mol. Biochem. Parasitol. 7: 63-74

Pays E, Lheureux M, Vervoort T, Steinert M (1981) Conservation of a variant-specific surface antigen gene in different trypanosome species and subspecies. Mol. Biochem. Parasitol. 4: 349-357

Pays E, Van Assel S, Laurent M, Darville M, Vervoort T, Van Meirvenne N, Steinert M (1983) Gene conversion as a mechanism for antigenic variation in trypanosomes. Cell 34: 371-381

Pays E, van Assel S, Laurent M, Dero B, Michiels F, Kronenberger P, Matthyssens G, Van Meirvenne N (1983) At least two transposed sequences are associated in the expression site of a surface antigen gene in different trypanosome clones. Cell 34: 359-369

Roth CW, Longacre S, Raibaud A, Baltz T, Eisen H (1986) The use of incomplete genes for the construction of a Trypanosoma equiperdum variant surface antigen gene. EMBO J 5: 1065-1070

Van der Ploeg LHT, Valerio D, De Lange T, Bernards A, Borst P, Grosveld FG (1982) An analysis of cosmid clones of nuclear DNA from Trypanosoma brucei shows that the genes for variant surface glycoproteins are clustered in the genome. Nucl. Acids Res. 10: 5905-5923

Vickerman K, Luckins AG (1969) Localization of variable antigens in the surface coat of Trypanosoma brucei using Ferritin conjugated antibody. Nature 224: 1125-1126

STRUCTURAL STUDIES ON THE GLYCOPHOSPHOLIPID MEMBRANE ANCHOR OF TRYPANOSOMA BRUCEI VARIANT SURFACE GLYCOPROTEIN

M.A.J. Ferguson, R.A. Dwek, S.W. Homans, and T.W. Rademacher

Oxford Oligosaccharide Group*, Department of Biochemistry, University of Oxford, South Parks Road, Oxford, England.

INTRODUCTION

The Trypanosome Surface Coat and its Membrane Anchor

In common with other African trypanosomes Trypanosoma brucei evades the mammalian host immune response by antigenic variation (reviewed recently by Boothroyd 1985). Antigenic variation involves the sequential expression of individual genes encoding immunologically distinct variant surface glycoproteins (VSGs) which form a dense cell surface coat. The current model of coat architecture is of a densely packed monolayer of VSG molecules covering the entire plamsa membrane of the organism. From a number of observations it would appear that the VSG coat acts as a permeability barrier to macromolecules, protecting the organism both from attack by the alternative complement pathway and from lysis mediated by antibodies directed against invariant membrane components. The survival of the parasite in the mammalian host is therefore entirely dependant on the integrity of the VSG cell surface coat. The coat is an extraordinarily specialized structure in molecular terms; all of the different VSGs must serve the same protective function whilst differing markedly in primary amino acid sequence to remain immunologically distinct. The only site of cross-reaction resides in a common carbohydrate epitope (Holder and Cross 1981) which forms part of the VSG membrane anchor.

The nature of VSG-plasma membrane anchoring remained an enigma for several years because early purification protocols produced VSG in a highly water soluble form (sVSG) with no physical evidence of a hydrophobic membrane-spanning domain. The complete amino acid sequence of sVSG (variant 117) confirmed the lack of such a polypeptide domain (Allen et al 1982). However when predicted amino acid sequences, from cDNA analyses, were compared with C-terminal amino acid sequences it became clear that primary VSG mRNA translation products possess C-terminal hydrophobic polypeptide extensions (of 17 or 23 amino acids, depending on the variant) which are absent from the isolated sVSG molecules (reviewed by Cross 1984). For a time it was considered that these extensions were the VSG membrane anchors and that during VSG purification a specific protease was responsible for their removal. This attractive model was short lived due to the important observation of Holder who showed that the α-COOH group of the sVSG C-terminal amino acid (Asp, Ser or Asn) was blocked by an amide linkage to ethanolamine, which was in turn linked to a novel carbohydrate moiety (reviewed by Holder 1985). This carbohydrate contains the only inter-VSG immunologically cross reacting determinant (CRD). The existence of this ethanolamine linkage absolutely

*Supported by Monsanto Company.

NATO ASI Series, Vol. H11
Host-Parasite Cellular and Molecular Interactions
in Protozoal Infections. Edited by K.-P. Chang and D. Snary

precludes the co-existance of the predicted hydrophobic peptide extensions and so the nature of VSG-membrane anchoring became obscure once again. A major breakthrough in this area was made by Cardosa de Almeida and Turner (1983) who showed that if Trypanosomes were solublized directly into boiling detergent buffers the VSG now behaved as an amphiphilic membrane protein with a hydrophobic membrane insertion site. Furthermore they showed that this membrane form of VSG (mfVSG) could be converted to sVSG by the action of an endogenous heat-sensitive trypanosome enzyme which they called 'enzyme x'. These studies suggested that mfVSG was anchored by some non-peptide hydrophobic domain (probably lipid in nature) and that conventional purification strategies involving trypanosome disruption at lower temperatures produced sVSG by the artifactual activation of 'enzyme x' (Cardosa de Almeida et al 1984). This paper reviews subsequent structural and biosynthetic studies on the membrane anchor of *T. brucei* mfVSG and presents recent data on what should become a general approach to obtain detailed fine-structure on this new class of membrane anchor.

RESULTS

Structural Studies

General features. Chemical and enzymatic analyses of the mfVSG membrane anchor have shown that it is a complex glycophospholipid, covalently linked to the C-terminus of the VSG polypeptide. The phospholipid portion of this structure is a dimyristyl-phosphatidylinositol as shown in Fig. 1. The presence of only a single species of fatty acid (myristic acid) in the structure was deduced by gas chromatography and gc-ms (Ferguson and Cross 1984). The association of the fatty acid with the C-terminus of the VSG polypeptide and with the CRD carbohydrate was established by compositional analysis of the reverse-phase HPLC purified myristylated Pronase fragment of mfVSG (i.e. the intact membrane anchor structure plus the VSG C-terminal amino acid) and comparison with the composition of the CRD containing Pronase fragment of sVSG (Ferguson et al 1985a). The analyses showed that mfVSG differed from sVSG by the presence of one glycerol and two myristic acid residues per mole of VSG, suggesting that the actual hydrophobic domain responsible for membrane insertion was a dimyristyl glycerol moiety. This was confirmed by studies on mfVSG which had been biosynthetically labelled with ^{3}H-myristic acid (^{3}H-mfVSG). Treatment of ^{3}H-mfVSG by acetolysis released the label as diglyceride acetate and digestion with a phospholipase A_2 released half of the label as free fatty acid. The stringent specificity of phospholipase A_2 further defined the structure as the sn-1,2-dimyristyl glycerol stereo-isomer with a phosphodiester bridge to the rest of the CRD at the glycerol 3-position.

It was shown by ^{1}H-NMR and chemical studies that the CRD glucosamine residue was not N-acetylated but existed in the free amino form (Strang et al 1986, Ferguson et al 1985b). This allowed specific degredation of the structure at the glucosamine residue by the action of nitrous acid. This deamination reaction converts glucosmine to 2,5-anhydro mannose with the simultaneous cleavage of the original glucosamine glycosidic bond. When this reaction was performed on ^{3}H-mfVSG, the label was released in a chloroform/methanol soluble form which by tlc co-migrated with phosphatidylinositol (see Fig. 1 for the site of cleavage). In addition the highly purified phosphatidylinositol-specific phospholipase C (PI-PLC) from S. aureus (Low 1981) was capable of rapidly releasing dimyristyl glycerol from ^{3}H-mfVSG. This enzyme will only recognize phospholipid structures with an inositol headgroup. These data predicted that sVSG should contain inositol phosphate linked to the CRD via the glucosamine residue (Fig. 1); this was confirmed by nitrous acid studies and gc-ms analysis (Ferguson *et al* 1985b). More recently the identification of dimyristyl phosphatidylinositol, released by nitrous acid, has been confirmed by fast atom bombardment-mass spectrometry (Schmitz *et al* 1986).

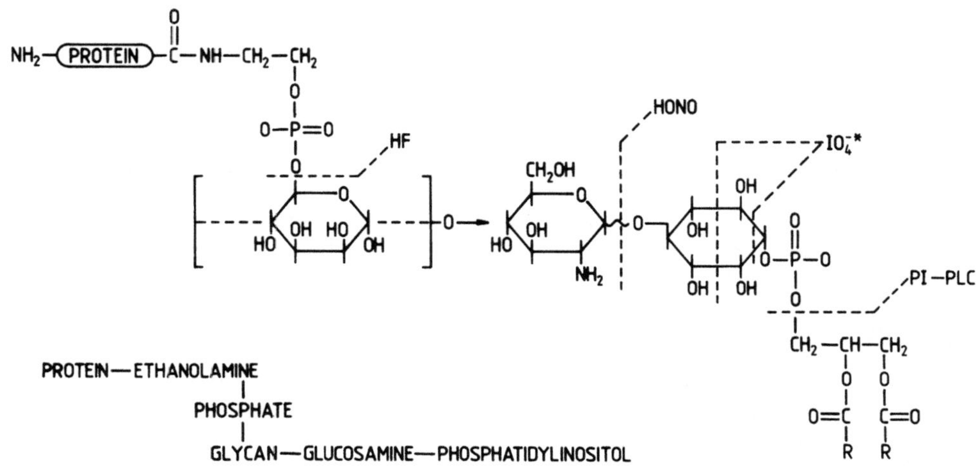

FIGURE 1. The partial structure of the glycosyl-*sn*-1,2-dimyristyl phosphatidylinositol membrane anchor of mfVSG (variant 117). The broken lines marked HF, HONO, IO_4^- and PI-PLC show the relevant cleavage sites of aqueous hydrogen fluoride, nitrous acid deamination, periodate oxidation and phosphatidylinositol-specific phospholipase C treatments respectively. The asterisk signifies tentative data discussed in the text.

The partial structure in Fig. 1 shows the substitution of the inositol ring by glucosamine at the 4- position. This is a tentative assignment made on the basis of periodate oxidation studies on ^{3}H-mfVSG. Following periodate treatment and reduction 70% of the label was recovered by chloroform/methanol extraction; 60% of this material migrated as phosphatidic acid, 25% as neutral lipid and 15% as an unidentified component by tlc (data not shown). Periodate oxidation of mono-substituted phosphoinositides causes the scission of C-C bonds between vicinal hydroxyls on the inositol ring and generates aldehyde groups. These can be reduced with $NaBH_4$ to produce polyols to which the phosphatide and substituent remain attached if the substituent is at the 2,3,5 or 6-position of the inositol ring (Hsieh *et al* 1978). If however the substituent is at the 4-position of the ring, scission between C2-C3 and C5-C6 will cause the phosphatide to be released entirely from the substituent (see Fig. 1). Theoretically, following reduction, the phosphatide (in this case dimyristyl glycerol-3-phosphate) should be linked to the 2-position of a glycerol residue derived from carbons 1,2 and 6 of the inositol ring. In this study the labelled phosphatide is indeed released from the substituent by periodate but tlc analysis shows that it is released as free dimyristyl glycerol-3-phosphate (phosphatidic acid). This phenomenon of the loss of the predicted glycerol moiety from the phosphate group has been observed previously for inositol 1-phosphate and inositol 1,4-bisphosphate where rapid over-oxidation to free phosphoric acid occurs (Tomlinson and Ballou 1961). By analogy with the above studies the periodate studies on ^{3}H-mfVSG are most consistent with a substitution at the 4-position of the inositol ring. This assignment however requires confirmation by alternative means such as methylation analysis.

Several lines of evidence suggest the existence of a phosphodiester bridge between the ethanolamine hydroxyl and the 6- position of one of the CRD mannose residues (Fig. 1). Firstly, the ability of acid hydrolysis to release ethanolamine from the CRD (Holder 1983a,b) precludes a direct ether linkage between the ethanolamine and the carbohydrate. Secondly, partial acid hydrolysis of the CRD results in the transient formation of small amounts of acid labile ethanolamine phosphate (identified by amino acid analyser, Ferguson et al 1985b) and thirdly, mannose-6-phosphate has now been directly identified as part of the structure by gc-ms sugar phosphate analysis of the CRD (details to be published elsewhere). Dephosphorylation of the CRD by 50% HF, 18 h, 0° increased the average yield of free mannose in the CRD from 2 to 3 moles per mole (as judged by gc-ms analysis before and after treatment) suggesting that Man-6-P is originally present at one mole per mole CRD.

Fine structure. The CRD containing Pronase fragment of sVSG (variant 117) contains all of the glycophospholipid membrane anchor (less the dimyristyl glycerol moiety) and the C-terminal aspartic acid residue of the polypeptide. This material was analysed by 1 and 2-dimensional (COSY) ^{1}H-NMR at 500 MHz which showed that the sample was highly heterogenous with respect to the carbohydrate (glycan) region of the structure (data not shown). Although some general conclusions could be drawn, such as the presence of two 6-O-substituted αGal residues in most of the structures, complete primary structure determination requires the purification and individual analysis of each form present in the total mixture. This type of heterogeneity in carbohydrate moieties is well known in glycoproteins where single asparagine glycosylation sites generally contain several different structures (see for example Parekh et al 1985).

To resolve this problem of heterogeneity in the glycophospholipid anchor we have developed a strategy to generate the various glycan structures in a neutral form free of the inositol phosphate and the C-terminal amino acid-ethanolamine-phosphate portion, (Fig. 2).

FIGURE 2. (Opposite) A new general scheme for obtaining neutralized glycan regions of glycophospholipid membrane anchors for sequencing. A general structure for this class of membrane anchor is shown at the top; AA, EtN and P are the C-terminal amino acid, ethanolamine and phosphate respectively. Two alternative routes can be taken to remove the polypeptide and the diacyl glycerol; (i) Pronase digestion, HPLC purification of the intact glycophospholipid then de-acylation with PI-PLC or (ii) PI-PLC treatment (in the case of sVSG this is performed by the endogenous trypanosome PLC during isolation) followed by Pronase digestion and purification of the same product. Nitrous acid deamination releases the inositol cyclic phosphate from the structure and subsequent reduction with NaB^3H_4 introduces a ^{3}H-label into the new 2,5-anhydro mannitol terminus. Aqueous HF treatment removes the phosphate from the structure to produce a neutral glycan suitable for further structural analyses, see text. An alternative to HF treatment involving hydrazinolysis, N-acetylation and methylation is shown. This approach may be used in future studies to retain the site of ethanolamine phosphate attachment to the glycan.

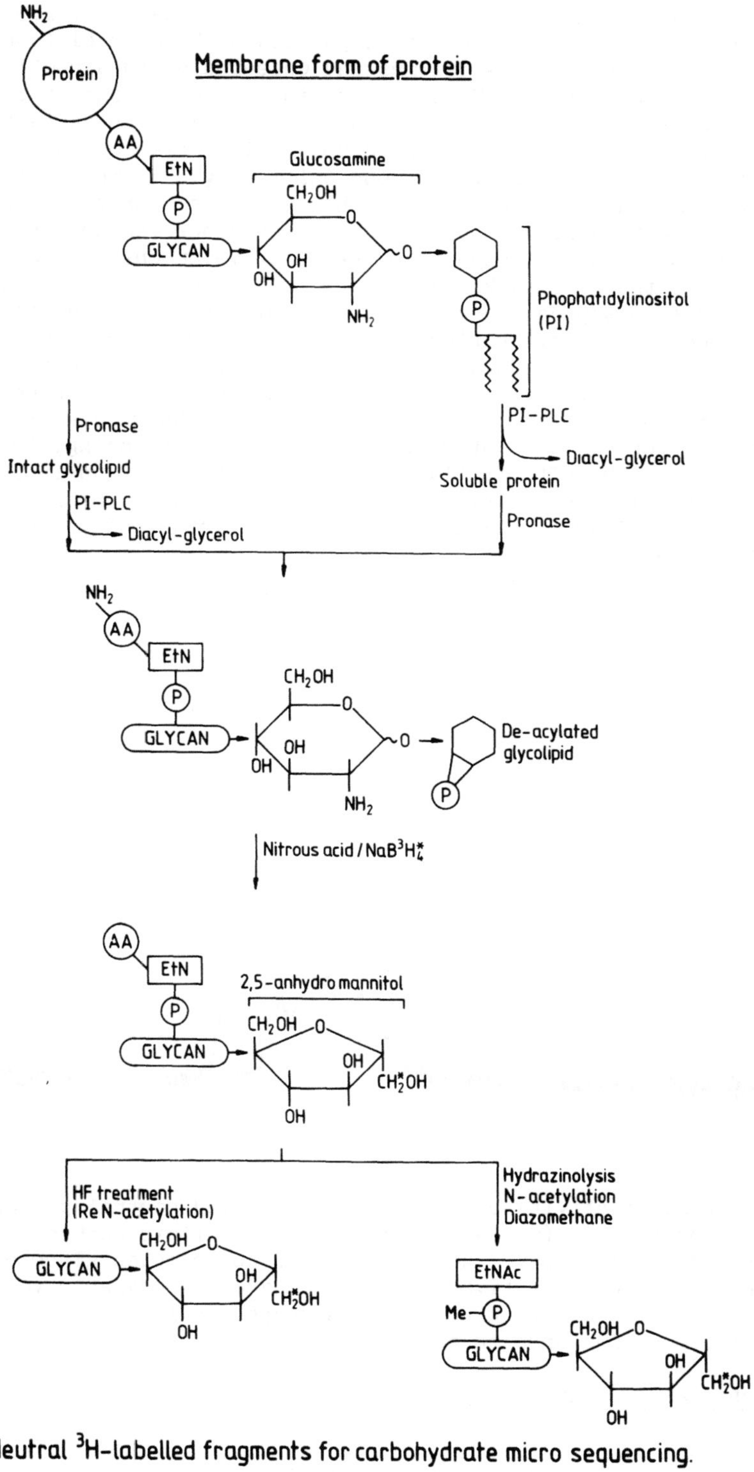

Neutral ^{3}H-labelled fragments for carbohydrate micro sequencing.

Once in this form, and conveniently labelled in the terminal 2,5-anhydro mannitol residue, the glycans can be separated by high-resolution gel-filtration on Bio-Gel P-4 columns and detected non-destructively by radioactivity flow monitors. This system rapidly establishes the size and degree of heterogeneity of the glycan structures. For VSG variant 117 the glycophospholipid anchor glycans are resolved into five obvious peaks ranging in size from 4.2 to 7.6 glucose units (Fig. 3). When this profile is optimally smoothed and resolution enhanced using Fourier transform image processing techniques there is strong evidence for the presence of another three minor peaks fused with the three major ones (data not shown). Each of the five peaks can now be subjected to high-voltage borate electrophoresis to separate any structures which co-elute on Bio-Gel P-4. This two dimensional purification system assures that each isolated structure is homogenous. Each unique structure will then be investigated by ^{1}H-NMR and methylation/gc-ms analysis to determine their fine structures and inter-relationships. In addition to the preparative power of high-resolution P-4 chromatography/radioactive monitoring, isolated glycan species can be efficiently sequenced using highly purified and defined exoglycosidases followed by re-analysis on the same system as the radiolabel is retained at all times at the undigested end of the product and recoveries are quantitative.

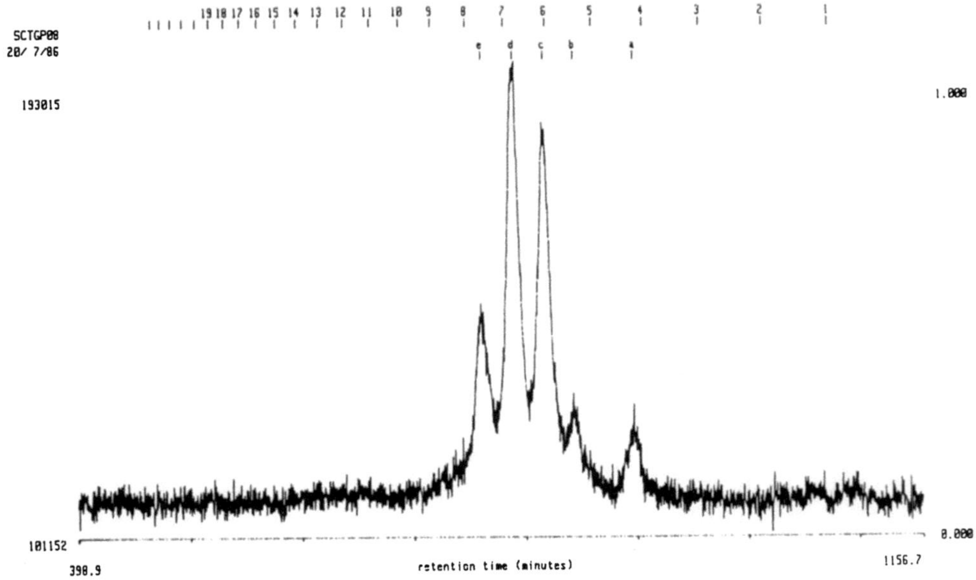

FIGURE 3. Radioactivity profile of the labelled neutral glycans generated from the VSG (variant 117) glycophospholipid anchor separated by Bio-Gel P-4 chromatography. The positions of co-injected glucose oligomers are shown at the top. The five peaks a - e have hydrodynamic volumes of 4.2, 5.4, 6.1, 6.8 and 7.6 glucose units respectively. This analytical scale analysis corresponds to about 5 nmoles (300 μg) of VSG.

Biosynthesis and Transport of mfVSG

Pulse-chase studies (Ferguson et al 1986) show that both the glycan (CRD) portion and the myristic acid moieties of the glycophospholipid anchor replace the predicted C-terminal hydrophobic peptide tails on newly synthesized VSG molecules within 1 minute of polypeptide synthesis. Similar results have also been reported by Bangs et al (1985, 1986). The rapid kinetics of this processing event suggest that the glycophospholipid is pre-assembled and transfered to the VSG en-bloc in the endoplasmic reticulum. Recently an elegant study by Krakow et al (1986) has provided strong evidence for the existence of a pre-assembled anchor species. Cryosection-immunogold labelling studies indicate that the glycophospholipid anchored mfVSG molecules traverse the Golgi stacks prior to their delivery to the plasma membrane (Duszenko et al 1986).

DISCUSSION

The covalent attachment of phosphatidylinositol, as a complex glycophospholipid, to protein represents a new class of protein-membrane anchor, reviewed by Low et al (1986). T. brucei mfVSG is the best characterized of this class of membrane protein in terms of its anchor structure and processing. The function of such an anchor instead of the usual transmembrane polypeptide sequence is still unclear, however the existence of an endogenous mfVSG→sVSG converting enzyme ('enzyme x') suggests that such anchors may be designed to be cleaved in response to specific cellular stimuli. The trypanosome enzyme has been identified as a phospholipase C (Ferguson et al 1985a) with specificity for the anchor structure rather than simple phospholipids or even phosphatidylinositol (Bülow and Overath 1986, Fox et al 1986). The precise location of the trypanosome enzyme is still unknown but pulse-chase studies show that the time taken for newly synthesized mfVSG to become accessible to the enzyme is similar to the transit time from the rough endoplasmic retention to the plasma membrane (Bangs et al 1986, Ferguson et al 1986). It is tempting to speculate that this enzyme activity resides in the outer leaflet of the plasma membrane such that upon some physiological stimulation the enzyme could release all or part of the VSG surface coat. Potentially this could be useful to the parasite during immune attack and/or VSG replacement during antigenic variation, although VSG release would need to be carefully controlled. The VSG coat has been observed to be shed from the plasma membrane during the in vitro differentiation of bloodstream forms to coatless procyclic forms (Overath et al 1983). This may be mediated by the phospholipase C. It may be significant that VSG release by the action of this enzyme would also result in the generation of biologically active diacyl glycerol in the plasma membrane which could stimulate a protein kinase C to mediate intracellular responses to the original coat releasing stimulus (Fig. 4). The ability to express an intrinsic membrane glycoprotein with a metastable membrane anchor might also be usefully exploited by other parasites. For instance cell surface glycoproteins involved in host cell recognition and/or penetration might be usefully released from the parasite surface during/after the invasion to allow the host and parasite plasma membranes to become separated. In this context the 63 Kd glycoprotein of Leishmania major, has been shown to be anchored by an analogous glycophospholipid membrane anchor (Etges et al 1986). Furthermore this molecule has been shown to be involved in macrophage recognition in L. mexicana mexicana (Russell and Wilhelm 1986) and L. m. amozonensis (Chang and Chang 1986). The 195 Kd merozoite surface glycoprotein of Plasmodium falciparum also contains a glycophospholipid moiety (Haldar et al 1985) although the function of this molecule is unknown. This new class of protein membrane anchor is not restricted to parasitic protozoa but appears to be widely distribtured occuring also in Torpedo acetylcholinesterase and mammalian acetylcholinesterase, akaline phosphatase, 5'-nucleotidase (reviewed by Low et al 1986) and rat thymocyte and brain Thy-1 antigen (Tse et al 1985).

We intend to undertake extensive structural analyses of the glycophospholipid anchors of both parasite and mamalian glycoproteins in order to assess the extent of common and species specific sub-structures. Some preliminary results on the anchor of VSG 117 are shown here. Compositional analysis of the CRD carbohydrate from several different VSGs (Holder 1985) suggests that they may all contain a common Man_2GlcN_1 'core' with varying amounts (0-8) of galactose. To assess their structural relationships the anchors of VSG 055 (8 galactose) and VSG 118 (0 galactose) will be analysed. Comparison of these VSG structures with those of other parasites and mammalian (host) structures using the high-sensitivity strategy outlined in Fig. 2 should allow us to identify any parasite-specific structural details which might be exploited in the development of chemotherapeutic agents. It is already clear from the composition of rat Thy-1 (Tse et al 1985) that the glycans of this mammalian anchor cannot be identical with those of VSG.

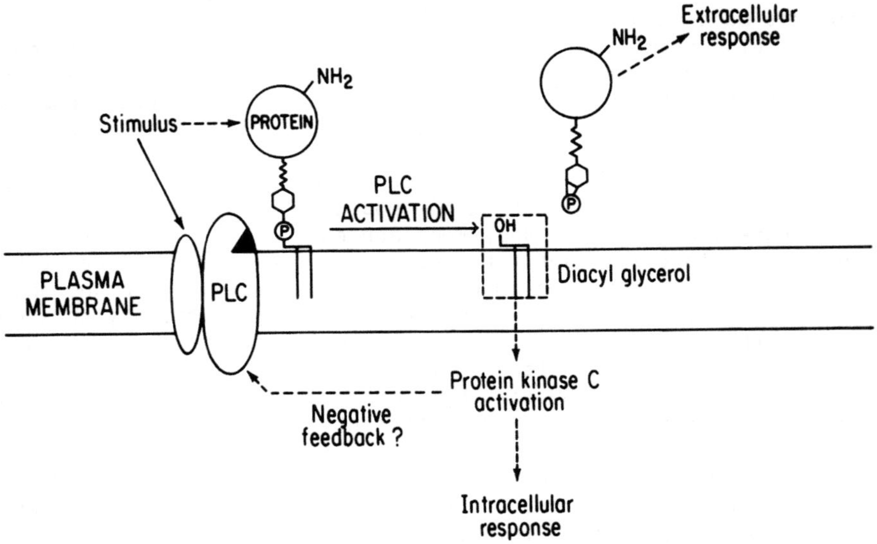

FIGURE 4. Hypothetical model of protein shedding mediated by an endogenous anchor-specific phospholipase C.

REFERENCES

Allen A, Gurnett LP and Cross GAM (1982) Complete amino acid sequence of a variant surface glycoprotein (VSG 117) from Trypanosoma brucei. J. Mol. Biol. 157, 527-546.

Bangs JD, Andrews NW, Hart GW and Englund PT (1986) Posttranslational modification and intracellular transport of a Trypanosome variant surface glycoprotein. J. Cell Biol. 103, 255-263.

Bangs JD, Hereld D, Krakow JL, Hart GW and Englund PT (1985) Rapid processing of the carboxyl terminus of a Trypanosome variant surface glycoprotein. Proc. Natl. Acad. Sci. 82, 3207-3211.

Boothroyd JC (1985) Antigenic variation in African Trypansomes. Ann. Rev. Microbiol. 39, 475-502.

Bülow R and Overath P (1985) Synthesis of a hydrolase for the membrane-form variant surface glycoprotein is repressed during transformation of Trypanosoma brucei. FEBS Lett. 187, 105-110.

Bülow R and Overath P (1986) Purification and characterization of the membrane-form variant surface glycoprotein hydrolase of Trypanosoma brucei. J. Biol. Chem. 261, 11918-11923.

Chang CS and Chang K-P (1986) Monoclonal antibody purification of a Leishmania membrane glycoprotein and its inhibition of Leishmania-macrophage binding. Proc. Natl. Acad. Sci. USA 83, 100-104

Cross GAM (1984) Structure of the variant glycoproteins and surface coat of Trypanosoma brucei. Phil. Trans. R. Soc. Lond. B 307, 3-12.

Cardoso de Almeida ML, Turner MJ (1983) The membrane form of variant surface glycoproteins of Trypanosoma brucei. Nature 302, 349-352.

Cardoso de Almeida ML, Allan LM, Turner MJ (1984) Purification and properties of the membrane form of variant surface glycoproteins (VSGs) from Trypanosoma brucei. J. Protozool. 31, 53-60.

Duszenko M, Ivanov IE, Ferguson MAJ, Plesken H, Cross GAM and Sabatini DD (1986) Intracellular transport of variant surface glycoprotein in Trypanosoma brucei. Manuscript submitted.

Etges R, Bouvier J and Bordier C (1986) The major surface protein of Leishmania promastigotes is anchored by a myristic acid labelled phospholipid. EMBO J. 5, 597-601.

Ferguson MAJ and Cross GAM (1984) Myristylation of the membrane form of a Trypanosoma brucei variant surface glycoprotein. J. Biol. Chem. 259, 3011-3015.

Ferguson MAJ, Haldar K and Cross GAM (1985a) Trypanosoma brucei variant surface glycoprotein has a sn-1,2-dimyristyl glycerol membrane anchor at its COOH terminus. J. Biol. Chem. 260, 4963-4968.

Ferguson MAJ, Low M G and Cross GAM (1985b) Glycosyl-sn-1,2-dimyristylphosphatidylinositol is covalently linked to Trypanosoma brucei variant surface glycoprotein. J. Biol. Chem. 260, 14547-14555.

Ferguson MAJ, Duszenko M, Lamont GS, Overath P and Cross GAM (1986) Biosynthesis of Trypanosoma brucei variant surface glycoproteins. J. Biol. Chem. 261, 356-362.

Fox JA, Duszenko M, Ferguson MAJ, Low MG and Cross GAM (1986) Purification and characterization of a novel glycan- phosphatidylinositol-specific phospholipase C from Trypanosoma brucei. J. Biol. Chem. - in press.

Haldar K, Ferguson MAJ and Cross GAM (1985) Acylation of a plasmodium falciparium merozoite surface antigen via sn-1,2-diacyl glycerol. J. Biol. Chem. 260, 4969-4974.

Holder AA (1985) Glycosylation of the variant surface antigens of Trypanosoma brucei. Current Topics in Microbiol. Immunol. 117, 57-74.

Holder AA and Cross GAM (1981) Glycopeptides from variant surface glycoproteins of Trypanosoma brucei. C-terminal location of antigenically cross-reacting carbohydrate moieties. Mol. Biochem. Parasitol. 2, 135-150.

Hsieh TC-Y, Kaul K, Laine RA and Lester RL (1978) Structure of a major glycophosphoceramide from tobacco leaves. Biochem. 17, 3575.

Krakow JL, Hereld D, Bangs JD, Hart GW and Englund PT (1986) Identification of a glycolipid precursor of the Trypanosoma brucei variant surface glycoprotein. J. Biol. Chem. 261, 12147-12153.

Low MG, (1981) Phosphatidylinositol-specific phospholipase C from Staphylococcus aureus. Meth. Enzymology 71, 741-746.

Low MG, Ferguson MAJ, Futerman AH and Silman I (1986) Covalently attached phosphatidylinositol as a hydrophobic anchor for membrane proteins. Trends in Biochemical Sciences 11, 212-215.

Overath P, Czichos J, Stock U and Nonnengaesser C (1983) Repression of glycoprotein synthesis and release of surface coat during transformation of trypanosoma brucei. EMBO 2, 1721-1728.

Parekh RB, Dwek RA, Sutton BJ, Ferandes DL, Leung A, Stanworth D, Rademacher TW, Mizuochi T, Taniguchi T, Matsuta K, Takeuchi F, Nagano Y, Miyamoto T and Kobata A. Association of rheumatoid arthritis and primary osteoarthritis with changes in the glycosylation pattern of total serum IgG. Nature 316, 452-457.

Russell DG and Wilhelm H (1986) The involvement of the major surface glycoprotein (gp63) of Leishmania promastigotes in attachment to macrophages. J. Immunol. 136, 2613-2620.

Schmitz B, Klein RA, Egge H and Peter-Katalinic J (1986) A study of the membrane attachment site of the membrane-form variant surface glycoprotein from Trypanosoma brucei brucei using lipid vesicles as a model of the plasma membrane. Mol. Biochem. Parasitol. 20, 191-197.

Strang A-M, Williams JM, Ferguson MAJ, Holder AA and Allen AK (1986) Trypanosoma brucei brucei variant surface glycoprotein contains non-N-acetylated glucosamine. Biochem. J. 234, 481-484.

Tomlinson RV and Ballou CE (1961) Complete characterization of the myo-inositol polyphosphates from beef brain phosphoinositide. J. Biol. Chem. 236, 1902-1911.

Tse AGD, Barclay N, Watts A and Williams AF (1985) A non-protein hydrophobic tail at the C-terminus of the Thy-1 glycoprotein of neurous and thymocytes. Science 230, 1003-1005.

METABOLISM OF TRYPANOTHIONE AND GLUTATHIONYLSPERMIDINE IN TRYPANOSOMATIDS

Alan H. Fairlamb and Graeme B. Henderson

Laboratory of Medical Biochemistry, Rockefeller University,
1230 York Avenue, New York, NY 10021, USA

INTRODUCTION

With few exceptions, both eukaryotic and prokaryotic cells contain high concentrations of the tripeptide γ-glutamyl-cysteinyl-glycine (glutathione; GSH) [1,2]. Glutathione plays a number of important roles in cellular metabolism, including maintenance of the correct intracellular thiol redox balance, the enzymatic removal of hydrogen peroxide and the scavenging of free radicals. Glutathione has also been implicated in the control of DNA and protein synthesis and regulation of enzyme activity [3]. The intracellular ratio of reduced to oxidized glutathione (GSH : GSSG) is kept high (GSSG < 1% of the total) due to the presence of the NADPH-dependent enzyme, glutathione reductase [E.C. 1.6.4.2].

Several previous studies have indicated that trypanosomatids contain glutathione, but reports of glutathione reductase activity were conspicuously lacking [4]. This situation was resolved with the discovery that dialyzed soluble extracts of _Trypanosoma brucei_, _T.cruzi_, _Leishmania mexicana_ and _Crithidia fasciculata_ are able to enzymatically reduce GSSG and other disulfides by NADPH, solely in the presence of a low molecular weight metabolite [6]. Significantly, this heat stable component was found to be uniquely present in extracts of trypanosomatids and not in a range of other organisms. Using dialyzed enzyme preparations from _T.brucei_, as a biological probe, this metabolite was isolated and purified from _C.fasciculata_. The structure was determined by amino acid analysis and mass-spectrometry and confirmed by chemical synthesis to be N^1,N^8-bis (L-γ-glutamyl-L-hemicystinyl-glycyl)-spermidine [5]. As this thiol-containing polyamine-peptide appears to be unique to trypanosomatids, the trivial name trypanothione was proposed. The structure is given in Fig. 1 [1].

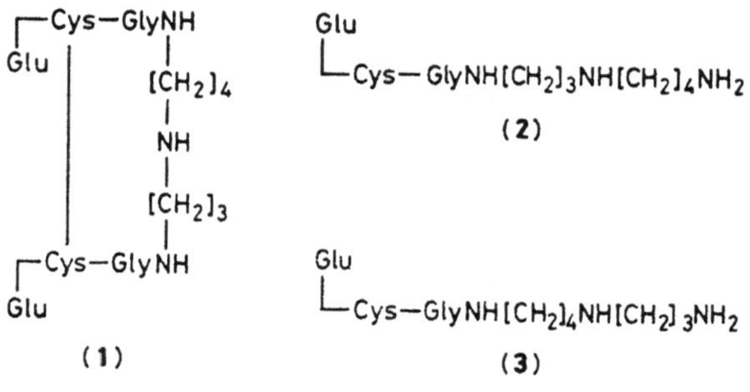

FIGURE 1. Structure of trypanothione disulfide [1]; N^1-mono- [2] and N^8-glutathionylspermidine [3].

NATO ASI Series, Vol. H11
Host-Parasite Cellular and Molecular Interactions
in Protozoal Infections. Edited by K.-P. Chang and D. Snary

Trypanothione disulfide ($T(S)_2$) is maintained in the cell in the reduced form (dihydrotrypanothione; $T(SH)_2$) by means of an NADPH-dependent flavoenzyme, trypanothione reductase. Thus, intracellular disulfides such as GSSG are reduced according to the following scheme:

(1)	$T(S)_2 + NADPH + H^+$	---->	$T(SH)_2 + NADP^+$
(2)	$T(SH)_2 + GSSG$	---->	$T(S)_2 + 2GSH$
Sum (3)	$GSSG + NADPH + H^+$	---->	$2GSH + NADP^+$

Reaction (1) is catalyzed by trypanothione reductase and reaction (2) can proceed non-enzymatically _in vitro_ at physiological pH. It is not known whether reaction (2) is enzyme catalyzed _in vivo_ or not. The sum of these reactions (3) yields the functional equivalent of the glutathione reductase--catalyzed reaction of other organisms.

In this article we will review our current understanding of the role of trypanothione in the cell, including the properties of trypanothione reductase, the discovery of glutathionylspermidine in trypanosomatids and the enzymatic removal of H_2O_2 by trypanothione peroxidase. The implications of these findings will be assessed with respect to the mode of action of certain existing drugs and the development of new chemotherapeutic compounds.

CHEMICAL SYNTHESIS AND PROPERTIES OF TRYPANOTHIONE

Trypanothione is not available in quantity from biological sources and, consequently, is prepared by chemical synthesis. The original method [6] has been modified [7] to allow the synthesis of analogous compounds, such as the isomeric N^1- and N^8-monoglutathionylspermidines (Figure 1 [2], [3], respectively). The method, which involves coupling the glycinyl-spermidine derivatives [15], [16] or [17] to a functionally protected γ-glutamylcysteine dipeptide [4] or [18], is outlined in Figure 2.

TABLE 1.
STANDARD REDUCTION POTENTIALS OF TRYPANOTHIONE AND OTHER BIOLOGICAL COMPOUNDS[a)]

Compound	E'_o, v
Dithiothreitol	-0.330
NADPH	-0.320
Lipoic acid	-0.288
Thioredoxin	-0.260
Trypanothione	-0.242 b)
Glutathione	-0.230

[a]Data from [8]. [b]Unpublished.

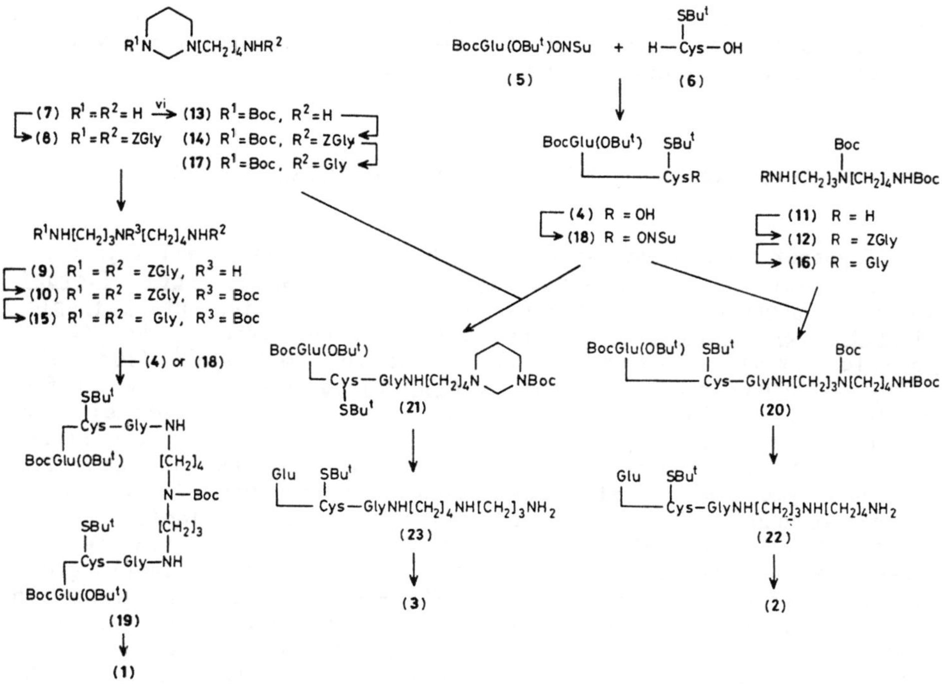

FIGURE 2. Synthesis of trypanothione disulfide and N^1-mono-and N^8-monoglutathionyl spermidine. From [7].

Space filling models of the 24-ring macrocycle, trypanothione disulfide, and the linear peptide, dihydrotrypanothione, indicate that both forms might assume the conformation of a ß-pleated sheat by the formation of three hydrogen bonds (between the carbonyl and amide functions of the peptide) and two salt bridges between the α-carboxyl and α-amino groups of each glutamate residue) [6]. However, NMR studies on trypanothione disulfide indicate that the molecule has considerable conformational flexibility in aqueous solution (Henderson, Fairlamb, Cerami and Cowburn, unpublished results).

Knowledge of the redox potential of trypanothione is important to an understanding of the role of this compound in the cell. The standard redox potential has been measured enzymatically by an equilibrium displacement technique similar to that used for the determination of the redox potential of thioredoxin [9]. From these experiments, a standard redox potential of $E_o' = -0.242 \pm 0.002$ v (n = 5) could be calculated (Fairlamb and Henderson, unpublished). This is marginally more negative than glutathione, but considerably less so than other physiological dithiols such as lipoic acid and thioredoxin (Table 1). Comparison with the redox potential (-0.330 v) of the biochemical reagent dithiothreitol (DTT) indicates that dihydrotrypanothione is a relatively poor reducing agent and thus can not simply be regarded as a 'biological DTT'.

Further research is necessary in this area, particularly to determine the non-enzymatic rates of thiol-disulfide exchange with other intracellular disulfides such as GSSG. This will be of value in predicting whether a dihydrotrypanothione : disulfide transferase is necessary for physiological rates of reduction of GSSG to occur in vivo.

PURIFICATION AND PROPERTIES OF TRYPANOTHIONE REDUCTASE

Trypanothione reductase activity has been found in all trypanosomatid species examined so far, including the parasites, T.brucei, T.cruzi and several Leishmania spp., and their free living counterparts C.fasciculata and Leptomonas seymouri. Depending on species and stage of the life cycle, specific activities range between 20 and 200 nmol min^{-1} $(10^8$ $cells)^{-1}$.

Trypanothione reductase from C. fasciculata has now been purified to homogeneity [10]. The enzyme is a disulfide-containing flavoprotein, similar in many respects to human glutathione reductase. Some of the structural and catalytic parameters of these enzymes are compared in Table 2. The oxidized enzyme shows a visible absorption spectrum (λ max = 464 nm) indicative of a flavoprotein. During reductive titration with dithionite, a new peak appears at 530 nm, indicating the formation of a charge-transfer complex between one of the newly reduced thiols and the FAD prosthetic group, as found for glutathione reductase. The NADPH-reduced enzyme is rapidly inactivated by the thiol alkylating agent, iodoacetamide, whereas the oxidized enzyme is not, suggesting that a redox-active cysteine is present in the active site. This was confirmed by labelling the active site of the NADPH--reduced enzyme with [^{14}C]iodoacetamide. A single trypsin fragment was found to have incorporated radiolabel and its amino acid sequence is shown in Figure 3. Strikingly, trypanothione reductase and human glutathione reductase show identical sequence homology through 14 residues. As in the case of human glutathione reductase, the cysteine close to the amino-terminus was exclusively labelled. The second cysteine close to the carboxyl terminus is not labelled, suggesting that it takes part in a charge-transfer complex with the FAD prosthetic group. Together, these data suggest a catalytic mechanism where electron flow from NADPH to the redox active cysteine occurs

TABLE 2.
PROPERTIES OF TRYPANOTHIONE REDUCTASE AND HUMAN GLUTATHIONE REDUCTASE

Property	Trypanothione reductase	Glutathione reductase
Flavin	FAD	FAD
pyridine nucleotide	NADPH	NADPH
M_r of monomer	53800	50000
redox active disulfide	yes	yes
charge transfer in EH_2	yes	yes
turnover no. (min^{-1})		
glutathione	3.1	12000
trypanothione	31000	9.6
active-site peptide	identical through 14 residues	

Reproduced from [10].

TR: H_3N^+----H H G P P H Y A A L G G T *C V N V G C V P K K Y----CO_2^-

GR: H_3N^+----R A A V V E S H K L G G T *C V N V G C V P K K V----CO_2^-

*Cysteine alkylated with [1-^{14}C]iodoacetamide

FIGURE 3. Amino acid sequence of active site peptides from trypanothione reductase (TR) and human glutathione reductase (GR). From [10].

via the FAD prosthetic group, as in glutathione reductase.

Although the properties of trypanothione and glutathione reductases are remarkably similar, each enzyme is specific for its respective disulfide substrate. Trypanothione reductase has an estimated turnover number of 31,000 min^{-1}, comparable to other disulfide flavoprotein dehydrogenases. Oxidized glutathione is an extremely poor substrate for trypanothione reductase, with a rate at least 10,000 times slower than that with trypanothione.

In the absence of either a full primary sequence or knowledge of the three-dimensional structure of the active site, further information can be obtained from kinetic studies using analogs of trypanothione. A series of such compounds have been synthesized and tested as substrates. (Henderson, Fairlamb, Ulrich and Cerami, unpublished). To briefly summarize these findings, it has been found that trypanothione reductase not only reduces cyclic disulfides such as trypanothione (K_m 51 µM), but will also use non-cyclic N^1-glutathionylspermidine disulphide (K_m 149 µM) and the mixed disulfide of glutathionylspermidine and glutathione (K_m 380 µM). Since these compounds could be formed _in vivo_, they may represent additional physiological substrates for the enzyme. Studies using other synthetic (non-physiological) disulfides highlight the importance of a basic amino-function on the substrate for binding to the enzyme. These findings are encouraging for the rational design and synthesis of specific inhibitors.

BIOSYNTHESIS OF TRYPANOTHIONE AND GLUTATHIONYLSPERMIDINE

Identification of glutathionylspermidine

When _C.fasciculata_ is cultured in the presence of [^{3}H]spermidine, about one third of the radiolabel is found on analysis by HPLC to be conjugated to sulfur-containing peptides (Figure 4, peaks [a], [b], [c] and [e]) and the remainder as spermidine (peak [d]). Compounds [a] and [b] could be identified as trypanothione disulfide and dihydrotrypanothione, respectively. From _in vitro_ double labelling experiments with [^{3}H]GSH and [^{14}C]spermidine, peaks [c] and [e] could be provisionally identified as glutathionylspermidine and glutathionylspermidine disulfide, respectively. This was confirmed by chemical synthesis using the method shown in Figure 2. The elution times of the N^1- and N^8- isomers were too close to unambiguously assign which isomer of glutathionylspermidine was present _in vivo_. Better resolution was obtained by separating the isomers as the dithionitrobenzoate (TNB) derivatives. Acid extracts of _C.fasciculata_ were derivatized with DTNB and analyzed by HPLC. Mixing experiments showed that N^1-glutathionylspermidine is the predominant isomer found in vivo [11].

Polyamine and thiol content of C.fasciculata

From the above radiolabelling experiments, the concentrations of trypanothione, glutathionylspermidine and total spermidine (after acid hydrolysis) can be calculated (Table 3). Note that approximately one third of the total spermidine is present in the cells as glutathionyl-spermidine conjugates. Using a more sensitive method in which extracts are derivatized with the fluorescent compound, mono-bromobimane [2], the concentrations of GSH, glutathionylspermidine and dihydrotrypanothione can be determined. In late stationary phase cells, glutathionylspermidine is the major intracellular thiol, in agreement with the above results. However, in log phase of growth, trypanothione is the predominant thiol species.

TABLE 3
THIOL AND POLYAMINE CONTENT OF C.FASCICULATA a)

Method	SPD b)	GSH	GSH-SPD	$T(SH)_2$	Growth phase
$[^3H]$spermidine	23.8	-	7.1	1.8	stationary c)
DTNB	-	-	7.4	-	stationary c)
mono-bromobimane	-	1.2	3.7	2.2	stationary d)
mono-bromobimane	-	1.3	2.5	4.2	log d)

a Concentrations in nmol min^{-1} $(10^8$ cells$)^{-1}$. b Total spermidine (SPD) after acid hydrolysis. c From [11]. d Unpublished results.

Glutathionylspermidine has previously been found in Escherichia coli [12], where it is barely detectable during the logarithmic phase of growth, but accumulates in stationary phase under acidic and anaerobic conditions, such that essentially all of the spermidine and a large proportion of the intracellular glutathione is converted to glutathionylspermidine. On dilution into fresh medium, glutathionylspermidine is hydrolyzed within 5 min to spermidine and glutathione. It has been suggested that glutathionylspermidine may serve as a means of sequestering these metabolites in an inactive form [12]. This would seem not to be so for C.fasciculata, since considerable amounts of glutathionylspermidine are present during all phases of growth (Table 3). However, the ratio of trypanothione to glutathionylspermidine in C.fasciculata does change appreciably during growth and thus may play an important regulatory role. Large quantities of glutathionylspermidine are also present in procyclic forms of T.brucei, but not in bloodstream forms, suggesting that glutathionylspermidine is associated with differential gene expression during the life cycle of this organism.

Biosynthetic pathway to trypanothione

Glutathione and spermidine can be converted in vitro by dialyzed extracts of C.fasciculata into N^1-glutathionylspermidine and trypanothione [11]. The reaction requires ATP and can not be replaced by GTP, CTP, UTP or ITP. Maximal incorporation occurs with phophoenolpyruvate and pyruvate kinase as ATP generating system. To test whether N^1- or N^8-glutathionylspermidine were intermediates of trypanothione biosynthesis, these compounds were substituted for spermidine in the incubation mixtures. Activity was measured by estimating the incorporation of [^{3}H]GSH into trypanothione and glutathionylspermidine by ion exchange chromatography and scintillation counting. Figure 5 shows that both N^1- and N^8-glutathionylspermidine will replace spermidine in the reaction and that ATP is essential for activity. Initial rates for incorporation are 2.2, 1.0 and 1.6 nmol min^{-1} mg^{-1} for spermidine, N^1- and N^8-glutathionylspermidine, respectively. The distribution of [^{3}H]GSH into products was measured by HPLC. With N^1- or N^8-glutathionylspermidine as substrates, 93 and 92 % of the products were found to be trypanothione, with the balance in N^1-glutathionylspermidine, in both cases. In contrast, with spermidine as substrate, 46 % of the product is N^1-glutathionylspermidine and 54 % trypanothione. Only trace amounts of the N^8-isomer of glutathionylspermidine become radiolabelled in these experiments, as found in vivo. It is not clear at present whether incorporation of [^{3}H]GSH into initially unlabelled N^1-glutathionylspermidine is due to exchange catalyzed by the 'trypanothione synthetase' enzyme(s) or whether hydrolysis, releasing free spermidine, followed by reincorporation has occurred. This question is best resolved by purification of the enzyme activity. Also, it is not known whether more than one enzyme is involved in the synthesis of N^1-glutathionylspermidine and trypanothione. However, partially purified preparations

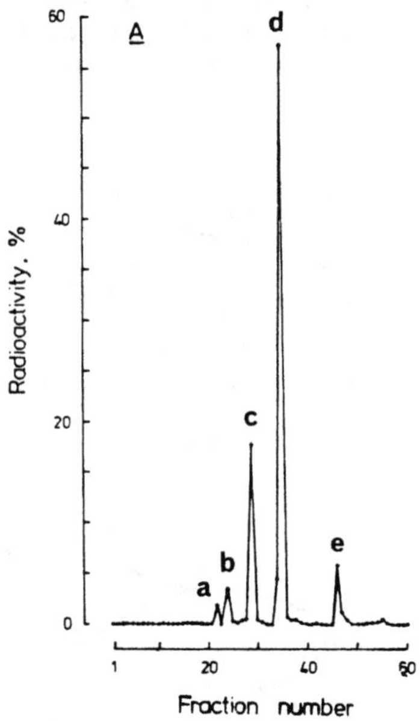

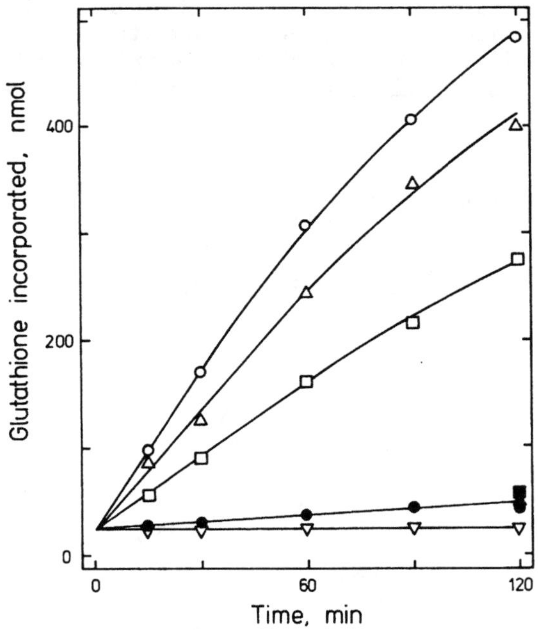

FIGURE 4. Distribution of radiolabel in extracts of C.fasciculata cultured with [^{3}H]spermidine. From [11].

FIGURE 5. Incorporation of [^{3}H]GSH into spermidine-peptides by extracts of C.fasciculata. Open symbols represent the complete assay system plus spermidine [o], N^1- [□] or N^8-glutathionylspermidine [△]. Closed symbols: ATP omitted. No incorporation was found if the enzyme extract was omitted [▽]. Reproduced from [11].

continue to catalyze the formation of both metabolites. The biosynthetic pathway to trypanothione is illustrated Figure 8 below.

PEROXIDE METABOLISM IN TRYPANOSOMATIDS

Previous studies have suggested that trypanosomes would appear to be deficient in the enzymatic defence mechanisms for the removal of H_2O_2, since glutathione peroxidase and catalase are either completely lacking or present only in trace amounts (reviewed in [4]). In mammalian cells, low intracellular concentrations of H_2O_2 are maintained by the glutathione peroxidase/glutathione reductase cycle (Figure 7). Since trypanosomatids have replaced glutathione reductase with trypanothione reductase, we investigated whether glutathione peroxidase of other organisms had been replaced with an analogous trypanothione peroxidase.

Peroxidase activity was measured in a coupled spectrophotometric assay, containing either glutathione reductase (GR) and GSH or trypanothione reductase (TR) and dihydrotrypanothione ($T(SH)_2$) (Figure 6). When dialyzed C.fasciculata extract was incubated with GR and GSH, no net increase in the rate of NADPH oxidation was observed on addition of t-butylhydroperoxide, indicating the absence of glutathione peroxidase (Figure 6, trace A). If trypanothione reductase and $T(SH)_2$ are substituted for GR and GSH, an apparent trypanothione peroxidase activity is observed (traces B, C and D). Trypanothione reductase itself contains no inherent peroxidase activity (trace B); only on addition of C.fasciculata extract is activity observed. Traces C and D show that

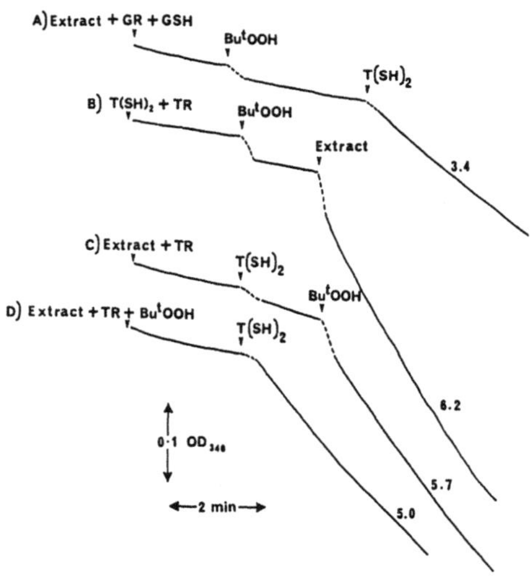

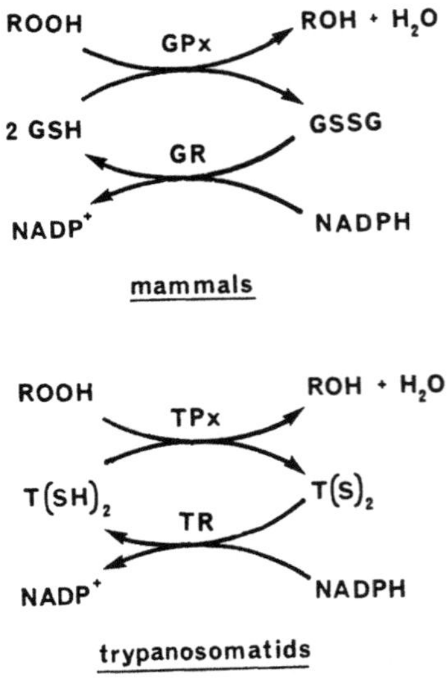

FIGURE 6. Trypanothione dependent peroxidase activity in C.fasciculata. Activity was measured spectrophotometrically at 340 nm as described in the text. The numbers refer to activity in nmol min^{-1}.

FIGURE 7. Removal of peroxides (ROOH) in mammalian cells and trypanosomatids

both dihydrotrypanothione and t-butylhydroperoxide are required for activity. Enzyme preparations from C.fasciculata will also use cumene hydroperoxide in place of t-butyl hydroperoxide, with specific activities of 4.0 and 5.2 nmol min^{-1} $(10^8\ cells)^{-1}$, respectively. Similar experiments have shown that bloodstream T.brucei contains trypanothione peroxidase with specific activities of 1.0 and 0.4 nmol min^{-1} $(10^8\ cells)^{-1}$ at 25 °C for t-butyl hydroperoxide and H_2O_2, respectively [Henderson, Fairlamb and Cerami, unpublished]. This is in contrast to a recent report stating that trypanothione-dependent peroxidase activity is absent from T.brucei [13]. The same authors note that intact bloodstream T.brucei are able to metabolize exogenously added H_2O_2 at a maximum rate of 2.6 nmol min^{-1} $(10^8\ cells)^{-1}$ at 37 °C. Allowing for the difference in temperature at which these studies were done, our results can therefore account for about one third of their reported activity. This may be an underestimate as we have not yet optimized our assay conditions.

These present findings now establish a definitive role for dihydrotrypanothione in the cell - protection against oxidative damage by removal of H_2O_2 by the trypanothione peroxidase / reductase cycle (Figure 7).

INTERACTION OF ARSENICALS WITH TRYPANOTHIONE

As early as 1909, Ehrlich proposed that trivalent arsenical compounds were selectively toxic to parasites by virtue of their ability to react with essential sulfhydryl groups [reviewed in 14]. Since that time, many studies have demonstrated that trivalent arsenicals form considerably more stable complexes with dithiols rather than with simple monothiols. Dihydrotrypanothione would therefore seem to be an ideal candidate for interaction with arsenical drugs. We have synthesized the adduct of dihydrotrypanothione and the trypanocidal drug, melarsen oxide. The purified derivative has the structure given in Figure 8, consistant with amino acid and mass spectrometric analysis [Fairlamb, Henderson and Cerami, unpublished]. On HPLC the derivative can be partially resolved into a pair of diastereoisomers, because of the asymmetry of the N^4-secondary amine in dihydrotrypanothione. The complex has a stability constant of 1.05 x 10^7 M^{-1}, and is more stable than those of glutathione or cysteine and melarsen oxide. When intact bloodstream forms of T.brucei are exposed to melarsen oxide, the derivative

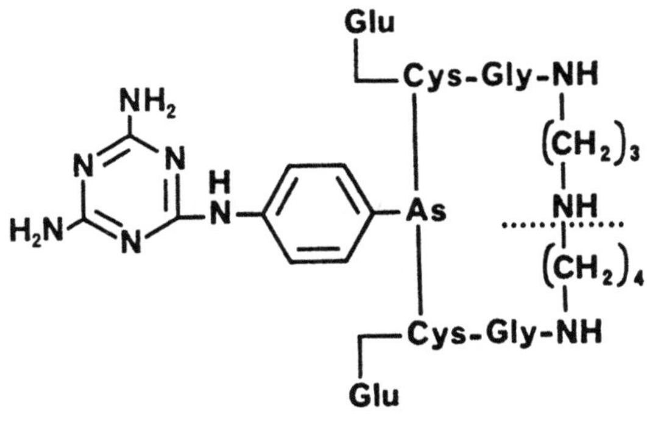

FIGURE 8. Structure of melarsen oxide derivative with dihydrotrypanothione.

is rapidly formed within the cell, reaching a maximum intracellular concentration of about 20 μM. No other forms of melarsen oxide are detectable in these extracts. Thus, the trypanothione derivative of melarsen oxide (MEL T) represents the principal acid-soluble form of the arsenical drug _in vivo_. MEL T is also an effective inhibitor of trypanothione reductase from _C.fasciculata_ (K_i = 28 μM) [Henderson, Fairlamb and Cerami, unpublished]. In the light of these findings, the mechanism by which trivalent arsenicals exert their selective effect requires reassessment.

TRYPANOTHIONE METABOLISM AS A TARGET FOR CHEMOTHERAPY

Our current understanding of the metabolism and functions of trypanothione is shown in Figure 9. The role of trypanothione in trypanosomatids would appear to mirror that of glutathione in other cells, in particular in maintaining intracellular thiol-redox balance and in eliminating peroxides. No doubt other functions of trypanothione remain to be discovered. Why these organisms should have evolved this appparently more elaborate metabolic

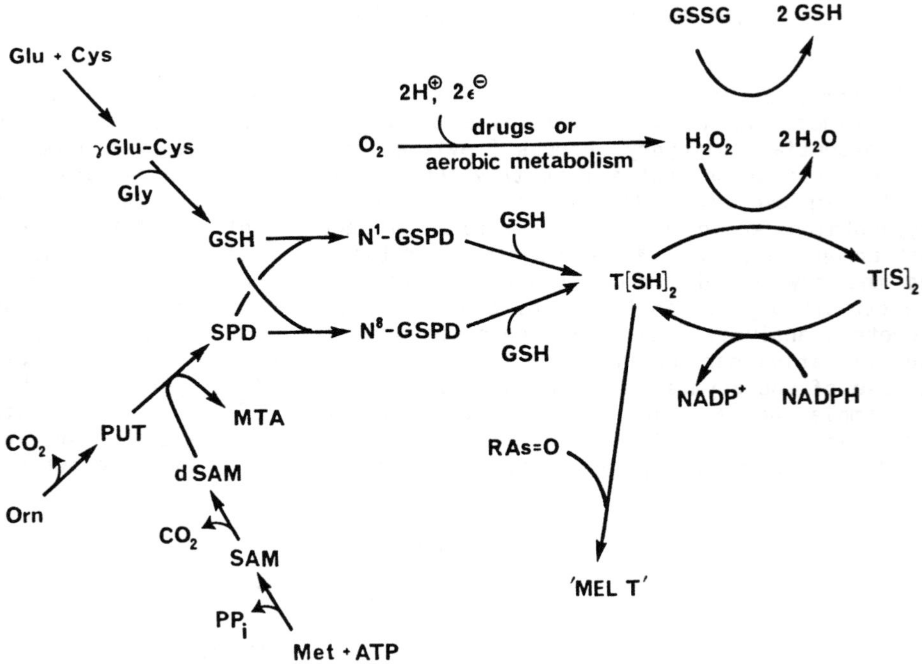

FIGURE 9. Metabolism and fuctions of trypanothione.

system is still a mystery. However, the trypanothione system presents a number of targets for 'rational' drug development. Three novel enzymes, discussed in this review, trypanothione reductase, trypanothione synthetase and trypanothione peroxidase are unique to the parasites. Thus, inhibition of any of these enzymes might be expected to compromise the parasites' ability to deal with oxidative stress, generated internally either as part of the organisms' own aerobic metabolism, or drug induced and generated externally as part of the host's immune response. Our present evidence, some of which is presented here, shows that two important drugs used in the treatment of African sleeping sickness (aromatic arsenicals and difluoromethylornithine; DFMO) clearly perturb trypanothione metabolism and function and that these effects may be involved in their selective toxicity. It may also offer an explanation for the synergistic effect of DFMO in combination with a wide range of apparently unrelated drugs [15].

In conclusion, it should be emphasised that there is an urgent need for new, effective drugs for the treatment of African sleeping sickness, Chagas' disease and leishmaniasis. We believe that the trypanothione system represents a particularly important target for further chemotherapeutic development, which could lead to a single drug active against all these diseases.

ACKNOWLEDGEMENTS

This work was supported by grants from the UNDP/World Bank/WHO Special Programme for Research and Training in Tropical Diseases, the National Institutes of Allergy and Infectious Diseases and the Rockefeller Foundation.

REFERENCES

1. Meister A and Anderson M (1983) Glutathione. Ann Rev Biochem 52:711-760
2. Fahey R C and Newton G L (1983) Occurrence of low molecular weight thiols in biological systems. In: Larsson A et al (eds) Functions of Glutathione: Biochemical, Physiological, Toxicological and Clinical Aspects. Raven Press New York, pp 251-260
3. Ziegler D M (1985) Role of reversible oxidation-reduction of enzyme thiols-disulfides in metabolic regulation. Ann Rev Biochem 54:305-329
4. Docampo R and Moreno S N J (1984) Free-radical intermediates in the antiparasitic action of drugs and phagocytic cells. In: Pryor W A (ed) Free Radicals in Biology. Academic Press New York, pp 243-288
5. Fairlamb A H and Cerami A (1985) Identification of a novel thiol-containing co-factor essential for glutathione reductase activity in trypanosomatids. Mol Biochem Parasitol 14:187-198
6. Fairlamb A H, Blackburn P, Ulrich P, Chait B T and Cerami A (1985) Trypanothione: a novel bis(glutathionyl)spermidine cofactor for glutathione reductase in trypanosomatids. Science 227:1485-1487
7. Henderson G B, Ulrich P, Fairlamb A H and Cerami A (1986) Synthesis of the trypanosomatid metabolites trypanothione, N^1-mono- and N^8-mono-glutathionylspermidine. J Chem Soc (Chem Commun) 593-594
8. Segel I H (1976) Biochemical Calculations, 2nd edn. Wiley New York, pp 414-415
9. Porque P G, Baldesten A and Reichard P (1970) Purification of a thioredoxin system from yeast. J Biol Chem 245:2363-2370
10. Shames S L, Fairlamb A H, Cerami A and Walsh C T (1986) Purification and characterization of trypanothione reductase from C.fasciculata: a new member of the family of disulfide-containing flavoprotein reductases. Biochemistry 25:3519-3526
11. Fairlamb A H, Henderson G B and Cerami A (1986) The biosynthesis of of trypanothione and N^1-glutathionylspermidine in C.fasciculata. Mol Biochem Parasitol: in press

12. Tabor H and Tabor C W (1975) Isolation, characterization and turnover of glutathionylspermidine from Escherichia coli. J Biol Chem 250:2648-2654
13. Penketh P G and Klein R A (1986) Hydrogen peroxide metabolism in Trypanosoma brucei. Mol Biochem Parasitol 20:111-121
14. Albert A (1973) Selective Toxicity. Wiley New York, pp 130-172
15. McCann P P, Bacchi C J, Nathan H C and Sjoerdsma A (1983) Difluoromethylornithine and the rational development of polyamine antagonists for the cure of protozoan infection. In: Singer T P and Ondarza R N (eds) Mechanisms of Drug Action. Academic Press, New York, pp 159-173

DIFFERENTIATION OF *Trypanosoma brucei* FROM BLOODSTREAM TO PROCYCLIC TRYPOMASTIGOTES

Peter Overath, Roland Bülow, Joachim Czichos and Bernhard Ehlers

Max-Planck-Institut für Biologie, Corrensstrasse 38, D7400 Tübingen, Federal Republic of Germany.

The life cycle of the unicellular flagellate, *Trypanosoma brucei*, in the mammalian host and the tsetse fly proceeds through a series of morphological states which are accompanied by profound alterations in ultrastructure and metabolism. The developmental step most extensively studied is the differentiation of bloodstream to procyclic trypamostigotes which occurs in the midgut of the fly after uptake with the blood meal. This transition, also termed transformation, is initiated in the mammal by the differentiation of dividing bloodstream forms with slender morphology to non-dividing, stumpy cells. The latter cells are considered to convert most readily to dividing procyclic cells in the midgut of the fly. Several culture media have been devised which allow a more or less synchronous transformation of either pleomorphic populations of bloodstream forms or monomorphic populations which have a uniformly slender morphology and arise from pleomorphic strains by frequent passage in rodents.

Among the many differences between bloodstream forms and procyclic cells two are particularly noticeable. First, the bloodstream forms are covered by a surface coat consisting of the variant surface glycoprotein (VSG). This coat is absent in procyclic cells. Second, in slender bloodstream forms metabolic energy is entirely derived from glycolysis and the mitochondrion is poorly developed. In stumpy forms, certain mitochondrial dehydrogenases can be detected and the presence of these enzymes is considered to be important for prompt transformation. Procyclic cells contain a fully developed mitochondrion and derive their energy mostly from oxidative phosphorylation.

In this article, three aspects of the differentiation of bloodstream forms to procyclic cells will be discussed. First, the signals triggering transformation are considered. In the second part, the regulation of VSG gene expression and the kinetics of mitochondrial development is examined. The third part concerns the possible mechanism which leads to the release of the surface coat during transformation.

The biology, molecular genetics, biochemistry as well as techniques of *in vitro* culture of African trypanosomes are covered in recent reviews by Vickerman (1985), Borst (1986), Fairlamb & Opperdoes (1986) and Brun & Jenni (1985).

NATO ASI Series, Vol. H11
Host-Parasite Cellular and Molecular Interactions
in Protozoal Infections. Edited by K.-P. Chang and D. Snary

Signalling transformation

Brun and Schönenberger (1981) discovered that intermediates of the citric acid cycle, notably citrate and *cis*-aconitate, dramatically stimulate the *in vitro* transformation of a pleomorphic population of bloodstream forms (Fig.1, upper part). These observations have been confirmed and extended

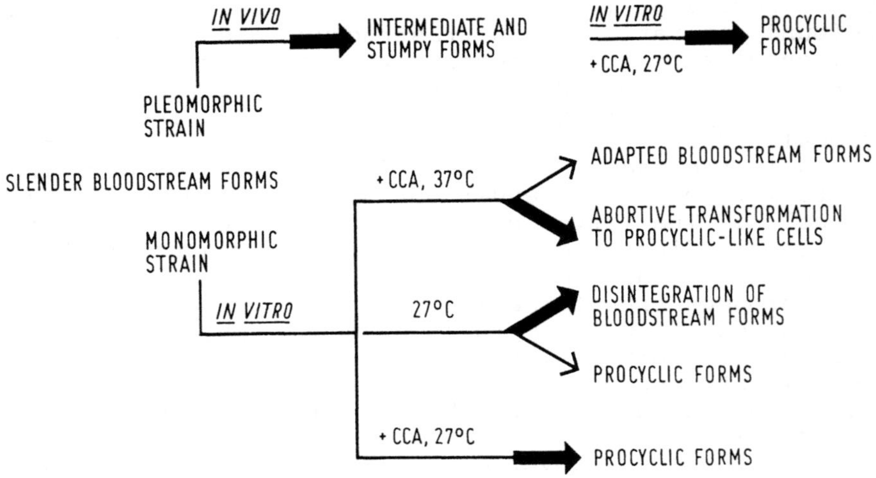

FIGURE 1. Citrate/*cis*-aconitate (CCA) and a temperature reduction as signals for differentiation. Bold arrows indicate major pathways. For further details see Overath *et al*., 1986.

using the monomorphic variant clone MITat 1.4 (117) of *T. brucei* stock 427 and a medium adequate for growth of both bloodstream forms and the newly transformed procyclic cells (Fig.1 lower part; Czichos *et al*., 1986; Overath *et al*., 1986). Synchronous transformation requires the combined action of two signals: the addition of *cis*-aconitate (6 mM) or CCA, a combination of citrate (3 mM) and *cis*-aconitate (3 mM), and a reduction in temperature from 37°C to 27°C. Under these conditions the cells transform within 24 h leading to an exponentially growing culture of procyclic cells. The response of the monomorphic parasite population to only one of the transformation signals is more complex. When exposed to CCA at 37°C the cells have the choice of adapting or undergoing an "abortive" transformation to non-dividing procyclic-like cells which subsequently disintegrate. CCA-adapted cells are indistinguishable in their growth characteristics and their respiratory properties from non-adapted bloodstream forms. Upon exposure to the other transformation signal, a reduction in temperature in the absence of CCA, the cells have

likewise the possibility to respond in two ways. The majority retain the morphology of bloodstream forms for several days, before finally disintegrating. In contrast, a small fraction transform successfully and then divide resulting eventually in an established culture of procyclic cells. The behavior at 27°C in the absence of CCA appears similar to that observed after uptake of clone MITat 1.4 with a bloodmeal into the midgut of the tsetse fly. Under these conditions most bloodstream forms disintegrate within the first hours after uptake and only a few cells transform successfully leading to a fulminating midgut infection within 10 days.

The mechanism by which the citric acid cycle intermediates stimulate the differentiation step remains unknown. Most likely, there exists no equivalent for this effective but artificial signal in the fly. In principle, a midgut infection can be initiated from a single transformed cell. Therefore, while synchronous differentiation is indispensable for biochemical studies *in vitro* it is not an essential requirement for cyclic transmission through the fly.

Repression of variant surface glycoprotein synthesis and mitochondrial development.

Rapid repression of VSG synthesis is a characteristic event when transformation is triggered by the addition of CCA and a reduction in temperature (Fig.2, Overath *et al.*, 1983; Ehlers *et al.*, 1986).[1] Approach to steady-state labeling and pulse-chase experiments using [^{3}H]adenine as the RNA precursor have shown that the mRNA coding for VSG is synthesized at a high

[1]The initial transformation protocol (Overath *et al.*, 1983) involved growth of bloodstream forms taken from mouse blood for 17 h at 37°C in MEM medium + 15% inactivated rabbit serum + CCA and subsequent transfer to medium SDM-79 + 10% inactivated bovine serum + CCA at 27°C. Thereafter, Czichos *et al.*, 1986, demonstrated that the entire experiment could be performed in MEM medium + 15% inactivated horse or fetal bovine serum and that CCA could be added at the start of the experiment or at the time of the reduction in temperature. Because CCA causes a variable repression of VSG synthesis (20-80%) during the incubation at 37°C (Overath *et al.*, 1986) the addition of the acids at the time of the temperature shift is the preferred protocol for studies on the kinetics of repression of the VSG gene (Ehlers *et al.*, 1986).

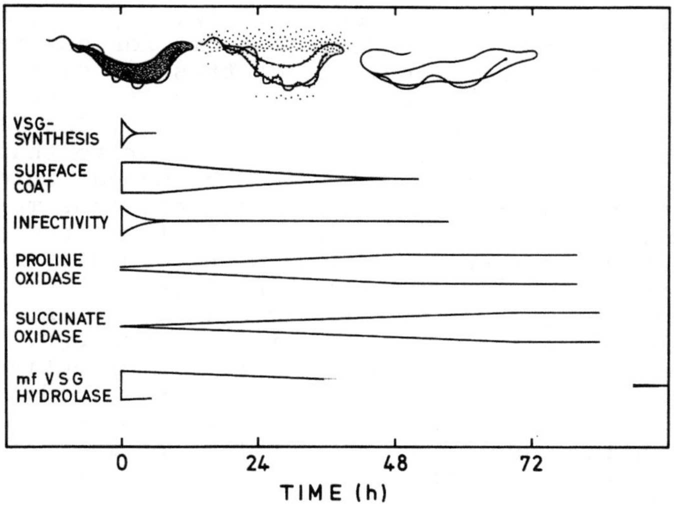

FIGURE 2. Developmental changes during transformation of bloodstream forms to procyclic cells. At time 0 h bloodstream forms grown overnight at 37°C in the presence of CCA were shifted to 27°C. For further details see Overath et al., 1983, 1986 and Bülow & Overath, 1985.

rate in bloodstream forms. VSG mRNA is moderately stable having a half-life, $t_{1/2}$ = 4.5 h (Fig.3A,D). Triggering transformation causes an immediate reduction in the rate of transcription by a factor of 20 and a 4-fold reduction in transcript stability ($t_{1/2}$ = 1.2 h, Fig.3B,D). One possible explanation for this shut-off is that CCA causes rapid changes in the concentration of intracellular ligands which modulate the activity of the VSG gene by interacting with putative transcription factors. The new VSG mRNA steady-state is 100-fold lower than in bloodstream forms. Within 24 h the VSG mRNA vanishes to the undetectably low level characteristic for procyclic cells (Fig.3C). Interestingly, cells treated with CCA at 27°C for up to 8 h can rapidly resume transcription and translation of the VSG gene upon removal of CCA and readjustment of the temperature to 37°C. Cells treated at 27°C with CCA for 30 h cannot reactivate VSG synthesis. At this time the cells are irreversibly committed to differentiation and the presence of CCA is no longer required. Therefore, transformation entails an initial reversible phase and a second phase in which return to the initial state is no longer possible (Czichos et al., 1986).

Whereas previous transformation protocols produced highly variable results regarding the kinetics of formation of mitochondrial functions (cf. summary in Fairlamb & Opperdoes, 1986), CCA-stimulated differentiation produces within 72 h a population of procyclic cells with properties very similar to

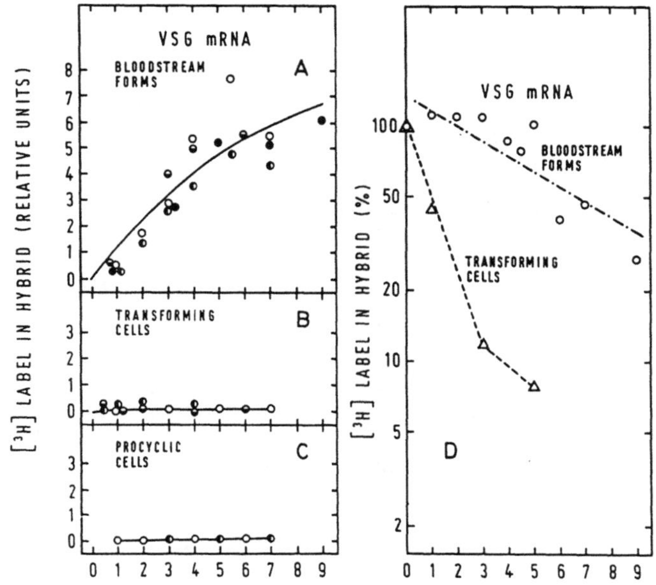

FIGURE 3. Kinetics of incorporation of [^{3}H]adenine into VSG mRNA in bloodstream forms at 37°C (A), in transforming cells at 27°C + CCA (B), and in procyclic cells at 27°C (C) Part D shows the decay of label in VSG mRNA in bloodstream forms at 37°C and in transforming cells at 27°C + CCA. For further details see Ehlers _et al._, 1986.

those of established procyclic cells. In particular, the cells rapidly acquire proline and succinate oxidase activity (Fig.2) and the rate of glucose oxidation declines to 20% of that observed in bloodstream forms. The development of a functional respiratory chain during transformation can be deduced from the increasing percentage of respiration which becomes sensitive to CN^- after inhibition of the glycerol-3-phosphate oxidase by salicylhydroxamic acid i.e. 49% at 24 h, 79% at 48 h, 75% at 72 h for glucose as subtrate. Under the same conditions 70% of glucose-stimulated respiration is CN^--sensitive in established procyclic cells. Interestingly, Haston (1975) demonstrated by microrespirometric studies on single transforming trypanosomes taken from the midgut of flies that changes in the rate of respiration of glucose, proline, and succinate occurred within 72 h after uptake with the bloodmeal. Therefore, mitochondrial development _in vitro_ and _in vivo_ occurs with similar kinetics.

Membrane-form VSG hydrolase and coat release

The variant surface glycoprotein is anchored in the cytoplasmic membrane of bloodstream forms by a 1,2-dimyristoylphosphatidylinositol residue linked _via_ a complex carbohydrate to the carboxy terminus of the polypeptide chain (Ferguson et al., 1985a). Upon cell rupture, this membrane-form VSG (mfVSG) is converted within minutes to a soluble form (sVSG) and dimyristoylglycerol by an endogenous phospholipase C-like hydrolase (Cardoso de Almeida & Turner, 1983; Ferguson et al., 1985b; Bülow & Overath, 1985). Possibly, this hydrolase plays a role in the release of the surface coat during transformation.

Using monoclonal antibodies as specific affinity reagents mfVSG hydrolase was purified to homogeneity from the membrane fraction of trypanosome lysates (Bülow & Overath, 1986). The enzyme (M_{SDS} = 39 000) is highly specific for mfVSG. Quantitative immunoprecipitation experiments demonstrate that a monoclonal antibody can precipitate all the mfVSG hydrolase activity present in detergent solubilized trypanosomes. Therefore, this enzyme appears to be the only activity in the cell which converts mfVSG to sVSG and diacylglycerol. Finally, reconstitution experiments into phospholipid vesicles show that the enzyme can hydrolyse mfVSG present in the same but not in a separate lipid vesicle. Most likely, the reaction can only occur when enzyme and substrate are exposed on the same face of the bilayer.

Several experiments indicate that the enzyme is located in intracellular vesicles rather than in the plasma membrane. First, cell fractionation studies performed in collaboration with F.R. Opperdoes (ICP, Brussels) demonstrate that the distribution of mfVSG hydrolase activity in sucrose gradients is different from that of typical plasma membrane enzymes. Second, treatment of bloodstream forms with trypsin causes essentially quantitative cleavage of mfVSG within 10 min and, subsequently, the rounding up of the cells in a way reminiscent of protoplast formation in bacteria. As shown in Fig. 4 enzyme activity is not affected by these conditions while it is rapidly inactivated after cell lysis. Therefore, the bulk of the hydrolase appears not accessible to trypsin cleavage at the cell surface. Third, immunofluorescence staining of cells briefly fixed with formaldehyde suggests that the enzyme resides in intracellular vesicles, many in apparently close association with the plasma membrane (Fig.5) although this result has yet to be confirmed by immunoelectronmicroscopy. In summary, the majority of the enzyme appears to reside in the cell rather than at the cell surface.

The following experimental observations are pertinent for the possible mechanisms of coat release during transformation. First, bloodstream forms growing exponentially *in vitro* slowly release VSG in the form of sVSG at a rate of about 1% per hour (R. Bülow, unpublished experiments). Second, pulse chase experiments on transforming cells of variant clones MITat 1.4 (Overath et al., 1983) and ILTat 1.3 (Shapiro, 1986) demonstrate that the coat is shed into the medium as sVSG and as a 6-10 000 Da smaller, proteolytic fragment lacking the cross-reacting determinant at the C-terminus. Third, the time course of the disappearance of the surface coat in a transforming trypanosome population (Fig.2) suggests that a single cell may require up to 24 h to strip its coat.

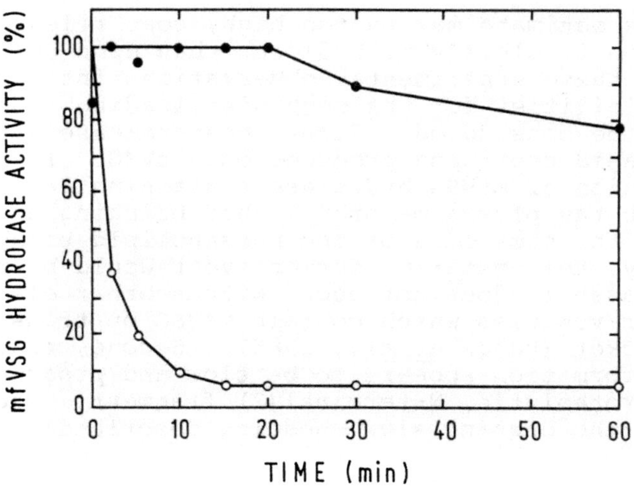

FIGURE 4. Treatment of living (●——●) or lysed MITat 1.2 (221) bloodstream forms (o——o) with trypsin (50 µg/ml) in Iscove's medium at 37°C. At various times samples were processed for the determination of mfVSG hydrolase activity (R. Bülow, unpublished experiments).

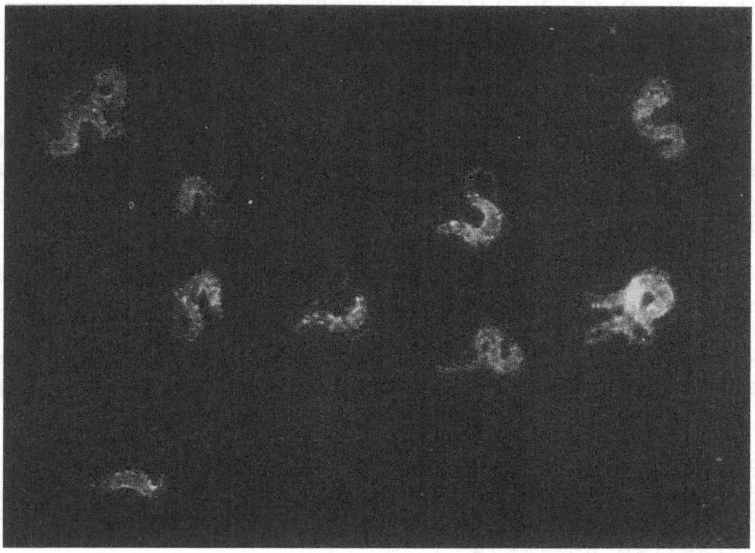

FIGURE 5. Localization of mfVSG hydrolase in bloodstream forms fixed for 10 min at 25°C with 2% formaldehyde using monoclonal antibody mAT3 and the biotin-streptavidin system (R. Bülow, unpublished results).

Although this estimate may be too high, coat release during transformation is clearly much slower than upon cell rupture.

In taking these experimental observations into account several possibilities for the mechanism leading to coat release must be considered. First, coat release after cell lysis is a rapid event and produces only sVSG. It may be caused by fusion of mfVSG hydrolase containing intracellular vesicles with the plasma membrane, thus bringing enzyme and substrate to the same face of the phospholipid bilayer. Interestingly, this putative fusion event would be rather specific because it does not occur with another class of intracellular vesicles which contain mfVSG on transit to the flagellar pocket (Bangs et al., 1986). Second, coat release during transformation appears to be slow and produces both sVSG and a proteolytic, N-terminal(?) fragment. The sVSG may be formed by the ongoing slow shedding described above for bloodstream forms which may involve a minor and as yet undetected fraction of mfVSG hydrolase at the surface. The proteolytic fragment could be formed inside the cell after endocytosis of mfVSG and then secreted via the flagellar pocket. Alternatively, a specific exocytotic event could expose a protease on the external face of the flagellar pocket.

In summary, there is little doubt that the mfVSG hydrolase is responsible for the surface coat release upon cell lysis. However, there is no convincing evidence as yet that this enzyme plays a specific role during differentiation of bloodstream to procyclic trypomastigotes.

REFERENCES

Bangs JD, Andrews NW, Hart GW, Englund PT (1986) Post translational modification and intracellular transport of a trypanosome variant surface glycoprotein. J Cell Biol 103:255-263.

Borst P (1986) Discontinuous transcription and antigenic variation in trypanosomes. Ann Rev Biochem 55:701-732.

Brun R, Jenni L (1985) Cultivation of African and South American trypanosomes of medical and veternary importance. Br Med Bull 41:122-129.

Brun R, Schönenberger M (1981) Stimulating effect of citrate and cis-aconitate on the transformation of Trypanosoma brucei bloodstream forms to procyclic forms in vitro. Z Parasitkd 66:17-24.

Bülow R, Overath P (1985) Synthesis of a hydrolase for the membrane-form variant surface glycoprotein is repressed during transformation of Trypanosoma brucei. FEBS Lett. 187:105-110.

Bülow, R, Overath P (1986) Purification and characterization of the membrane-form variant surface glycoprotein hydrolase of *Trypanosoma brucei*. J Biol Chem 261:11918-11923.

Cardoso de Almeida ML, Turner MJ (1983) The membrane form of variant surface glycoproteins of *Trypanosoma brucei*. Nature 302:349-352.

Czichos J, Nonnengaesser C, Overath P (1986) *Trypanosoma brucei*: *cis*-Aconitate and temperature reduction as triggers of synchronous transformation of bloodstream to procyclic trypomastigotes *in vitro*. Exp Parasitol 62:283-291.

Ehlers B, Czichos J, Overath P (1986) RNA turnover in *Trypanosoma brucei*. Mol Cell Biol submitted for publication.

Fairlamb AH, Opperdoes FR (1986) Cabohydrate metabolism in African trypanosomes, with special reference to the glycosome. In: Morgan MJ (ed) Carbohydrate metabolism in cultured cells. Plenum Publishing Corporation, New York.

Ferguson MAJ, Low MG, Cross GAM (1985a) Glycosyl-*sn*-1,2-dimyristylphosphatidylinositol is covalently linked to *Trypanosoma brucei* variant surface glycoprotein. J Biol Chem 260:14547-14555.

Ferguson MAJ, Haldar K, Cross GAM (1985b) *Trypanosoma brucei* variant surface glycoprotein has an sn-1,2-dimyristylglycerol membrane anchor at its COOH terminus. J Biol Chem 260:4963-4968.

Haston WS (1975) Studies on the early development of *Trypanosoma brucei* in the vector *Glossina morsitans*. PhD thesis, Glasgow University.

Overath P, Czichos J, Stock U, Nonnengaesser C (1983) Repression of glycoprotein synthesis and release of surface coat during transformation of *Trypanosoma brucei*. EMBO J 2:1721-1728.

Overath P, Czichos J, Haas C (1986) The effect of citrate/*cis*-aconitate on oxidative metabolism during transformation of *Trypanosoma brucei*. Eur J Biochem 160:175-182.

Shapiro SJ (1986) *Trypanosoma brucei*: Release of variant surface glycoprotein during the parasite life cycle. Exp Parasitol 61:432-437.

Vickerman K (1985) Developmental cycles and biology of pathogenic trypanosomes. Br Med Bull 41:105-114.

DIGESTIVE ENZYMES, RECEPTOR-MEDIATED ENDOCYTOSIS AND THEIR ROLE IN THE NUTRITION OF THE BLOODSTREAM-FORM TRYPANOSOME.

Fred R. Opperdoes*, Isabelle Coppens*+ and Pierre Baudhuin*+

*International Institute of Cellular and Molecular Pathology and +Laboratoire de Chimie Physiologique, Université Catholique de Louvain, B-1200 Brussels, Belgium

INTRODUCTION

Many types of eukaryotic cells are capable of taking up macromolecules from the external environment. Pinocytosis as well as receptor-mediated endocytosis lead to the invagination of the plasma membrane and the formation of endocytotic vesicles. The latter then fuse with primary lysosomes, resulting in the formation of phagolysosomes or digestive organelles.

Electron-microscopic studies on several representatives of the Trypanosomatidae have shown that these protozoan parasites are also capable of the uptake of macromolecules. Ingestion of protein from a ferritin-containing medium was first demonstrated for Trypanosoma mega (Steinert and Novikoff, 1960) and similar experiments subsequently demonstrated the existence of pinocytosis in the case of T. rhodesiense (Brown et al., 1965), T. brucei (Jadin, 1971) and T. cruzi (Jadin, 1971). Langreth and Balber (1975) have carried out detailed ultrastructural studies on T. brucei and have shown that macromolecular material first accumulates in the flagellar pocket where it enters large spiny-coated vesicles. Subsequently such vesicles become continuous with a collecting membrane system and eventually the foreign material ends up in large digestive vacuoles (Langreth and Balber, 1975).

NATO ASI Series, Vol. H11
Host-Parasite Cellular and Molecular Interactions
in Protozoal Infections. Edited by K.-P. Chang and D. Snary

Biochemical evidence in support of such a scenario is also available. Several acidic hydrolases, associated with membrane-bounded organelles, resembling the lysosomes of other organisms, have been described in the case of T. brucei (Steiger et al, 1980). Fairlamb and Bowman (1980a) described the uptake of the inert macromolecular compound polyvinyl pyrrolidone (PVP) by T. brucei at a rate compatible with fluid endocytosis, while the lysosomotropic trypanocidal drug Suramin was taken up at an 18-fold higher rate, suggestive of a more specific type of uptake (Fairlamb and Bowman, 1980b). Finally, Opperdoes and Van Roy (1982b) have shown that T. brucei takes up the non-degradable macromolecular detergent Triton WR-1339 and accumulates it into phagolysosomes containing the hydrolases acid proteinase and α-D-mannosidase.

In this paper we summarize data, partly already published, partly new, that together provide information on the utilization of substrates by the African trypanosome T. brucei and its capability of uptake and digestion of macromolecular material through a mechanism of receptor-mediated endocytosis.

THE PLASMA MEMBRANE

The plasma membrane of the Trypanosomatidae is thightly associated with an array of sub-pellicular microtubules. Contrary to the situation with the mammalian microtubules the sub-pellicular microtubules of the Trypanosomatidae are stable and do not collapse upon cell rupture and subsequent cell-fractionation experiments. Since the association of the plasma membrane with microtubules gives it a high density (1.20 - 1.25 $g.cm^{-3}$) in sucrose gradients (Steiger et al, 1980; Opperdoes and Steiger, 1981), this has allowed us to unambiguously identify the plasma membrane in subcellular fractions of T. brucei and to separate it from the flagellar pocket membrane.

α-D-glucosidase

This enzyme has been used by us as an unambiguous surface marker in subcellular fractionation experiments (Steiger et al., 1980). It is highly accessible by its substrate in intact bloodstream forms, virtually non-latent in cytoplasmic extracts and the majority remains membrane-bound and is sedimentable. The enzyme is easily solubilized after treatment with Triton X-100, suggesting a loose attachement to the surface. α-D-glucosidase has a neutral pH optimum. In the insect stage of *T. brucei* the enzyme is equally active (0.5 nmole min^{-1} mg^{-1}), but, although still associated with the plasma membrane, completely latent when measured on intact-cell preparations, while freely accessible in cytoplasmic extracts (Opperdoes and Steiger, 1981). These observations are compatible with an association with the outer face of the pellicular membrane in the bloodstream form and with the inner face in the insect stage. It was suggested that, in view of its alternating between the two faces of the plasma membrane and the fact that the enzyme is not found in the monoxeneous insect stage Trypanosomatid *Crithidia luciliae*, α-glucosidase might fulfil a role in the turn--over of the surface-coat glycoprotein (Opperdoes and Steiger, 1981).

3'-Nucleotidase

This enzyme has extensively been described in the case of *Leishmania donovani* (Gottlieb and Dwyer, 1983), *Crithidia fasciculata* (Gottlieb, 1985) and the African trypanosomes (McLaughlin, 1982; Gottlieb et al., 1986; Gbenle et al. 1986). It is a plasma-membrane enzyme and cytochemical techniques have shown that in *Leishmania spp.* the enzyme is exposed on the outside of the cells. In *T. brucei* the enzyme has a V_{max} of 82 nmol min^{-1}(mg protein)$^{-1}$ with a K_m for 3'-AMP of

0.45 mM. The enzyme is strongly inhibited by the trypanocidal drug Suramin (Gbenle et al. 1986). During our studies on this enzyme in T. brucei we have noticed that it shares a great number of properties, including its sensitivity towards Suramin, with another enzyme recently described for this organism, Ca^{+2}-dependent alkaline endoribonuclease (Gbenle and Akinrimisi, 1982, Gbenle et al, 1986). Indeed our work and that of Gottlieb et al. (1986) has shown that 3'-nucleotidase in addition to its action on nucleotides has nuclease activity and that both enzymes have an identical localization. Both activities, therefore, probably belong to one and the same enzyme.

An interesting observation, however, is that while 3'-nucleotidase of insect stages is fully exposed on the outside of the cell, the enzyme of bloodstream forms is completely latent. Apparently the integration of proteins in the plasma membrane is subject to drastic changes upon the transformation from bloodstream form to insect stage. It is not clear whether such changes in composition of the plasma membrane are the result of the presence or absence of a surface coat or that they reflect differences in the availability of substrates in the extracellular environment of the two life-cycle stages.

Other Plasma-Membrane Proteins

A protein kinase activity, which was first described in the case of Trypanosoma gambiense (Walter, 1978), has been found to be associated with the plasma membrane of T. brucei (Walter and Opperdoes, 1982). This enzyme has a preference for phosvitin as the acceptor protein, is inhibited by micromolar concentrations of Suramin and is probably associated with the inner face of the plasma membrane since it cannot be measured in intact cell preparations.

Also in T. gambiense, a phosphoprotein phosphatase has been demonstrated which has a broad substrate specificity in that it acts on histones, protamine and phosvitin (Walter, 1982).

This enzyme is also present in T. brucei, where it is mainly soluble although a minor but significant part was found to be associated with the plasma-membrane fraction (Walter and Opperdoes, 1982).

Typical plasma-membrane enzymes such as 5'-nucleotidase and ouabain-sensitive ATPase, described for numerous eukaryotic cell types, and sometimes claimed to be present also in T. brucei, are, however, in our experience absent from this organism (Steiger et al, 1980).

THE FLAGELLAR POCKET

The flagellar pocket membrane is continuous with the plasma membrane and electron microscopy suggests that the pocket membrane might be covered with surface coat at the same tight packing as is seen in the plasma membrane. The flagellar pocket is the only part of the trypanosome's surface which is not supported by subpellicular microtubules and, therefore, is the only area on the cell's surface which has enough flexibility to allow invagination and thus endocytosis.

Several enzymes have been localized in the flagellar pocket membrane of T. brucei. These are acid phosphatase, acid pyrophosphatase and adenylate cyclase (Walter and Opperdoes, 1982).

Acid Phosphatase

This enzyme which covers the entire surface of some Trypanosomatidae, such as Leishmania donovani (Dwyer and Gottlieb, 1984), can only be detected in the flagellar pocket membrane in T. brucei by cytochemical techniques (Langreth and Balber, 1975), where it is thought to cover the lining of the membrane exposed toward the lumen of the pocket (Steiger et al, 1980). This observation has been important in identifying flagellar pocket membrane fractions in cell-fractionation experiments and was confirmed by the fact that no acid phosphatase was found in fractions enriched in plasma membranes. The enzyme in T. brucei, although exposed towards the external medium has an acidic pH optimum. Unlike the lysosomal phosphatases of mammalian cells this enzyme is less sensitive to inhibition by tartrate or fluoride (Steiger et al., 1980).

Studies on the substrate specificity of the enzyme have revealed considerable differences in affinity of acid phosphatase for the various substrates. The enzyme has a K_m for methylumbelliferyl phosphate (MUP) of 0.055 mM, for p-nitrophenylphosphate (PNP) of 0.23 mM and for glycerol 2-phosphate (GP) of 7.84 mM. The V_{max} in a crude homogenate was highest with the latter substrate (100 nmol min^{-1} mg $protein^{-1}$) and only 50% of that with the other two substrates.

The enzyme appears to be extremely sensitive to Suramin. The Ki in the presence of either GP or PNP is 5 μM while a value of 8 μM is found with MUP as substrate. The corresponding activity of the host (rat-liver phosphatase) is only slightly inhibited by Suramin.

Acid Pyrophosphatase

Another acid hydrolase, pyrophosphatase, has been found to be associated with the flagellar pocket membrane and possibly the lysosomal fraction of T. brucei (unpublished experiments). This enzyme, which has a pH optimum of 3.9, has a K_m for pyrophosphate of 0.29 mM. The enzyme is strongly inhibited by Suramin with a Ki = 1 μM.

LYSOSOMES

Morphometry of T. brucei bloodstream forms has revealed that the vacuolar system characterized by small vesicles resembling lysosomes does not represent more than 0.9% of the total cell volume (Coppens, I., unpublished), while specialized blood cells, such as macrophages, may have as much as 5% of lysosomal volume. These vesicles are heterogeneous in size and sometimes tubular in shape. Langreth and Balber (1975), by using cytochemical techniques, have shown that such vesicles do contain acid phosphatase activity, although this could not be confirmed by us using subcellular fractionation techniques (Steiger and Opperdoes, 1980). Four other important acid hydrolases have been localized in lysosomes: proteinase, α-D-mannosidase, RNAse and phospholipase.

Acid proteinase

This enzyme, previously incorrectly referred to as cathepsin D (Steiger et al., 1979), is not sensitive to pepstatin, and is completely inhibited by 1 mM p-chloromercuribenzoate. This proteinase has a pH optimum of 3.6 when acting on haemoglobin (Steiger et al., 1980), while with casein as substrate its pH optimum is close to neutral (6.5) (Opperdoes and Steiger, 1981).

α-D-Mannosidase

Mannosidase has been reported for *T. cruzi* (Steiger et al., 1979; Avila et al., 1979) as well as for *T. brucei* (Steiger et al., 1979; 1980). The *T. brucei* enzyme has a pH optimum of 4.6 and is not stimulated by divalent cations as reported in the case of the *T. cruzi* enzyme.

Other Enzymes

Lysosomes contain also phospholipase A_1 activity which displays a pH optimum of 5.2 with phosphatidyl choline as substrate. This enzyme is inhibited by diisopropyl fluorophosphate and differs from the, soluble phospholipase A_1 described by Hambrey et al. (1981) (cf Opperdoes and Van Roy, 1982a). The lysosomal enzyme is slightly inhibited by Suramin, while the soluble phospholipase A_1 is much more sensitive. Although the total phospholipase activity of T. brucei is subject to drastic changes going from bloodstream form to procyclic insect stage, the lysosomal activity remains unaltered.

Cell-fractionation experiments have provided evidence for the presence of an acid RNAse activity in the lysosomes of the T. brucei bloodstream form, in addition to other cell compartments (Steiger et al, 1980). This activity, which has not been further characterized, has a pH optimum of 5.6 when acting on total yeast RNA. It is quite distinct from the alkaline RNAse activity described earlier and which is thought to be identical to the 3'-nucleotidase of the plasma membrane.

FLUID-PHASE ENDOCYTOSIS

We have studied the uptake of serum proteins by T. brucei at 30°C to get an idea of the rate of fluid endocytosis. Total serum proteins were taken up by the trypanosomes at a rate similar to that of an abundant protein such as albumin. At high albumin concentration (>1 mg ml^{-1}) an average clearance rate of 0.08 µl (mg protein)$^{-1}$ h^{-1}, or 1.4 % of the total cell volume per hour, was measured. This value is similar to that of Fairlamb and Bowman (1980a) determined with polyvinyl pyrollidone. At low albumin concentrations (< 5 µg ml^{-1}), however, the apparent clearance rate increased to 2.1 µl (mg protein)$^{-1}$ h^{-1}, suggesting a significant contribution of adsorptive endocytosis.

RECEPTOR-MEDIATED ENDOCYTOSIS

When low-density lipoproteins (LDL) were used instead of total plasma proteins or albumin, a much higher clearance rate was observed. This uptake was (1) strongly stimulated by the presence of calcium, (2) temperature dependent and (3) independent of the concentration of the extracellular protein used. The highest clearance was observed with rabbit LDL (195 µl mg^{-1} h^{-1}), while the clearance for total lipoproteins as well as for high-density lipoproteins (HDL) was considerably less.

Similar results were obtained with the iron-transport protein transferrin. This protein was taken up with a clearance rate of 19.5 µl mg^{-1} h^{-1}, while the apotransferrin, lacking the iron, was taken up at a rate of only 0.31 µl mg^{-1} h^{-1}.

In the case of LDL its uptake was not inhibited by a large excess of either albumin or transferrin, while the addition of excess unlabelled LDL drastically reduced the rate of clearance. The reciprocal experiment carried out on transferrin gave similar results.

When trypanosomes were preincubated with trypsin at 30°C the uptake of LDL was strongly inhibited while that of albumin was not affected at all.

Although these experiments together demonstrate that the bloodstream form of *T. brucei* contains two types of receptor, one specific for low-density lipoproteins and the other for transferrin, they do not tell us anything about their localization on the cell surface.

MORPHOLOGICAL DEMONSTRATION OF ENDOCYTOSIS

To study protein uptake by trypanosomes at the level of the electron microscope various proteins such as albumin, LDL and transferrin were labelled by absorbing them to colloidal gold

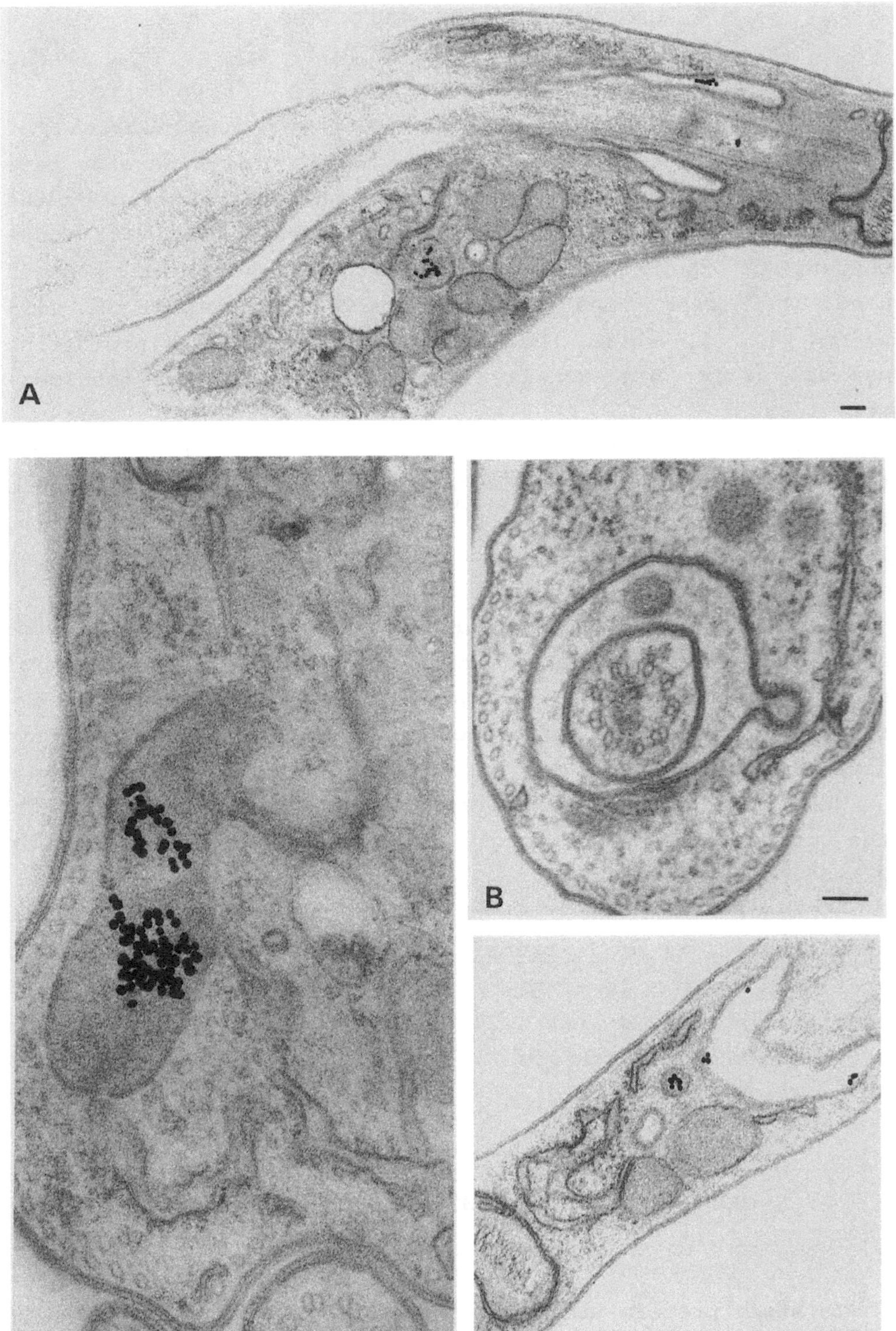
A
D
B
C
0.1

particles. With albumin very few gold particles could be identified inside the cells after an incubation of 60 min at 30°C. With LDL and transferrin, however, an entirely different situation was encountered. After incubation at 4°C numerous particles were found covering the lining of the flagellar pocket, while particles were never seen attached to the plasma membrane. When the temperature was increased to 30°C gold particles were found in vesicles in the cytosol in close proximity to the flagellar pocket, as well as concentrated in intracellular lysosome-like structures and digestive tubules. These observations suggest that the site of entry for macromolecular material is the flagellar pocket and that, therefore, the receptors for both LDL and transferrin must be localized on the membrane of this pocket.

DISCUSSION

Bloodstream trypanosomes circulate in an extremely rich environment which contains all components essential for their growth. Most of these components, such as free amino acids and carbohydrates, penetrate the trypanosomal membrane either freely, or by facilitated diffusion, or by active transport. Some essential components, however, do not circulate in a free form, but are part of macromolecules that have to be digested first, before they can be utilized by the trypanosome. Examples of such components are the nucleosides that are liberated by the action of plasma-membrane associated 3'-nucleotidase on

---

Figure 1 : Uptake of LDL by bloodstream form trypanosomes. Cells were incubated for 1 h at 30°C in the presence of 3.2 µg LDL adsorbed on gold per ml. A, LDL in the lumen of the flagellar pocket and in a lysosome-like structure. B, formation of a spiny-coated vesicle. C, LDL at the lining of the flagellar pocket and in an endocytotic vesicle. D, digestive structure loaded with LDL particles. Bar represents 0.1 µm.

RNA (Gbenle et al., 1986) or the sugars made available from oligosaccharides or glycoproteins by the action of plasma-membrane glycosidases.

Other essential components circulate while tightly bound to proteins, such as cholesterol that is associated with LDL particles and iron which is tightly bound to the carrier transferrin. Both cholesterol and iron are deeply buried inside the interior of their respective carriers and, therefore, can only be made available to the parasite after interiorization of the carrier protein followed by its digestion.

With the data presented in this paper we have shown that the bloodstream form of T. brucei is well capable of interiorizing macromolecular material. This either occurs by fluid endocytosis as is the case with an abundant protein, such as albumin, or by receptor-mediated endocytosis as is the case for two less abundant proteins, LDL and transferrin. In each case the proteins enter the flagellar pocket first. No evidence has been found for any interaction of proteins with the plasma membrane of the trypanosome. Once inside the flagellar pocket the serum protein either remains in the fluid phase where it may have weak interaction with components of the pocket, or binds to specific receptors present on the pocket membrane. In both cases invagination of the membrane then leads to the formation of endocytotic vesicles and the capture of the protein. Such vesicles then fuse with primary lysosomes, the interiorized material is digested and the essential nutrient liberated.

It is of interest to note that a primitive protozoan parasite utilizes the same carrier proteins as does the mammalian the host for its supply of essential nutrients. It also has at its disposal similar mechanisms of receptor mediated-endocytosis. An intriguing question is whether the trypanosome LDL and transferrin receptors are evolutionary related to their mammalian homologues and whether such proteins were already present before the Trypanosomatidae branched off from the main stream of eukaryotic evolution, some one billion years ago (Opperdoes, 1985), or that these receptors represent recent adaptations to the parasitic style of live in the mammalian bloodstream.

An answer to this question will soon come from the cloning and the sequencing of the genes that code for these trypanosome receptor proteins.

Fairlamb and Bowman (1980b) have shown that Suramin is taken up by trypanosomes at a rate 18-fold higher than that of fluid endocytosis. They, therefore, have proposed that this drug might enter the trypanosome in combination with a protein and they suggested that this protein could be albumin to which it thightly binds. Since we have shown that the trypanosomes lack a specific receptor for albumin, the increased uptake rate of Suramin is most likely due to its association with proteins other than albumin for which there exist specific receptors. In this respect it is of interest that several of the enzymes that Suramin encounters on its way from extracellular fluid to secondary lysosomes are strongly inhibited by Suramin. They, therefore, must have a high affinity for the drug and be capable of displacing it from its putative carrier. In view of the fact that Suramin is becoming concentrated on its way to the lysosomes, each of the enzymes mentioned above should be considered as a possible target for the drug. Since in the case of Suramin the mode of action is not well understood it can not be excluded that the interference with uptake and digestion of macromolecular material is one of the first effects exerted by this drug.

ACKNOWLEDGEMENTS

This work received the financial support from the UNDP/World Bank/WHO Special programme for Research and Training in Tropical Diseases and from the "Institut pour l'Encouragement de la Recherche Scientifique dans l'Industrie et l'Agriculture".

REFERENCES

Avila JL, Casanova MA, Avila A, Bretana A (1979) Acid and neutral hydrolases in Trypanosoma cruzi. Characterization and assay. J Protozool 26:304-311

Brown KN, Armstrong JA, Valentine RC (1965) The ingestion of protein molecules by blood forms of Trypanosoma rhodesiense. Exp Cell Res 39:129-135

Dwyer DM, Gottlieb M (1984) Surface membrane localization of 3'- and 5'-nucleotidase activities in Leishmania donovani promastigotes. Mol Biochem Parasitol 5:139-150

Fairlamb AH, Bowman IBR (1980a) Uptake of the trypanocidal drug Suramin by bloodstream forms of Trypanosoma brucei and its effects on respiration and growth rate in vivo. Mol Biochem Parasitol 1:315-333

Fairlamb AH, Bowman IBR (1980b) Trypanosoma brucei : maintenance of concentrated suspensions of bloodstream trypomastigotes in vitro using continuous dialysis for measurement of endocytosis. Exp Parasitol 49:366-380

Gbenle GO, Akinrimisi EO (1982) Studies on the catalytic properties of the calcium-dependent endoribonuclease of Trypanosoma brucei cytoplasm. Mol Biochem Parasitol 5:213-220.

Gbenle GO, Opperdoes FR, Van Roy J (1986) Possible identity of 3'-nucleotidase and calcium-dependent endoribonuclease of T. brucei and their localization in the plasma membrane. Acta Trop, in press.

Gottlieb M, Dwyer DM (1983) Evidence for distinct 5'- and 3'-nucleotidase activities in the surface membrane fraction of Leishmania donovani promastigotes. Mol Biochem Parasitol 7:303-317

Gottlieb M (1985) Enzyme regulation in a Trypanosomatid: effect of purine starvation on levels of 3'-nucleotidase activity. Science 227:72-74.

Gottlieb M, Gardiner PR, Dwyer DM (1986) 3'-nucleotidase activity in procyclic and bloodstream stages of Trypanosoma rhodesiense. Comp Biochem Physiol 83B:63-69

Jadin JM (1971) Cytologie et cytophysiologie des Trypanosomatidae. Acta Zool Pathol 53:1-168

Hambrey PN, Mellors A, Tizard IR (1981) The phospholipases of pathogenic and non-pathogenic Trypanosoma species. Mol Biochem Parasitol 2, 177-186

Langreth SG, Balber AE (1975) Protein uptake and digestion in bloodstream and culture forms of Trypanosoma brucei. J Protozool 22:40-53

McLaughlin J (1982) Subcellular distribution of particle-associated antigens in Trypanosoma rhodesiense. J Immunol 128:1656-2263

Opperdoes FR, Steiger RF (1981) Localization of hydrolases in cultured procyclic trypomastigotes of Trypanosoma brucei. Mol Biochem Parasitol 4:311-323

Opperdoes FR, Van Roy J (1982a) The phospholipases of Trypanosoma brucei bloodstream forms and cultured procyclics. Mol Biochem Parasitol 5:309-319

Opperdoes FR, Van Roy J (1982b) Involvement of lysosomes in the uptake of macromolecular material by bloodstream forms of Trypanosoma brucei. Mol Biochem Parasitol 6:181-190

Opperdoes FR (1985) Biochemical peculiarities of trypanosomes, African and South American. Br Med Bull 41:130-136.

Steiger RF, Van Hoof F, Bontemps J, Nyssens-Jadin M, Druetz JE (1979) Acid hydrolases in trypanosomatid flagellates. Acta Trop 36:335-341.

Steiger RF, Opperdoes FR, Bontemps J (1980) Subcellular fractionation of Trypanosoma brucei bloodstream forms with special reference to hydrolases. Eur J Biochem 105:163-175

Steinert M, Novikoff AB (1960) The existence of a cytostome and the occurrence of pinocytosis in the trypanosome Trypanosoma mega. Biophys Biochem Cytol 8:563-569

Walter RD (1978) Multiple protein kinases from Trypanosoma gambiense. Hoppe Seyler's Z Physiol Chem 359:601-606

Walter RD (1982) Interaction of Suramin and berenil with phosphorylation and dephosphorylation reactions in Trypanosoma gambiense. In : Nuclear techniques in the study and control of parasitic diseases of man and animals, Proceedings series, Int. Atomic Energy Agency

Walter RD, Opperdoes FR (1982) Subcellular distribution of adenylate cyclase, cyclic AMP phosphodiesterase, protein kinases and phosphoprotein phosphatase in Trypanosoma brucei. Mol Biochem Parasitol 6, 287-295.

THE CYTOSKELETON OF *TRYPANOSOMA BRUCEI*: GENES AND PROTEINS

R. Braun, T. Seebeck, K. Behrens, M. Imboden, A. Schneider and V. Küng

Institute of General Microbiology, University of Bern, 3012 Bern, Switzerland

INTRODUCTION

The cytoskeleton of eukaryotic cells is in general made up of three components: microfilaments, intermediary filaments and microtubules. In mastigophores such as trypanosomes the predominant and most striking cytoskeletal element are the microtubules.

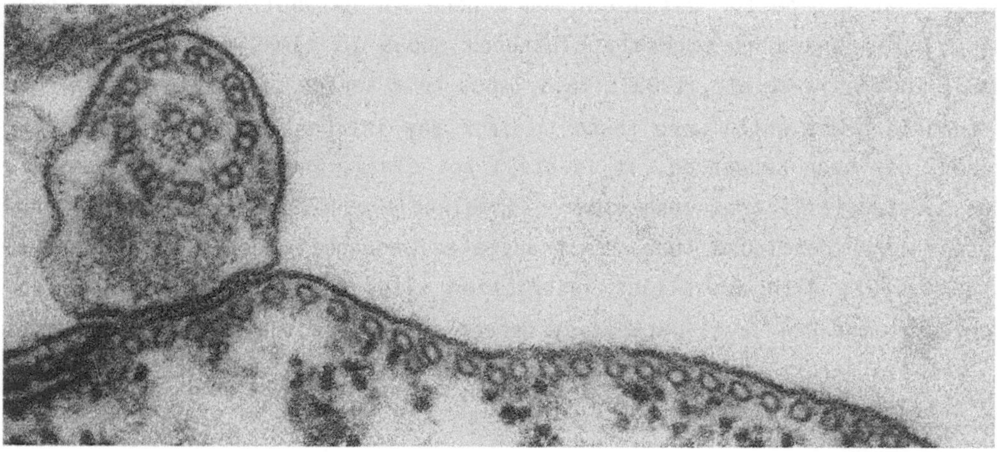

Figure 1. Transverse section through a *Trypanosoma brucei* cell. Lower half: cell body, top left: flagellum. x 220'000.

Two types of microtubules can be seen: grouped microtubules in the flagellum, and individual, subpellicular microtubules in the cytoplasm, just inside of the cell membrane. The bulk of the microtubular mass in both systems is made up of α- and β-tubulin with further proteins contributing to the specialized structures and functions of flagellar and subpellicular microtubules. Note also in the flagellum, below the (9x2)+2 microtubules a highly ordered fibrillar network, the paraflagellar rod (PFR).

NATO ASI Series, Vol. H11
Host-Parasite Cellular and Molecular Interactions
in Protozoal Infections. Edited by K.-P. Chang and D. Snary

RESULTS

Over the last few years our laboratory and others have analyzed the genes coding for tubulin and their transcription as well as the biochemistry of tubulin proteins.

1. Organization of the tubulin genes

Thanks to the high degree of sequence conservation of tubulin in widely different organisms, it was possible to clone the tubulin genes of T.brucei with the help of a mammalian tubulin gene. From genomic clones it was deduced that a gene for α-tubulin and a gene for β-tubulin made up a repeat of 3.7 kb, which is tandemly clustered about 10 times (Seebeck et al., 1983; Thomashow et al., 1983). This leads to a series of alternating α- and β-tubulin genes which were found to lack any introns. Although one 3.7 kb repeat has been sequenced, it is still not clear, whether the repeats differ substantially from each other (Kimmel et al., 1985). From protein analyses it is concluded that the translated sequences are nearly identical (see below), although slight restriction site polymorphism has been observed at the DNA level between different repeats of the cluster.

2. Transcription of the tubulin genes

Since tubulin are the single most abundant proteins in T.brucei, the mRNAs for these two proteins is also present in large amounts in the cytoplasm. Detailed structural analysis revealed that the mRNA for α- and β-tubulin are 1950 and 2300 nucleotides long (Imboden et al., 1986).

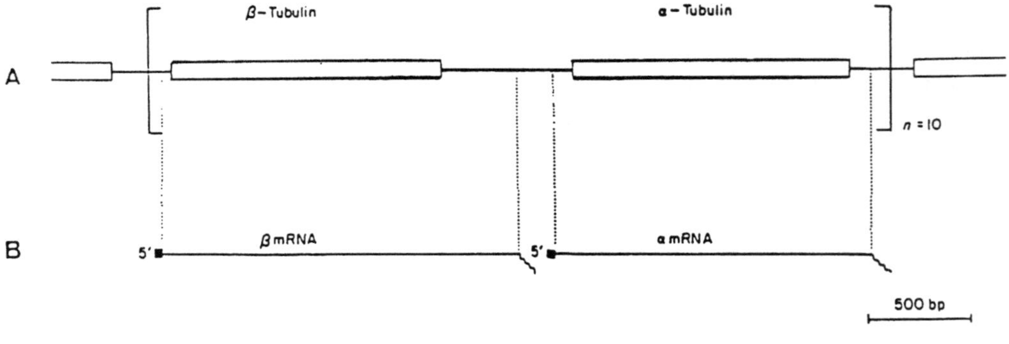

Figure 2. Structure and expression of a tubulin repeat. In (A) a single repeat is marked by brackets. The open boxes indicate the coding regions. In (B) the dotted lines indicate colinearity between genes and transcripts. Dark squares: miniexons; wavy lines: poly A tails of the mRNAs.

It is interesting to note that tubulin mRNAs as well as all other trypanosome mRNAs so far analyzed, carry a 35 nucleotide miniexon at their 5'-end (Sather and Agabian, 1985). All miniexons of a given species have the same nucleotide sequence and they are added to the mRNA after transcription is completed, probably by trans-splicing (for review see Braun, 1986). Miniexons and the enzymes adding them to mRNA could serve as targets for chemotherapy, since mammalian mRNA is not capped by miniexons. No promotors, enhancers or other regulatory signals of transcription have so far been found in trypanosomes: in some instances the primary transcripts seem to be extremely large, implying processing of a huge precursor. (Swinkels et al., 1986; for review see Borst, 1986). The pre-mRNAs for tubulin have not yet been characterized in detail, though it is clear that only about 150 nucleotides separate the two stable transcripts on the DNA map.

3. Tubulin proteins

In vitro, in the presence of taxol, trypanosomal tubulins can be polymerized to structures which can, by electron microscopy, be identified as microtubules (Stieger et al. 1984). As expected, α- and ß-tubulin of T.brucei are similar to the homologous proteins from mammals, however, some minor differences have anyway been found.

In two dimensional gel electrophoresis the ß-tubulin of T.brucei appears as a single spot, indicating that only one single polypeptide is present. For α-tubulin two spots can be discerned, α_1 and α_3 (see figure 3).

In vitro translation of mRNA showed α_1 to be the primary product, while α_3 is a tubulin which has been modified through acetylation (Gull et al., 1986). Both α_1 and α_3 can undergo a second modification, namely tyrosinylation at the C terminus of the polypeptide chain. The two types of modification have been studied in a rather detailed manner, both with regard to temporal and also spatial distribution of the modified forms. The α_1 and α_3 iso-forms are both in the flagellar and in the subpellicular compartment, whereby the flagella contain mostly α_3 (Schneider et al., 1986). The soluble pool has only α_1, presumably newly synthesized. Tyrosinolated tubulin occurs in all above mentioned cellular compartments, though with some temporal restrictions (Sherwin et al., 1986). This modified tubulin is found predominantly in newly made microtubules and not in older ones. Taken together, these experiments show a rather complex distribution pattern of modified tubulins even in a rather simple unicellular organism.

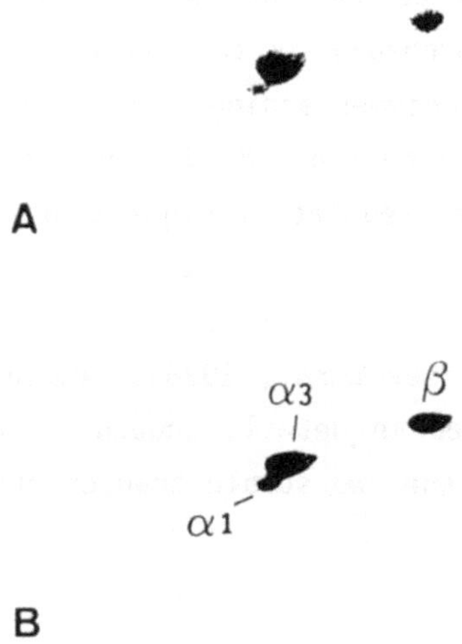

Figure 3. Two dimensional gel electrophoresis of a whole cell lysate. A. Coomassie-stain. B. Immundetection with a mixture of antibodies against α-tubulin and β-tubulin.

4. Further cytoskeletal proteins

In the mitotic spindle, the flagellum and the subpellicular layer microtubules clearly have different functions. They are most probably only able to fullfil these diverse functions if they are associated with further proteins. Presently studies are underway to identify and characterize such proteins and to isolate their genes. In particular we are interested in the proteins of the PFR and in proteins which anchor the subpellicular microtubules to the membrane.

DISCUSSION

Thanks to the new procedures of gene technology and immunology it has become possible to study the cytoskeleton of trypanosomes in a detailed manner. This is of interest in itself, since here microtubules are the dominant structural element, in contrast to the situation with mammalian cells. The microtubule-membrane association may well be decisive for cell shape and motility. In addition, the cytoskeleton may be a possible target for chemotherapy, for which potent drugs are sorely lacking.

ACKNOWLEDGEMENT

Supported by grant 3.650-0.84 of the Swiss National Science Foundation.

REFERENCES

Borst P (1986) Discontinuous Transcription and Antigenic Variation in Trypanosomes. Ann Rev Biochem 55, 701-732

Braun R (1986) Discontinuous RNA Synthesis through Trans-Splicing. Bio Essays, in press

Gull K, Hussey PJ, Sasse R, Schneider A, Seebeck T and Sherwin T (1986) Tubulin Isotypes: Generation of Diversity in Cells and Microtubular Organelles. J Cell Sci Suppl 5, in press

Imboden M, Blum B, DeLange T, Braun R and Seebeck T (1986) Tubulin mRNAs of Trypanosoma brucei. J Mol Biol 188, 393-402

Kimmel BE, Samson S, Wu J, Hirschberg R and Yarbrough LR (1985) Tubulin genes of the African trypanosome Trypanosoma brucei rhodesiense: nucleotide sequence of a 3.7-kb fragment containing genes for alpha and beta tubulins. Gene 35, 237-248

Sather S and Agabian N (1985) A 5' spliced leader is added in trans to both α- and β-tubulin transcripts in Trypanosoma brucei. Proc Natl Acad Sci USA 82, 5695-5699

Schneider A, Sherwin T, Russel DG, Gull K and Seebeck T. Subpellicular and Flagellar Microtubules of Trypanosoma brucei brucei contain the same Alpha-Tubulin Isotypes. In preparation

Seebeck T, Whittaker PA, Imboden M, Hardman N and Braun R (1983) Tubulin genes of Trypanosoma brucei: A tightly clustered family of alternating genes. Proc Natl Acad Sci USA 80, 4634-4638

Shervin T, Schneider A, Sasse R, Seebeck T and Gull K. Distinct Localization and Cell Cycle Dependence of C-Terminally Tyrosinloated Alpha-Tubulin in the Microtubules of Trypanosoma brucei brucei. In preparation

Stieger J, Wyler T and Seebeck T (1984) Partial Purification and Characterization of Microtubular Protein from Trypanosoma brucei. J Biol Chem 259, 4596-4602

Swinkels BW, Gibson WC, Osinga KA, Kramer R, Veeneman GH, Van Boom JH and Borst P (1986) Characterization of the gene for the microbody (glycosomal) triosephosphate isomerase of Trypanosoma brucei. EMBO J 5, 1291-1298

Thomashow, LS, Milhausen M, Rütter WJ and Agabian N (1983) Tubulin Genes are Tandemly linked and clustered in the Genome of Trypanosoma brucei. Cell 32, 35-43

GENETIC RE-ASSORTMENT OF AFRICAN TRYPANOSOMES IN GLOSSINA

L. Jenni and R. Brun
Swiss Tropical Institute
Basel, Switzerland

The first direct demonstration of hybrid formation between trypanosomes in the laboratory has been recently reported. Two clones of Trypanosoma (T.) brucei have been simultaneously transmitted through Glossina m. morsitans. After full cyclical development of the trypanosomes in the flies, hybrid trypanosomes have been identified by electrophoretic analysis of several enzymes and with DNA probes for single copy and surface antigen genes. These results demonstrate that genetic exchange can take place in African trypanosomes.

INTRODUCTION

Various species of trypanosomes occur in man and animals. In Africa, trypanosomes causing sleeping sickness in man (Trypanosoma brucei gambiense and T.b. rhodesiense) are cyclically transmitted by tsetse flies (genus Glossina). There have been repeated speculations regarding the occurrence of mating and sexually differentiated forms in these trypanosomes.

NATO ASI Series, Vol. H11
Host-Parasite Cellular and Molecular Interactions
in Protozoal Infections. Edited by K.-P. Chang and D. Snary

PREVIOUS WORK

In general trypanosomes reproduce by binary fission in the blood and other tissues of infected mammalian hosts, and also in the intestinal tract and salivary glands of Glossina. Sexual processes which involve genomic re-assortment in parasitic protozoa are well known among sporozoans such as Plasmodia and Coccidia. Meiotic recombination and gene re-assortment associated with sexual conjugation are important and powerful factors in eukaryotic evolution and has been investigated in African trypanosomes since the beginning of this century (Wenyon 1926).

Fiennes (1945) reported sexual phenomena for Trypanosoma congolense and Fairbairn et al. (1946) published their findings on syngamy in Trypanosoma rhodesiense and T. simiae.

Some indirect evidence for genetic exchange between trypanosomes in the tsetse fly was provided by Culwick et al. (1951), based on retention or loss of infectivity to man, and on morphological changes resulting from the first simultaneous transmission of different trypanosome stocks of Trypanozoon subspecies. However, the lack of well-defined genetic markers, coupled with the use of uncloned populations makes it difficult to interpret some of this earlier work on the mixed transmission of trypanosome isolates. Clear evidence for mating in trypanosomes would advance our knowledge of the basic biology of these parasites and assist the interpretation of the steadily accumulating epidemiological observations (Tait 1983).

STATE OF KNOWLEDGE

Tait (1983) has reviewed the evidence provided by various experimental approaches which have been recently applied to the long-standing question of kinetoplastid sexuality. Increasing evidence for genetic exchange emerged from analyses of the distribution of enzyme variants in African trypanosome stocks (Trypanosoma (T.) brucei and Trypanosoma (N.) congolense) isolated from different areas in Africa during the last 10 years (Gibson et al. 1980; Richner and Jenni 1986; Tait 1980; Young and Godfrey 1983). These authors found many enzymes which showed single and multiple-banded patterns after electrophoresis, depending on the trypanosome population that was analysed. Organisms showing one band were assumed to be homozygous at the respective locus whereas those showing multiple bands were assumed to be heterozygous. By applying the Hardy-Weinberg law, Tait (1980) found that the frequencies of the various forms of each enzyme in the parasite populations provided good evidence that mating between trypanosome was taking place.

The data obtained by Culwick et al. (1951) and Letch (1984) indicated that the sexual processes of Trypanosoma (T.) brucei probably occurred during the cyclical development in the tsetse fly vector.

Based on these data a series of mixed transmission experiments have been carried out in Basel using cloned and well characterized trypanosome populations which differed from each other in several aspects (Jenni et al. 1984). The prerequisites for these simultaneous transmission studies with respect to different markers were that both parental clones had:

- to be homozygous single banded for distinct isoenzymes with different electrophoretic mobilities.
- to be transmissible with comparable speed and mature infection rates through the same tsetse species.
- to show distinct DNA hybridization patterns with specific DNA probes.
- to belong to distinct serodemes.

In one series of experiments cloned progeny trypanosome populations showed "hybrid" properties for most of the above markers. The parental and progeny clones have been extensively analysed in collaboration with several research groups and the results have been recently published (Jenni et al. 1986). The enzyme phenotypes of the stocks derived from the mixed transmissions were found to differ in 4 enzymes from those of either parental clone. They also appeared to be heterozygous for the variant alleles for which the parental clones were homozygous. Comparable results were obtained with DNA probes showing different hybridization patterns between the parental clones, thus demonstrating that the "offspring" were hybrid not only at the level of gene products but also at the level of the genes examined. In addition, the analysis of inheritance of size variation of the intermediate class of chromosomes separated by pulse field gel electrophoresis and restriction site and length polymorphisms of kinetoplast DNA showed an uniparentally inheritance (Sternberg et al., in press). These experiments have clearly demonstrated that gene exchange can occur between African trypanosomes of the subgenus Trypanosoma (T.) brucei and that it most probably occurs at specific stages during cyclical development in the fly.

FUTURE RESEARCH

It is clear from the results obtained so far that further data are required in order to establish the basic features of this genetic system, namely to answer the following questions:

- What are the prospects of finding other compatible trypanosome clones which show a detectable degree of gene re-assortment?

- What is the frequency of hybrid formation in a given pair of compatible trypanosome clones in the laboratory and how do these data relate to events occurring in the field?
- Where does hybrid formation occur in the fly?
- Are there specific mating types in trypanosomes?
- Are trypanosomes homo- or heterothallic?

Regarding the question of meiosis and syngamy involved in hybrid formation, basic data on ploidy are essential to the analysis of a system of gene exchange. Independent groups of experiments have revealed that T. brucei ssp are diploid for most genes as shown by measurements of DNA content and complexity of the genome (Borst et al. 1982), analysis of restriction enzyme length polymorphisms in distinct glycolytic enzyme genes (Gibson et al. 1985), and analysis of enzyme electrophoretic banding patterns (Gibson et al. 1980; Tait 1980).

Most recently fluctuations in DNA content, determined by microfluorometry, were investigated during cyclical transmission of one of the parental clones used in the above mixed infection experiments. Results showed that the metacyclic forms contained half the amount of DNA when compared to bloodstream forms (Zampetti-Bosseler et al., in press). This indicates that meiosis may occur in the fly and that syngamy occurs shortly after cyclical transmission.

The regulatory mechanisms underlying and controlling cell physiology, genetic exchange and differentiation leading to hybrid formation in African trypanosomes are not yet known. We are a long way from understanding the possible molecular mechanisms (e.g. macromolecular, electrical and chemical) controlled and regulated by the trypanosome itself and the part played by the tsetse fly as the insect vector.

REFERENCES

Borst P, Van der Ploeg M, Van Hoek JFM, James J. (1982). On the DNA content and ploidy of trypanosomes. Mol Biochem Parasitol 6: 13-23

Culwick AT, Fairbairn H, Culwick RE (1951) The genectic relationship of the polymorphic trypanosomes and its practical implications. Ann Trop Med Parasitol 45: 11-29

Fairbairn H, Culwick AT, Gee FL (1946) A new approach to Trypanosomiasis. Ann Trop Med Parasitol 40: 421-452

Fiennes RN (1945) A sexual reproduction cycle of T. congolense. Nature 156: 390

Gibson WC, Marshall TF de C, Godfrey DG (1980) Numercial analysis of enzyme polymorphism. A new approach to the epidemiology and taxonomy of trypanosomes of the sub-genus Trypanozoon. Adv Parasitol 18: 175-245

Gibson WC, Osinga KA, Michels PAM, Borst P (1985) Trypanosomes of subgenus Trypanozoon are diploid for houskeeping genes. Mol Biochem Parasitol 16: 231-242

Jenni L, Marti S, Betschart B (1984) Simultaneous cyclical transmission of Trypanosoma brucei gambiense and Trypanosoma brucei brucei. In: Tümbay E, Yasarol S, Oezcel MA (eds) Fourth European Multicolloquium of Parasitology, Izmir 14-19 Oct 1984, 97, Bilgeham Publ Izmir.

Jenni L, Marti S, Schweizer J, Betschart B, Le Page RWF, Wells JM., Tait A, Paindavoine P, Pays E, Steinert M (1986) Hybrid formation between African trypanosomes during cyclical transmission. Nature 322: 173-175

Letch CA (1984) A mixed population of Trypanozoon in Glossina palpalis palpalis from Ivory Coast. Trans Roy Soc Trop Med Hyg 78: 627-630

Richner D, Jenni L (1986) Characterization of cyclically transmitted Trypanosoma (T.) brucei isolates from man. Acta Tropica 43: 21-29

Sternberg JM, Tait A, Haley S, Schweizer J, Jenni L (in press) The inheritance of chromosomal and kinetoplast variants in T. brucei spp. In: Agabian N, Goodmann H, Noguiera N, (eds), Molecular strategies of parasitic invasion, UCLA Symposia on Molecular and Cellular Biology, New Series, 42

Tait A (1980) Evidence for diploidy and mating in trypanosomes. Nature 287: 536-538

Tait A (1983) Sexual processes in the kinetoplastida. Parasitology 86: 29-57

Wenyon CM (1926) Protozoology. Vol 1. London, Baillière, Tindall & Cox

Young C, Godfrey DG (1983) Enzyme polymorphism and the distribution of Trypanosoma congolense isolates. Ann Trop Med Parasitol 77: 467-481

Zampetti-Bosseler F, Schweizer J, Pays E, Jenni L, Steinert M (in press) Evidence for haploidy in metacyclic forms of Trypanosoma brucei. Proc Nat Acad Sci USA

CELL SURFACE GLYCOPROTEINS OF TRYPANOSOMA CRUZI

David Snary[1], Michael A.J.Ferguson[2], Anthony K.Allen[3], Michael A.Miles[4] and Alan Sher[5].

1. Wellcome Biotechnology, Beckenham, Kent, U.K., 2 Dept. of Biochemistry, University of Oxford, U.K., 3 Charing Cross and Westminster Hospital Medical School, London, U.K., 4 London School of Hygiene and Tropical Medicine, London, U.K., 5 NIAID, Bethesda, Maryland, U.S.A.

INTRODUCTION

The oblagative intracellular parasitic protozoa Trypanosoma cruzi is the causative agent for Chagas disease or South American trypanosomiasis. This parasite has a complex life cycle and is transmitted by reduviid bugs by contaminative means, the parasite has a non-dividing infectious trypomastigote stage which is found in both the blood of infected mammals and in the insect hind-gut, although morphologically similar the two trypomastigote forms are distinguishable by other criteria. The parasite multiplies in the insect gut in the flagellated epimastigote form and intracellularly within the mammalian host in a rounded amastigote form (Brener 1973). There has been considerable interest in cell surface molecules of this parasite because of their potential both for highly discriminatory diagnostic tests (Schechter et al, 1983) and as vaccines (Snary 1983). A variety of methods have been used to isolate these surface antigens and most have been found to be glycosylated (reviewed by Snary 1985a), these include a 90 000 mol. wt. protective glycoprotein isolated by lectin affinity chromatography, an 85 000 mol. wt. blood trypomastigote specific glycoprotein the gene for which has recently been cloned (Peterson et al, 1986), the LPPG which is a complex molecule containing lipid, phosphate, peptide and carbohydrate which was isolated by phenol extraction,

NATO ASI Series, Vol. H11
Host-Parasite Cellular and Molecular Interactions in Protozoal Infections. Edited by K.-P. Chang and D. Snary

and a 72 000 mol.wt. glycoprotein specific to those stages of the life cycle found in the insect vector. This latter glycoprotein is the most characterized of all the T.cruzi glycoproteins and was the subject of the studies to be discussed here.

MATERIALS AND METHODS

The sources of the parasites and the methods used have essentially been described in previous publications (Snary et al, 1981, Ferguson et al, 1983, Snary 1983) except for the gas chromatography-mass spectrometry (gc-ms) analyses. A Hewlett-Packard 5996A gc-ms system with an HP1000 computer system was used. Methanolysis/TMS analysis was performed by a modified version of Chaplin (1982) using a 25x0.32mm fused silica capillary column with bonded SE30 phase (Alltech RSL 150) and a temperature program of 140°(2 min) to 250° at 6°/min. Alditol acetates were prepared following hydrolysis in 100ul 0.25M H_2SO_4 in 93% aqueous acetic acid (2h at RT followed by 2h at 80°). Sulphuric acid was removed on a column (0.6ml) of Dowex AG3 (acetate form) eluted with five column volumes of 50% methanol. The dried eluate was reduced with 0.3ml 10mg/ml $NaBH_4$ for 3h, acidified with acetic acid and evaporated three times with methanol to remove boric acid. The dried residue was acetylated with acetic anhydride (4h at 100°). Following the removal of acetic anhydride under vacuum the products were partitioned between water and dichloromethane. The alditol acetates recovered in the organic phase were analysed using a 30x0.32mm Supelcowax 10 capillary column and a temperature program of 90°(1min) to 230° at 20°/min. In both systems on-column injection was used with the carrier gas at 25psi. The column eluates were split using an open/split interface and analysed by both mass spectrometry for identification and flame ionisation detection (FID) for quantitation. Mass spectra were collected using electron impact at 70eV with source pressure of $2x10^{-5}$Torr and temperature of 150°. Transfer lines were held at 280 and the FID temperature was 300.

RESULTS AND DISCUSSION

The 72 000 mol.wt.glycoprotein (Gp 72) was originally identified using a monoclonal antibody (WIC 29.26), the specificity of the antibody clearly demonstrated that this glycoprotein, or at least the epitope defined by the monoclonal antibody, was not present on blood trypomastigotes or amastigotes, but was present on epimastigotes and reduviid derived metacyclic trypomastigotes (Snary et al. 1981, Snary 1983). Immunoprecipitation from ^{125}I radiolabelled cell surface proteins with this antibody followed by SDS PAGE gave a broad band of around 72 000 mol. wt in the Y strain of the parasite, but the apparent molecular weight of this protein varied from strain to strain within a range of 69 000 to 76 000 (Chapman et al. 1984). The monoclonal antibody originally used to identify GP 72 and further monoclonal antibodies which were shown by competitive radioimmune assays to bind to the same or closely related epitopes did not react with all isolates of T.cruzi equally, the parasite has been divided into broad groups or zymodemes based on their isoenzyme patterns, and the monclonal antibodies failed to react to or reacted only weakly to Z1 isolates (Chapman et al. 1984). However, the ability of oligosaccharide side chains released from the glycoprotein by alkali digestion to inhibit monoclonal antibody binding in RIA, and the periodate sensitivity of the epitope indicated that the monoclonal antibodies were binding to the carbohydrate side chains of the glycoprotein (Ferguson et al. 1983). Z1 isolates could therefore have lower levels of the glycoprotein, different patterns of glycosylation or some other factor such as conformation of the protein could be influencing the accessibility of the determinant. Two further antisera were therefore used to probe for the presence of Gp 72 in apparently negative isolates, one was a polyclonal rabbit serum raised against Gp 72 purified by antibody affinity chromatography, and the other was a monoclonal antibody generated from mice which had been immunised with Gp 72 in which the epitope identified by the original monoclonal had been destroyed by periodate oxidation. Both antisera bound to isolates which had previously been negative (Kirchhoff et al. 1984), an observation which suggested that the zymodemes can differ from each other at the level of their glycosylation enzymes, and since carbohydrate has been implicated in cell surface functions of many different cells including T.cruzi (Snary 1985b), differences of this type could influence the properties of the parasite, for example their tissue tropism.

The composition of Gp 72 is unusual in that it contains high levels of phosphate as well as atypical sugars (Ferguson et al. 1983). Release of the carbohydrate side chains from the protein by alkali digestion (β elimination) and their fractionation by ion exchange chromatography revealed firstly that the phosphate was covalently linked to the carbohydrate and secondly that there were two classes of side chain. The majority of the side chains were phosphorylated, contained the pentose and 6-deoxyhexoses, and were linked through xylose and fucose probably to serine or threonine in the polypeptide backbone. A minority of the side chains were most likely of the high mannose type, did not contain any pentose or 6-deoxyhexoses and were probably linked through glucosamine to asparagine residues in the polypeptide, some of these chains were phosphorylated but at a level lower than that found for the major complex structures. Previous analyses of the carbohydrate composition of Gp 72 had suggested that ribose was present, the identification of this sugar being based on gc analysis of the glycoprotein by the methanolysis/trimethylsilylation (TMS) procedure (Chambers & Clamp 1971). This analytical method produces mixed anomers of the released methyl glycoside such that each monosaccharide yields a characteristic pattern of peaks by gc. The pentose ribose produces two peaks Rib 1 and Rib 2, 20% and 80% of the total area respectively. Peaks with the correct relative retention times and relative areas were observed in all Gp 72 analyses and consequently identified as being ribose (Ferguson et al. 1983). Recent analysis of Gp 72 and its carbohydrate fractions by gc-ms using a high-resolution capillary column still produced peak patterns consistent with the presence of ribose. However, the mass spectra of the peaks revealed that the Rib 1 position peak did not belong to a sugar derivative and that the Rib 2 position peak had a mass spectrum consistant with a 6-deoxyhexose rather than a pentose derivative, suggesting the presence of rhamnose. To test this a sample of alkali released oligosaccharide (type IIIb, Ferguson et al. 1983) was reanalysed by the alditol acetate procedure which gives good resolution of all pentose and deoxyhexose derivatives. The chromatogram in Figure 1A

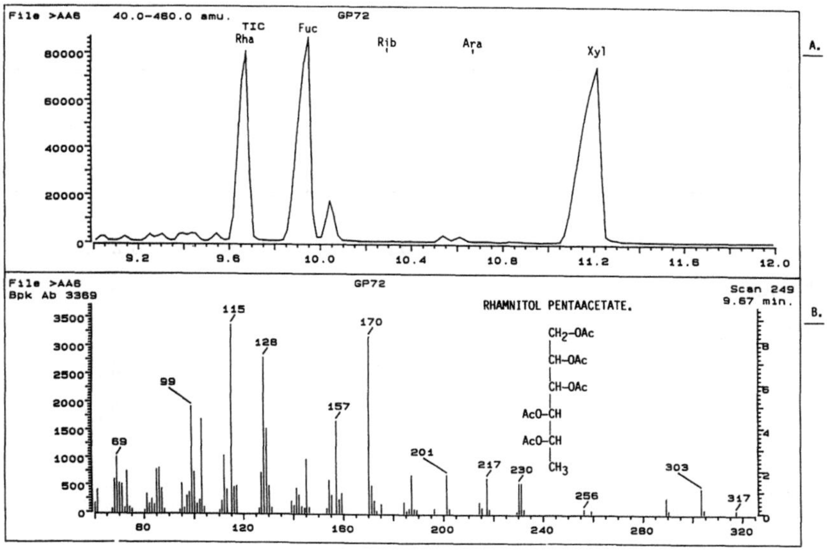

Figure 1. A, Detail of the pentose, 6-deoxyhexose region of the gc-ms chromatogram (total ion current). Rha, Fuc, Rib, Ara, Xyl show the relative positions of the rhamnose, fucose, arabinose and xylose alditol derivatives respectively. B, Mass spectrum of the peak co-eluting with authentic rhamnose derivative in Fig.1A. The mass spectrum is identical to that of rhamnose pentaacetate.

shows the presence of the previously identified fucose and xylose, the absence of any ribose and the presence of a peak with the identical retention time and mass spectrum (Fig 1B) to an authentic rhamnose derivative (rhamnitol pentaacetate). These data provide unequivocal evidence for the presence of rhamnose rather than ribose in GP 72 and confirm all other previous identifications. The new analyses (by bulk methanolysis/TMS and alditol acetate methods) agree well quantitatively with published values and a revised composition for Gp 72 is shown in Table 1. The presence of ribose in this glycoprotein would have been unique and totally unexpected, and although rhamnose has been found in plant and bacterial glycoconjugates its occurrence in higher eukaryotic glycoproteins has not previously been reported. Its presence is evidence for unique structures and with the greater sensitivity of modern carbohydrate analytical

Table 1. Composition of Gp 72

	% by weight	Moles/100 amino acids
Mannose	8.2	12.3
Galactose	15.6	23.3
Glucose	1.3	2.0
Glucosamine	2.9	4.4
Fucose	6.5	10.8
Xylose	10.6	19.1
Rhamnose	4.3	7.7
Protein (total amino acids)	44	
Phosphate	10	26.0

methods we hope to obtain more detailed information on the fine structure of the oligosaccharides of this unusual glycoprotein.

In concert with the extensive structural data which has revealed unusual molecular properties of Gp 72, it is probable that the biological properties of the glycoprotein have also been discovered. Clearly the glycoprotein was found on those stages of the organism present in the insect vector and presumably its prime function must relate to this phase of the life cycle. In fact the monoclonal antibody to the carbohydrate epitope has been shown to block the ability of the epimastigote to transform to the trypomastigote form. The transformation was induced by implanting epimastigotes in Millipore chambers into mice, for complete transformation implantation for 24h was needed, but the parasite could be removed from the chambers after 8h when it was still in the epimastigote form, but on subsequent culture *in vitro* for 16h, transformation to the trypomastigote form was completed (Sher *et al*. 1983). It was this latter phase of the transformation *in vitro* which was inhibited by inclusion of monoclonal antibody in the culture medium (Sher & Snary, 1982). The inhibition required divalent antibody and Fab fragments were not effective, but only very low levels of antibody were needed and over 50% maximal inhibition was given by a serum dilution of $1:10^{4}$. This promoted the idea that the monoclonal antibody was mimicing a gut lectin which could cross-link the surface of the parasite and fix it both in the dividing epimastigote form, and attach it to the gut wall retaining the infection within the insect vector.

Gp 72 has also been implicated in the immunology of *T.cruzi* both as a protective antigen and as an activator of complement. Epimastigotes are immediately lysed in fresh serum by the activation of the alternate complement pathway, this has been shown to take place because C3 binds very specifically to Gp 72 (Joiner *et al*, 1983). However, although Gp 72 is present on metacyclic trypomastigotes, this stage does not activate C3, Sher _et al_. (1986) have therefore suggested either that the structure of Gp 72 or its membrane microenvironment may be modified on metacyclic trypomastigotes. Although immunisation with Gp 72 has been shown to protect mice against a metacyclic challenge, the quality of the protection determined both by the level of the maximum blood parasitaemia attained and the duration of readily detectable blood parasites was not as good as the protection given by an alternate glycoprotein fraction. This was almost certainly related to the stage specificity of the two glycoproteins, whereas Gp 72 induced protection could only function by killing the invading parasite and reducing the effective infecting dose, the alternate glycoprotein fraction would have given an immune response potentially able to act on the mammalian stages of the parasite as well.

In conclusion a phosphorylated glycoprotein Gp 72 with a unique sugar composition has been shown to be present on the insect stages of _T.cruzi_, this glycoprotein is believed to be involved in the control of morphogenesis of the parasite, in its complement sensitivity and can induce a degree of protective immunity against infection.

REFERENCES

Brener Z (1973) Biology of _Trypanosoma cruzi_. Ann Rev Microbiol 27: 347-382.

Chapman MD, Snary D, Miles MA (1984) Quantitative differences in the expression of a 72000 molecular weight cell surface glycoprotein (GP 72) in _Trypanosoma cruzi_ zymodemes. J Immunol 132: 3149-3153.

Chambers RE, Clamp JR (1971) An assessment of methanolysis and other factors used in the analysis of carbohydrate containing materials. Biochem J 125: 1009-1018.

Chaplin MF (1982) A rapid sensitive method for the analysis of carbohydrate components in glycoproteins using gas-chromatography. Anal Biochem 123: 336-341.

Ferguson MAJ, Allen AK, Snary D (1983) Studies on the structure of a phosphoglycoprotein from the parasitic protozoan _Trypanosoma cruzi_. Biochem J 213: 313-319.

Kirchhoff LV, Hieny S, Shriver GM, Snary D (1984) Cryptic epitope explains the failure of a monoclonal antibody to bind to certain isolates of Trypanosoma cruzi. J. Immunol 133: 2731-2735.

Joiner KA, Hieny S, Kirchhoff LV, Sher A (1985) GP72, the 72 kilodalton glycoprotein, is the membrane acceptor site for C3 on Trypanosoma cruzi. J Exp Med 161: 1196-1212.

Peterson DS, Wrightsman RA, Manning JE (1985) Cloning of a major surface antigen gene of Trypanosoma cruzi: identification of a nonapeptide repeat. Nature 322: 566-568.

Schechter M, Flint JE, Voller A, Guhl F, Markinkelle CJ, Miles MA (1983) Purified Trypanosoma cruzi specific glycoprotein for discriminative serological diagnosis of South American trypanosomiasis (Chagas disease). Lancet II: 939-941.

Sher A, Snary D (1982) Specific inhibition of the morphogenesis of Trypanosoma cruzi by a monoclonal antibody. Nature 300: 639-640.

Sher A, Crane MSJ, Kirchhoff LV (1983) Incubation in mice provides a signal for differentiation of Trypanosoma cruzi epimastigotes to trypomastigotes. J Protozool 30: 278-283.

Snary D (1983) Cell surface glycoproteins of Trypanosoma cruzi: protective immunity in mice and antibody levels in human chagasic sera. Trans R Soc Trop Med Hyg 77: 126-129.

Snary D (1985a) The cell surface of Trypanosoma cruzi. Current Topics Microbiol Immunol 117: 75-92.

Snary D (1985b) Receptors and recognition mechanisms of Trypanosoma cruzi. Trans R Soc Trop Med Hyg 79: 587-590.

Snary D, Ferguson MAJ, Scott MT, Allen AK (1981) Cell surface antigens of Trypanosoma cruzi: use of monoclonal antibodies to identify and isolate an epimastigote specific glycoprotein. Mol Biochem Parasitol 3: 343-356.

A SPECIFIC INHIBITOR OF TRYPANOSOMA CRUZI NEURAMINIDASE FOUND IN HUMAN PLASMA

R.P. Prioli, I. Rosenberg and M.E.A. Pereira

New England Medical Center Hospitals
Department of Geographic Medicine and Infectious Diseases
750 Washington Street, P.O. Box 041
Boston, MA 02111
U.S.A.

INTRODUCTION

American trypanosomiasis or Chagas' disease is caused by the flagellate protrozoan Trypanosoma cruzi and represents a major health problem in Central and South America affecting over 10 million people (Brener, 1973; Hudson and Hindmarsh, 1985). Carlos Chagas first documented the disease in 1909. In his 78 works published between 1903 and 1935 he described the disease, its etiology, its vectors and its reservoirs. Chagas' disease predominantly striking the rural population, still represents an exciting challenge to the scientific community because the molecular mechanisms of its pathogenicity have yet to be unravelled.

T. cruzi has a complex biphasic life cycle comprising infection of vertebrate hosts and transmission by arthropod vectors. It has three distinct stages in its biological cycle. The amastigote or reproductive form is present intracellularly in mammals and multiplies by binary fission. After extensive multiplication inside the host cells, amastigotes transform into non-dividing trypomastigotes approximately 24 hours prior to cell lysis and parasite release (Zingales and Collins, 1985). Once in the bloodstream trypomastigotes circulate for a variable length of time before penetrating other cells (particularly muscle) where they transform to the amastigote form and resume the cycle. It is during the trypomastigote stage that transmission to the vector occurs. Circulating trypomastigotes are ingested as part of the insect's blood meal. In the acute phase of the disease, tissue and bloodstream forms are abundant, whereas in the chronic stage parasites are rarely seen in blood smears and are only detectable using more sensitive techniques such as Xenodiagnosis, which consists of feeding clean vectors on patients. Once in the reduviid bug, trypomastigotes transform into epimastigotes in the lumen of the insect midgut where they further multiply by binary fission. At the distal end of the insect gastrointestinal tract, epimastigotes transform once again into trypomastigotes (metacyclic) which are then excreted in the feces of the insect during a blood meal. These metacyclic forms are highly infective and gain access to the human bloodstream through the mucosa or via skin abrasions caused by scratching the bitten area.

NATO ASI Series, Vol. H11
Host-Parasite Cellular and Molecular Interactions
in Protozoal Infections. Edited by K.-P. Chang and D. Snary

Recently, Pereira (1983a) demonstrated the presence of neuraminidase (NA) activity in trypomastigotes. This enzyme activity was found to be developmentally regulated, displaying the highest activity in trypomastigotes, low activity in epimastigotes and is undetectable in amastigotes. Furthermore, it has been shown that live intact trypomastigotes are capable of desialylating mammalian cells and glycoconjugates during infection in vivo and in vitro (Pereira 1983a; Libby et al, 1986). Neuraminidase was found to be heterogeneously distributed in T. cruzi strains and clonal isolates and may possibly be associated with parasite myotropism (Pereira and Hoff, 1986). T. cruzi NA has generated a great deal of interest because it is thought that it might be involved in adherence of the trypomastigote to the host cell and to play a role in its subsequent internalization.

Neuraminidases or sialidases are widely distributed, being produced by many microbes, mammalian and avian tissues. They catalyze the hydrolysis of sialic acid from the cell surface and from soluble glycoconjugates (Rosenberg and Schengrund, 1976).

The wide occurence of sialic acids in the more exposed positions of molecules and their strong negative charge (pKa=2.6) suggest an involvement in various activities affecting cellular behaviour or survival. These activities include regulation of the longevity of erythrocytes which are removed by the spleen and liver upon loss of their sialic acid moiety. Similar clearance can be artificially induced by treating erythrocytes with neuraminidase. Removal of sialic acid from vascular endothelium has been shown to increase adhesion of circulating platelets (Gorog et al, 1972) and supressess the intravascular adherence of granulocytes (Gorog et al, 1980), whereas desialylation of cardiac cells enhances Ca^{2+} flux (Frank et al, 1977). Sialic acid in neurons, however, has been found to be involved in the release of neurotransmitters (Suzuki et al, 1980). In addition, lymphocytes treated with neuraminidase show altered functional properties. These include changes in the homing pattern, enhanced responses to several antigens and mitogens, and increased stimulation of mixed lymphocyte culture (Lundgren and Simons, 1971; Alnaes and Rahaminoff, 1975). The latter may merely reflect an increase in antigenicity since certain antigens on the surface of intact cells are more extensively exposed following sialic acid hydrolysis (Bagshawe and Currie, 1968). Further involvement of sialic acid in immunological functions, namely the regulation of the alternative complement pathway, has also been reported (Fearon, 1978).

Many of the biological activities mentioned above involve a surface recognition mechanism which is apparently triggered by the removal of the sialic acid moiety from the cell surface. It would, therefore, be of great interest to determine whether the T. cruzi neuramindase is involved in a similar manner in the parasite's association with host cells.

Cruzin

1. Identification:

T. cruzi neuraminidase inhibitory activity was first detected in sera from patients with Chagas' disease and was originally thought to be anti-T. cruzi antibody. However, it soon became evident that this was not the case since control sera from individuals who had never been exposed to T. cruzi possessed levels of cruzin equivalent to those in which the inhibitor was first detected. In addition, IgG absorbed from a pool of chagasic sera had very low inhibitory titre and sera devoid of either IgM, IgG, IgD, IgE, lambda or kappa chains retained inhibitory activity. Further conclusive evidence was provided when the presence of a T. cruzi neuraminidase inhibitor was confirmed in agammaglobulinemic sera. From these experiments it was concluded that a proteinaceous component of human plasma could specifically inhibit T. cruzi neuraminidase. This protein was therefore named cruzin. Since many human plasma proteins have been identified and characterized, it is possible that cruzin has already been designated a physiological role. In this context, several plasma proteins have been analyzed to determine if they possessed inhibitory activity. To date, all purified plasma proteins tested did not inhibit T. cruzi neuraminidase. Among those proteins tested were alpha 2 macroglobulin (protease inhibitor), orosomucoid, amyloid p-substance, C-reactive protein (acute phase proteins), IgG1, 2, 3 and 4, IgM, IgA, IgD, complement component C_3 and albumin.

Cruzin is not found exclusively in man. Different strains of mice can be ranked according to the amount of inhibitor detected in their plasma. C57BL/10 had 8 fold less cruzin than C_3H/eb although other strains tested had intermediate amounts. Interestingly, the two strains mentioned differ greatly in their susceptibility to infection with T. cruzi (Trischmann and Bloom 1982). It is feasible that, at least in mice, genetic predisposition to acquiring T. cruzi infection may be partially related to cruzin.

2. Detection:

Detection of cruzin was achieved using the inhibition of a neuraminidase assay based on peanut lectin agglutination (PNA) (Pereira, 1983b). Briefly, human erythrocytes were desialylated by T. cruzi neuraminidase exposing beta galactosyl residues to which PNA binds avidly, causing the red blood cells to agglutinate. Inhibition of T. cruzi neuraminidase was obtained as a function of the cruzin concentration. In addition, the amount of sialic acid released correlated with the lectin hemagglutination titre. One inhibitory unit was defined as the amount of cruzin required to account for a 50% inhibition of neuraminidase activity.

3. Purification:

Purification of cruzin from human plasma was achieved by sequential fractionation with polyethylene glycol (PEG), ion exchange and gel filtration. The isolation scheme was initiated by fractionating whole plasma with PEG, followed by application of the fraction containing the majority of the inhibitory activity, the 20% PEG precipitate, to a CM-Affi- Gel-Blue ion exchange column. Most contaminating proteins were eluted with high concentrations of sodium chloride (3.0 M) followed by guanidine hydrochloride (2.0M). Greater than 80% of the inhibitory activity was recovered by eluting with potassium thiocyanate (1.5M). Fractions containing the inhibitor were pooled, dialyzed against PBS, concentrated and applied to a pre-calibrated Ultrogel AcA 34 gel column.

The final preparation obtained from the latter column, as a symmetrical protein peak, contained highly purified inhibitor. The yield and increase in specific activity for each step of the purification is shown in Table 1.

TABLE 1
PURIFICATION OF CRUZIN FROM PLASMA

Steps	Total protein mg	Total activity units	Specific activity units/mg	Purification fold	Yield %
1. Plasma	1930	16,000	8.2	1	100
2. PEG 20% precipitate	181	2,640	14.5	2	16.5
3. CM Affi-Gel	6.2	2,000	322.5	39	12.3
4. AcA 34	1.6	1,504	940	114	9.4

Isolation of cruzin proved to be a straightforward endeavor. The accessibility of large quantities of plasma and the rapid, efficient fractionation of major plasma proteins by standard chromatographic procedures allowed the purification scheme to be carried out in 2 days.

4. Characterization:

Molecular characterization of cruzin was initiated by analyzing the purified material on SDS-PAGE. Under both reducing and non- reducing conditions cruzin migrated as a single sharp band of molecular weight, 25,000±1,000 dl. In its native state cruzin was shown to be 246,000 ±20,000 dl, following chromatography on Ultrogel AcA 34. Cruzin proved to be heat-labile, losing >90% of its activity after a 10 minute

incubation at 100°C. Trypsin treatment resulted in a dose dependent decrease in activity. However, even at the highest concentration of trypsin used, the activity could not be completely abolished. Inhibitory activity was preserved in preparations containing up to 1% SDS as well as after repeated freezing and thawing. Cruzin showed no protease or protease inhibitory activity.

5. Kinetics of Inhibition:

The kinetics of inhibition was studied to determine the manner by which cruzin inhibits T. cruzi neuraminidase. These studies indicated a reversible and possible non-competitive mechanism. Reversibility of total inhibition over a wide range of cruzin concentrations was achieved as a function of time. Human erythrocytes were selected as substrate to investigate the kinetics of cruzin inhibition because (as will be described below) such inhibition was contingent upon the physical size of the substrate. The suitability of this substrate was confirmed using Vibrio cholera neuraminidase and 2-deoxy-2,3-dehydro N-acetyl-neuraminic acid as an inhibitor. Using either erythrocytes or fetuin (Meindl et al, 1974), V. cholera neuraminidase was competitively inhibited by the latter inhibitor. The inhibition constant (K_i) for cruzin was 5.9×10^{-9}M and the K_m for the substrate was 4.1×10^{8} erythrocytes/ml.

Inhibition of _T. cruzi_ neuraminidase was dependent upon the actual size of the substrate. One inhibition unit was obtained with at least 100 fold less inhibitor using human erythrocytes or Burkitt's lymphoma cells as compared to substrates such as 4-methylumbelliferyl-N-acetyl-α-D-neuramide ammonium salt and N-acetyl neuramin lactose. It is beyond the scope of this chapter to elucidate the mechanism of inhibition by cruzin. Nonetheless the fact that the enzyme's active site is stoichiometrically hindered from binding only to larger substrates, may be indicative of a trapping mechanism. If this proves to be the case, it will not be without precedent since alpha-2-macroglobulin, a protease inhibitor found in human plasma, has a similar mechanism of inhibition (Feldman et al, 1985).

6. _Specificity_:

One of the most remarkable features of cruzin is its specificity for _T. cruzi_ neuraminidase. It was ineffective against several neuraminidases including those from _Vibrio cholera_, _Clostridium perfringens_, influenza virus and _Arthrobacter urefaciens_. Surprisingly, even the neuraminidase from the closely related trypanosome, _T. rangeli_, was not inhibited by cruzin. This peculiar characteristic may enable cruzin to be of immediate use as a diagnostic tool in epidemiological surveys in areas where _T. cruzi_ and _T. rangeli_ concomitantly infect man.

CONCLUSION

The involvement of _T. cruzi_ neuraminidase in the pathogenesis of Chagas' disease has yet to be elucidated. One current working hypothesis is that the hematological changes observed in laboratory animals (Cardoso and Brener, 1980) and in cattle (Esievo et al, 1982) infected with trypanosomes are due to the trypanosome neuraminidase (Pereira, 1983a). Our recent findings have indicated that the enzyme may also play a role in adsorption to and penetration of the trypanosome into host cells (Prioli et al, manuscript in preparation). _T. cruzi_ strains with high neuraminidase activity were found to be less virulent than those with low enzyme activity (Pereira and Hoff, 1986). Furthermore, the neutralization of _T. cruzi_ neuraminidase activity by cruzin may influence the exiting of parasites from infected cells in a manner similar to that described for myxovirus (Schulman, 1970). The studies mentioned above strongly suggest an important role for _T. cruzi_ neuraminidase at the host-parasite interface.

ACKNOWLEDGEMENTS

This work was supported by N.I.H. grant AI 18102 and by UNDP/World Bank/WHO Special Programme for Research and Training in Tropical Diseases. We thank Ms. Irene Doucette for excellent secretarial assistance.

REFERENCES

Alnaes E, Rahaminoff R (1975) On the role of mitochondria in transmitter release from motor nerve terminals. J. Physiol. 248:285.

Bagshawe KD, Currie GA (1968) Immunogenicity of L 1210 murine Leukemia cells after treatment with neuraminidase. Nature 218:1254.

Brener Z (1973) The biology of Trypanosoma cruzi. Annal. Rev. Microbiol.27:349.

Cardoso JE, Brener Z (1980) Hematological changes in mice experimentally infected with T. cruzi. Mem. Inst. Oswaldo Cruz 75:97.

Esievo KAN, Saror DI, Ilemogbabe I, Hallaway K (1982) Variation in erythrocytes surface and free sialic acid concentrations during experimental Trypanosoma vivax infection in cattle. Res. Vet. Sci. 32:1.

Frank JS, Langer GA, Nudd LM, Seradarian D (1977) The myocardial cell surface, its biochemistry and the effect of sialic acid and calcium removal on its structure and cellular ionic exchange. Circ. Resp. 41:702.

Fearon DT (1978) Regulation of membrane sialic acid of beta 1-H dependent deacy dissociation of amplification C3 convertase of the alternative complement pathway. Proc. Natl. Acad. Sci. USA 75:1971.

Feldman SR, Gonias SL, Pizzo SV (1985) Model for α-2 macroglobulin structure and function. Proc. Natl. Acad. Sci. USA 82:5700.

Gorog P, Kovacs IB, Born GVR (1980) Suppression of the intravascular adherence of granulocytes by N-acetyl neuraminidase (sialic) acid. Brit. J. Exp. Pathol. 61:490.

Gorog P, Schrausfstatter I, Born GVR (1972) Effect of removing sialic acid from endotheliam on the adherence of circulating platelets in artheries in vivo. Proc. R. Soc. Lond. (Biol) 214:471.

Hudson L, Hindmarsh PJ (1985) The relationship between autoimmunity and Chagas' disease: causal or coincidental? Curr. Top. Microbiol. Immunol. 117:167.

Libby P, Elroy J, Pereira MEA (1986) A neuraminidase from Trypanosoma cruzi removes sialic acid from the surface of mammalian myocardial and endothelial cells. J. Clin. Invest. 77:127.

Lundgren G, Simons SL (1971) Effect of neuraminidase on the stimulatory capacity of cells in human mixed lymphocyte culture. Clin. Exp. Immunol. 9:915.

Meindl P, Bodo P, Palese P, Schulman J, Toppy H (1974) Inhibition of neuraminidase activity by derivatives of 2-deoxy-2,3-dehydro-N-acetyl neuraminic acid. Virology 58:457.

Pereira MEA (1983a) A developmentally regulated neuraminidase activity in Trypanosoma cruzi. Science 219:1444.

Pereira MEA (1983b) A rapid and sensitive assay for neuraminidase using peanut lectin hemagglutination: Application to Vibrio cholera and Trypanosoma cruzi. J. Immunol. Meth. 63:25.

Pereira MEA, Hoff R (1986) Heterogenous distribution of neuraminidase activity in strains and clones of Trypanosoma cruzi and its possible association with parasite myotropism. Mol. Biochem. Parasitol. 20:183-189.

Rosenberg A, Schengrund CL (1976) Sialidases. In Biological Roles of Sialic Acid. Rosenberg and Schengrund (eds), pp. 295; Plenum Press, New York.

Schulman JL (1970) Effects of immunity on transmission of influenza: Experimental studies. Progr. Med. Virol. 12:128.

Suzuki T, Takauji M, Nagai T (1980) Distribution of sialic acid in frog skeletal muscle and effect of neuraminidase on Ca uptake and ATPase activity of sarcoplasmic reticulum. Jpn. J. Physiol. 30:61.

Trischmann TM, Bloom BR (1982) Genetics of murine resistance to Trypanosoma cruzi. Infec. Immun. 35:546.

Zingales B, Collins W (1985) Trypanosoma cruzi: interaction with host cells. Curr. Top. Microbiol. Immunol. 117:129.

DEFINITION OF TRYPANOSOMA CRUZI ANTIGENS BY ANTIBODIES AND T CELLS

L.Hudson[1] & F. Guhl[2]

[1]Department of Immunology, St. George's Hospital Medical School, London SW17 ORE, UK and [2] Department of Biological Sciences,
University of the Andes, Bogota, Colombia.

In the majority of protozoal diseases one may argue with certainty that a 1- or 2- log reduction in parasite load, achieved by immunoprophylaxis or chemotherapy, would benefit the host and might avoid symptomatic disease. Recent studies on the nature of the immune response elicited by Trypanosoma cruzi suggests that a similar beneficial association cannot be axiomatically assumed for South American trypanosomiasis and Chagas' disease. The trypanosome undoubtedly initiates the disease and, during the early parasitaemic phase, the severity of acute stage disease is dependent upon the intensity of infection, thereafter the location of the parasite and its total numbers become less important. Other processes, notably those dependent on the nature of the immune response elicited by infection, serve to exacerbate the pathogenesis which results in chronic stage cardiopathy and digestive manifestiations.

THE PROBLEMS STATED

The immune response elicited by T.cruzi infection is sufficient in its quality and quantity to virtually eliminate the parasite a few months after initial infection. Provided the immune response remains intact, no recrudescence of parasitaemia is ever seen; the immunological check on parasite proliferation and dissemination is powerful and effective. Acute stage disease may be observed, but it is usually benign and, in any case, resolves with the termination of the parasitaemic phase. Even so a variable proportion of infected patients progress to develop chronic stage disease in the virtual absence of parasites.

The nature of the chronic disease is of great interest in the present context of antigen selection for immunoprophylaxis: (i) the proportion of infected patients showing disease shows great geographical variation, for example, in some parts of Brazil a large proportion of patients develop overt chronic stage disease

NATO ASI Series, Vol. H11
Host-Parasite Cellular and Molecular Interactions
in Protozoal Infections. Edited by K.-P. Chang and D. Snary

whereas in Colombia the vast majority of patients are asymptomatic, (ii) the type and severity of disease shows a similar geographical variation, in Venezuela symptomatology is limited to a mild cardiopathy whereas in Brazil the heart and digestive tract may show severe megalic formations, (iii) a recrudescence of chronic stage parasitaemia following immunosuppression (Barousse et al 1980) has no influence on the severity or progression of chronic disease and (iv) cross-reactive autoimmunity, involving both T and B lymphocyte mediated responses, develops in experimental and clinical infections (reviewed by Hudson and Britten 1985). Under these conditions, it has been argued that only complete sterilising immunity would be sufficient to achieve clinical benefit; for escape of only a few parasites might elicit an autoimmune pathogenesis.

The existence of autoimmunity in Chagas' disease is not open to doubt (Hudson 1985), however, its role in disease pathogenesis has not been established and might be unimportant (Kierszenbaum 1985). In addition, Köberle (1974) has provided evidence suggesting that acute phase destruction of nervous tissue might modulate or contribute to the severity of chronic stage disease, possibly via the anti-parasite immune response acting upon passively acquired parasite antigens (Ribeiro dos Santos and Hudson 1980). Consequently, a reduction in the availability of these antigens, by even a partial reduction in parasitaemia, could be expected to ameliorate acute phase damage and so lessen chronic phase disease.

These crucial observations give cause for optimism and provide a valuable guide to the types of antigens to be considered for clinical use: (i) selective immunisation will be essential, thus avoiding those antigens that might give rise to autoimmune cross reactions (Teixeira et al 1975) or other pathologic consequences and (ii) one should use immunoprophylaxis to achieve the maximal resistance to infection, but need not set the goal unrealistically high. Incomplete but significant protection could bring desirable clinical benefit.

ANTIGENS WITH IMMUNOPROPHYLACTIC AND PATHOLOGIC POTENTIAL.

Antigens associated with resistance to infection.

In common with many other micro-organisms, active infection with T.cruzi achieves a high level of resistance to secondary infection; a level of resistance much greater than that achieved by immunisation with killed parasites or subfractions thereof (Krettli and Brener 1982). Resistance was correlated with a special functional class of antibodies able to react with

the surface of living trypomastigotes and so produce complement mediated lysis (CML) or viable immunofluorescence (VIF). Intriguingly, this antibody activity is lost when patients are given aggressive long term chemotherapy (Krettli et al 1982) and is never developed by mice immunised with fixed trypomastigotes.
In both instances, however, sera are positive for those antigens detected by immunofluorescence or ELISA in conventional diagnostic serology.

Comparison of CML/VIF positive and negative sera from man and mouse, using immunoprecipitation of polypeptides from 125-I labelled trypomastigotes, showed that the loss of CML/VIF reactivity was associated with the loss of several of the high molecular weight components in the immunoprecipitation profile (app Mr 120-160 kDa - Martins et al 1985). A similar, but less complex difference was seen when using sera from mice infected with live trypomastigotes or immunised with formaldehyde fixed trypomastigotes. The latter group produced sera which were CML/VIF negative even though they showed high titres on conventional assays. Significantly, immunoprecipitation analysis revealed a difference of only one major polypeptide at app Mr 160k Da, thus identifying it as an antigen with potential for immunoprophylaxis.

Monoclonal antibodies reacting with the surface of viable trypomastigotes have been isolated in this laboratory and tested for their potential to protect mice by the protocol shown below:

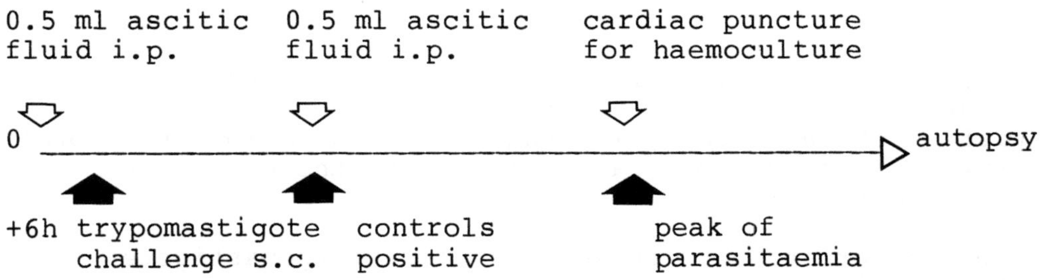

Five monoclonal antibodies with VIF activity have been isolated and tested for their protective effect. Two of these antibodies, EE1.E3 and NN1.D8 lengthened the time to patent parasitaemia and reduced both the peak and duration of the parasitaemic wave, as shown in the tables overpage:

Antibody	Reaction by immunofluorescence with: Live parasites			Fixed parasites		
	*T	A	E	T	A	E
2A2.C4	-	-	-	+	+	+
EE1.E3	++	+	-	+	+	-
NN1.D8	-	++	-	ND	ND	-

*T,A,E - trypomastigote, amastigote, epimastigote. ND - not determined

Passive antibody	Peak no's	Percent of control	Parasitaemia peak day	Parasitaemia duration
none	$7.0x10^4$	-	10	7-12
2A2.C4	$5.0x10^4$	93	10	7-12
EE1.E3	$1.1x10^3$	1.6	11	11-13
NN1.D8	$1.0x10^3$	1.4	11	11-13

Mice were challenged with $5x10^2$ Y strain trypomastigotes.

The monoclonal antibody 2A2.C4 was raised against epimastigotes of T.cruzi and does not recognise live parasites of any life cycle stage. Administration of a total of 1ml of this antibody failed to modify the course of parasitaemia.

Further experiments showed that anti-trypomastigote neutralisation was completely effective and so escape *in vivo* was due to the persistence of intracellular amastigotes (Heath, A and Hudson, L. to be published). Effective immunity against the intracellular parasites will probably require the participation of T cells. Experiments to isolate these antigens by immune affinity chromatography and gene cloning are now in progress.

Interaction between T.cruzi and T.rangeli.

Patients from Choachi and Tibu in Colombia have been followed as part of a longitudinal study of the pattern of T.cruzi polypeptides recognised during the evolution of Chagas' disease. Patients and controls were selected by standard ELISA and immunofluorescence techniques for the diagnosis of T.cruzi infection, and have now been followed for 6 years. Sera from patients were screened by immunoprecipitation using Renex 30 solubilised 125-I labelled trypomastigotes and, on SDS-PAGE analysis, sera from the majority of patients recognised a profile of polypeptides very similar to those reported previously from Brazil (Zingales et al 1984; Martins et al 1985).

Two classes of patients showed marked differences in their precipitation profile (i) those recognising additional or completely different polypeptides or (ii) by a complete absence of T.cruzi specific immunoprecipitation. More detailed serological analysis showed that the latter patients, previously diagnosed as having T.cruzi infection, were in fact infected with an apparently non-pathogenic trypanosome, T.rangeli (Guhl, Hudson, Marinkelle, Jaramillo and Bridge, submitted for publication). Three groups of patients were finally identified; either uncomplicated T.cruzi or T.rangeli infections or a mixed infection. Intriguingly, both immune precipitation and immunoblotting showed that patients with mixed infection consistently produced a stronger anti-T.cruzi response than patients with an infection of either parasite alone.

T.rangeli shares the same insect vectors with T.cruzi over an appreciable part much of its range and in principle should be transmitted more efficiently as infection is gained by innoculation rather than contamination (T.cruzi). It is interesting to note that Chagas' disease assumes its most benign character in those countries (Venezuela and Colombia) where high vector infection rates with T.rangeli have been reported. Conversely, in Brazil and Chile, where T.rangeli is only infrequently reported, Chagas' disease is severe. It is conceivable that this might be the result of an interaction between the two parasites; an early acquired T.rangeli infection does not prevent T.cruzi infection but might protect against disease pathogenesis. In our present series of 200 patients, 2 have converted from T.rangeli infection alone to a mixed infection during the period of surveillance (Hudson, L.& Guhl, F. to be published) and so now are receiving special attention.

On present evidence, the nature of the interaction between T.rangeli and T.cruzi is not clear. Experimental studies in mice have shown that the GP72 glycoproteins on epimastigotes of both parasites cross react strongly. However, immunisation of mice with T.rangeli trypomastigotes, with up to 6 booster injections fails to produce an antiserum which will recognise T.rangeli trypomastigote polypeptides either by immunoprecipitation or immunoblotting (Guhl et al 1985). These observations, and the fact that T.rangeli attains a barely detectable parasitaemia even in mice immunosuppressed by γ-irradiation, suggests that the adaptive immune response (B or T cell mediated) cannot be an important barrier to the uninhibited proliferation of T.rangeli trypomastigotes in vivo.

Antigens associated with pathogenesis

Two classes of T.cruzi antigen may play an important role in the development of Chagas' disease:(i) parasite derived antigens with high affinity for host cell surfaces and (ii) host-parasite cross-reactive antigens.

The phenomenon of parasite antigen adsorption to infection and normal cells in vitro and in vivo may explain the change in the nature of the early inflammatory infiltrate. Initially inflammatory cell are found around "nests" of intracellular parasites, but inflammation becomes more diffuse late in the acute phase (Andrade, 1983). Acquisition of parasite antigens renders the modified host cells susceptible to the host's own anti-parasite immune response (Araujo, 1985) and could explain the intense denervation observed during the resolution of parasitaemia (Ribeiro dos Santos and Hudson 1981). In clinical situations intense tissue destruction by this mechanism could lead to the generation of anti-laminin antibodies (Szarfman et al 1982) and might release internal components of neuronal and muscle cells which cross-react with T.cruzi (Wood et al 1982, Snary et al 1983) to initiate a self sustaining cycle of autoimmunisation and tissue destruction in the chronic disease.

This hypothesis provides early markers of disease (e.g. denervation) as a basis for testing candidate vaccines both for their in vivo efficacy in protection and their potential for pathogenesis. To date the GP90 glycoprotein (see Snary - this volume) has been tested in this way using marmosets with partial protection and no pathology (Scott et al 1985).

T CELL LINES AND HYBRIDOMAS.

Antigen-specific T cell lines and clones with a helper/inducer phenotype have been isolated from chagasic patients (Britten and Hudson, 1985). However, little work has been published on antigen characterisation because of the problems associated with in vitro maintenance, in particular, the supply of peripheral blood mononuclear cells for antigen presentation and "feeders". T cell hybridomas have been prepared from the lymph nodes of T.cruzi infected mice (Britten V. to be published) using the AKR thymoma BW5147. Ag8. TIB 48. In contrast to the human situation where only helper/inducer lines were routinely isolated (Britten and Hudson 1985), supernatants from the majority of the mouse hybridomas isolated express an antigen non-specific suppressor function. Although this function is constitutively expressed, the hybridomas have an unstable karyotype and require repeated cloning for the maintenance of function.

The recent combination of antigen fractionation by SDS-PAGE and Western blotting for identification of specific antigens responsible for T cell stimulation (Lamb and Young, 1986) should have a major application in T.cruzi infection where many of the remaining problems of immunological protection centre on the need to erradicate the persistent intracellular infection and to identify the cross reactive antigens recognised by T cells during chronic disease. In this way not only might it be possible to prevent infection in the unaffected population but also, using the approach of "anti-idiotype" immune regulation similar to that used in the control of experimental autoimmune encephalomyelitis (Ben-Nun et al 1981), it might be possible to halt the pathogenic process in those 20 million individuals already infected and in danger of developing overt disease.

CONCLUSIONS AND PROSPECTS

The interplay between T. cruzi and its host is very complex and many of the factors which control infection or the expression of disease remain to be identified. Even so, one may recognise significant advances in several areas which have direct bearing on these problems.

The association of specific anti-living trypanosome antibodies with resistance to infection might provide a unique route to the identification of protective antigens, in a manner analogous to the that used in the study of malaria. Similarly, the identification of a possible interaction between T. cruzi and T. rangeli in promoting resistance, not to infection but to disease, should provide novel avenues for exploration. Clearly, it is only with a fundamental understanding of the parasite's interaction with the host's immune system, and its potential for disease pathogenesis, that the final goal - prevention or cure of infection and avoidance of disease - will be achieved.

ACKNOWLEDGEMENTS.

We are extremely grateful to our colleagues, identified in the text, for permission to cite their unpublished data and to Dr Veronica Britten for her help in manuscript preparation. Work from both our laboratories was supported by contracts with the Commission of the European Economic Community.

REFERENCES

Andrade Z. 1983 Mechanisms of myocardial damage in Trypanosoma cruzi infection. In: "Cytopathology of parasitic disease." Ciba Foundation Symposium 99. pp214-233. Pitman, London.

Araujo FG. 1985 Trypanosoma cruzi: expression of antigens on the membrane surface of parasitized cells. J. Immunol. 135:4149-4154.

Barousse AP, Costa JA, Epasto U, La Plume H, and Segura EL. 1980 Chagas' disease and immunosuppression (in Spanish) Medicina 40: 17-26.

Ben-Nun A, Wekerle H and Cohen IR. 1981 Vaccination against autoimmune encephalomyelitis with T-lymphocyte line cells reactive against myelin basic protein. Nature 292:60-61.

Britten V and Hudson L. 1985 Isolation and characterisation of human T-cell lines from a patient with Chagas' disease. The Lancet, September 21:637-639

Guhl F, Hudson L, Marinkelle CJ, Morgan SJ and Jaramillo C. 1985 Antibody response to experimental Trypanosoma rangeli infection and its implications for immunodiagnosis of South American trypanosomiasis. Acta Tropica 42:311-318.

Hudson L. 1985 Autoimmune phenomena in chronic chagasic cardiopathy. Parasitology Today. 1:6-9.

Hudson L and Britten V. 1985 Immune response to South American trypanosomiasis and its relationship to Chagas' disease. Brit. Med. Bull. 41: 175-180.

Kierszenbaum F. 1985 Is there autoimmunity in Chagas' disease? Parasitology Today. 1:4-6.

Köberle F. 1974 Pathogenesis of Chagas' disease. In: "Trypanosomiasis and Leishmaniasis with special reference to Chagas' disease." Ciba Foundation Symposium No 20 pp 137-158 Elsevier, North Holland.

Krettli AU and Brener Z. 1982 Resistance against Trypanosoma cruzi associated to anti-living trypomastigote antibodies. J. Immunol. 128: 2009-2015.

Krettli AU, Cançado JR and Brener Z 1982 Effect of specific chemotherapy on the levels of lytic antibody in Chagas' disease. Trans. Roy. Soc. Trop. Med. Hyg. 76: 334-339.

Lamb, DB and Young JR. 1986 A novel approach to identification of T cell epitopes in Mycobacterium tuberculosis using human T lymphocyte clones Immunol. (In press).

Martins MS, Hudson L, Krettli AU, Cançado JR and Brener Z. 1985 Human and mouse sera recognise the same polypeptide associated with immunological resistance to Trypanosoma cruzi infection. Clin. Exp. Immunol 61: 343-350.

Ribeiro dos Santos R and Hudson L 1980 Trypanosoma cruzi: immunological consequences of parasite modification of host cells. Clin. Exp. Immunol. 40: 36-41.

Ribeiro dos Santos R and Hudson L. 1981 Denervation and the immune response in mice infected with Trypanosoma cruzi. Clin. Exp. Immunol. 44:349-354.

Scott MT, Neal RA and Woods NC. 1985. Immunisation of marmosets with Trypanosoma cruzi cell surface glycoprotein (GP90). Trans. Roy. Soc. Trop. Med. Hyg. 79:451-454.

Snary D, Flint JE, Wood JN, Jessell TM and Miles MA. 1983. A monoclonal antibody with specificity for Trypanosoma cruzi, central and peripheral neurones and glia. Clin. Exp. Immunol. 54:617-624.

Szarfman A, Terranova VP, Rennard SI et al. 1982 Antibodies to laminin in Chagas' disease. J. Exp. Med. 155:1161-1171.

Teixeira ARL, Teixeira ML and Santos-Buch CA 1975 The immunology of experimental Chagas' disease. IV Production of lesions in rabbits similar to those of chronic Chagas' disease in man. Am. J. Pathol. 80: 163-180.

Wood JN, Hudson L, Jessell TM and Yamamoto M. 1982. A monoclonal antibody defining antigenic determinants on subpopulations of mammalian neurones and Trypanosoma cruzi parasites. Nature 296:34-38.

Zingales B, Abuin G, Romanaha AJ, Chiari E and Colli W. 1984. Surface antigens of stocks and clones of Trypanosoma cruzi isolated in humans. Acta Tropica 41:5-16.

MOLECULAR PATHOGENICITY OF LEISHMANIA IN INTRACELLULAR PARASITISM OF MACROPHAGES: ASSOCIATION OF VIRULENT PHENOTYPE WITH INCREASED EXPRESSION OF A MAJOR MEMBRANE GLYCOPROTEIN AND MICROSOMAL N-ACETYLGLUCOSAMINYL-TRANSFERASE

K.-P. Chang and J. A. Kink
Dept. Microbiology, UHS/Chicago Medical School, N. Chicago, IL 60064, USA

INTRODUCTION

Molecular mechanism of virulence in protozoal infection remains largely unknown. Of particular relevance are glycoproteins, especially those on the parasite surface, as they have been often implicated in the evasion of host immunity or invasion of host cells. Since glycosylation of proteins is known to change their properties and possibly their functions (Schwarz & Datema 1982), molecular regulation of this pathway may bear on the expression of surface glycoproteins as virulent determinants.

A useful model to examine the possibilities raised is experimental leishmaniasis. The causative agents are trypanosomatids of the genus, *Leishmania*, which live extracellularly as promastigotes in the sandfly vector and intracellularly as amastigotes in the macrophages of the mammalian host (see Chang, Fong & Bray 1985). Promastigotes not only infect macrophages exclusively as their host cells but also survive, differentiate and multiply in their phagolysosomes - the very intracellular site where other microorganisms are usually killed (see Chang 1983). Leishmania surface glycoproteins are apparent virulent factors, as they appear to play important roles in the sequential events of host-parasite cellular interactions to establish intracellular parasitism of macrophages (Chang & Fong 1983). This view is strengthened by finding the association of active protease activity with a major leishmania membrane glycoprotein of 63 kDa (gp63) (Etges, Bovier & Bordier 1986; Chaudhuri & Chang 1987).

Here, we show that the virulent phenotype of *L. m. amazonensis* is accompanied by an increased expression of gp63. Its role as an important virulent determinant is suggested. Leishmania virulence is also associated with an elevated level of N-acetylglucosaminyltransferase and N-glycosylation of proteins. We propose that molecular regulation of these events is involved in the control of leishmanial virulence.

MATERIALS AND METHODS

The source of *L. m. amazonensis* (LV78) and J774G8 macrophages, methods of their cultivation and leishmanial infection of BALB/c mice were described previously (Chang 1980; see Chang & Hendricks 1985). [^{35}S]methionine incorporation, SDS-PAGE and monoclonal affinity purification of gp63 followed established methods (Chang & Chang 1986; Chang et al. 1986). Assays for N-acetylglucosaminyltransferase, [^{3}H]mannose incorporation, DNA isolation and related techniques are described elsewhere (Kink & Chang 1987). Promastigotes of avirulent phenotype were those passaged weekly for prolonged periods *in vitro*. Promastigotes of virulent phenotype used were of two different types: those passaged weekly for less than three months and cells made resistant to 10 ug/ml of tunicamycin selected by stepwise increases of drug concentrations (Kink & Chang 1987).

NATO ASI Series, Vol. H11
Host-Parasite Cellular and Molecular Interactions
in Protozoal Infections. Edited by K.-P. Chang and D. Snary

RESULTS AND DISCUSSION

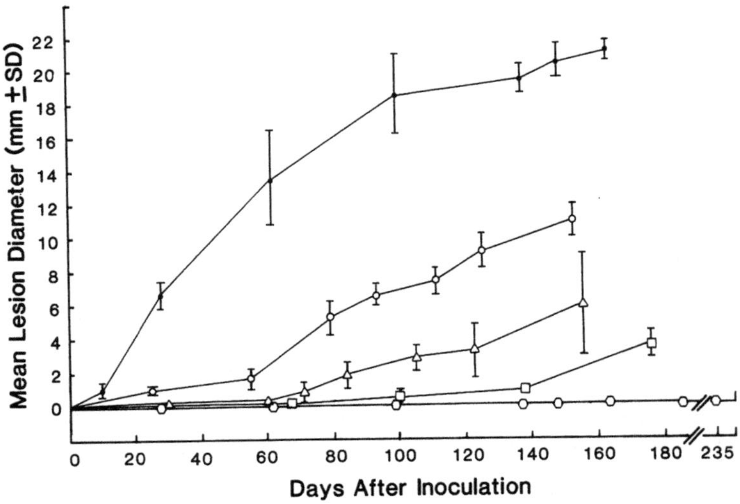

FIGURE 1. Progressive loss of leishmania virulence in vivo with periods of in vitro cultivation of L. m. amazonensis promastigotes in vitro. Amastigotes isolated from mouse lesions were allowed to differentiate into promastigotes and the latter were subsequently subcultured weekly (=5-6 generations) in Hepes-buffered Medium 199 plus 10% heat-inactivated fetal bovine serum at 27 °C. A group of 4-6 BALB/c mice were each inoculated at the tail base with 5 X 10^6 promastigotes at late-log to stationary phase of growth after subcultivation in vitro for the following periods in month: 1.5 (●), 6 (o), 14 (△), 19 (□) and 21 ( ). The size of lesions was measured by using a caliper at different post-infection periods as indicated.

Virulent Phenotype of Leishmanial Promastigotes: Reduction, maintenance and Induction in vitro

Virulence of promastigotes is defined here as their ability to produce lesions at the tail base of BALB/c mice and to parasitize macrophages in vitro. Assessment of virulence is based on the development of the lesion and its size and the increase in the number of intracellular leishmanias during a given period of infection with a specific number of promastigotes. The loss of leishmanial virulence as defined has long been noted for all species to accompany repeated passages in vitro as promastigotes (see Chang 1983; Chang & Hendricks 1985 for review). The generation of avirulent cells under such conditions varie in its rapidity and completness with species of Leishmania and media used.

Fig. 1 shows that the promastigotes of L. m. amazonensis lose their virulenc to mice gradually during their in vitro cultivation for periodsof up to about two years, as determined by the size of the lesions and their development with time of infection. The largest and most rapidly developing lesions seen are those produced by promastigotes passaged for less than eight weeks (about 50 generations). With increasing periods of their in vitro cultivation, promastigotes produced lesions, which are progressively reduced in size and delayed in the onset of their development. Those grown in vitro for 21 months or more

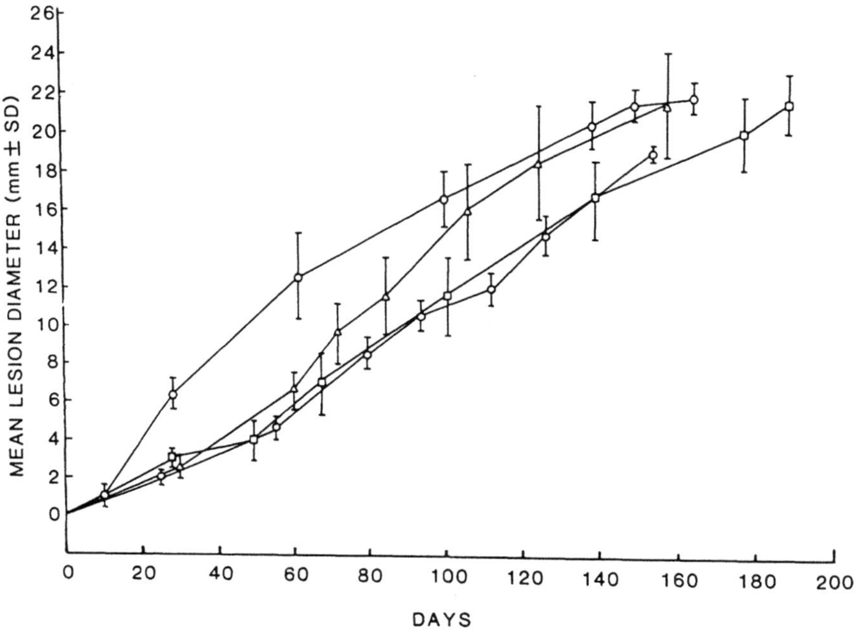

FIGURE 2. Maintenance of leishmanial virulence in vivo of tunicamycin-resistant L. m. amazonensis promastigotes independent of their period of in vitro cultivation. Promastigotes grown in vitro for about three months were made resistant to tunicamycin at 10 ug/ml as described (Kink & Chang 1987) and subsequently subcultured in drug-containing medium as described in the legend to Fig. 1. The growth rate of tunicamycin-resistant cells is very comparable to that of the parental wild type. Group of mice were each infected with 5 X 10^6 tunicamycin-resistant promastigotes after cultivation for 5 (O), 16 (△), 19 (□) and 21 (o) months.

(more than 500 generations) apparently become avirulent, as they produced no papable lesions in mice after infection for eight months or more. Pretreatment of virulent wild type with tunicamycin at 10 ug/ml at 27°C for 10 hours also renders them non-infective to mice, although cells so treated are not killed and can recover fully when released from the treatment.

In contrast to the wild type promastigotes, those made resistant to tunicamycin sustain prolonged in vitro cultivation without losing their virulence to mice (Fig. 2). Tunicamycin-resistant promastigotes grown in vitro for five months and those for up to 21 months produced lesions of comparable size and similar in the course of development. This maintenance of virulence by tunicamycin-resistant variants is their inherent property. It is not due to the effect of the drug carried over with the inocula against host immune system, as they were passaged in drug-free medium for 4-5 times in a period of two weeks and then washed thrice before use for inoculation.

When made resistant to tunicamycin, avirulent promastigotes obtained by prolonged in vitro cultivation regain their virulence to mice (not shown). This recovery of virulence is however partial. Recently, Katakura (1986) has reported his success in recovering the virulence of L. donovani by repeated infection of BALB/c mice for 15 times or more. These findings indicate that leishmanial virulence is a phenotype inducible under appropriate selective pressures.

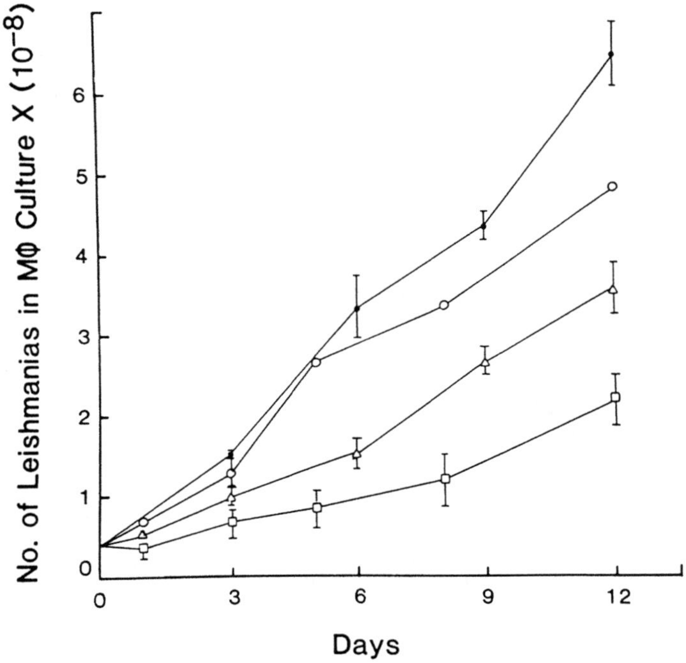

FIGURE 3. Progressive loss of infectivity to macrophages of L. m. amazonensis with periods of its cultivation as promastigotes in vitro. Infection was done at 35°C in 25 cm^2 TC-flasks, each of which received 4 X 10^6 J774G8 cells and 20 X 10^6 promastigotes in four ml Hepes-buffered Medium RPMI 1640 with 20% heat-inactivated fetal bovine serum. The total number of intracellular leishmanias per flask was estimated by: The total number of macrophages X average number of leishmanias per cell X % infected macrophages at different periods of post-infection as indicated. Promastigotes used as inocula were those subcultured at 27°C for the following periods in months: 2 (●), 6.5 (o), 14 (▲) and 20 (□).

Interaction of Promastigotes with Macrophages as the Cellular Basis of Leishmania virulence

Infection of J774G8 macrophages with the same group of virulent and avirulent cells in vitro yielded the same results. Promastigotes grown for shorter periods in vitro gave better infection than those grown for longer periods (Fig. 3). Tunicamycin-resistant promastigotes behave like the former, irrespective of their in vitro culture periods, while virulent wild type pretreated with tunicamycin fail to establish intracellular parasitism (not shown). Infection of macrophages with virulent promastigotes in vitro is marked by a higher yield of intracellular leishmanias during a period of up to 12 days. There is no significant difference in the number of intracellular leishmanias seen for all groups of promastigotes during the early period of infection for up to 10 hours. Thus, steps beyond the initial host-parasite recognition for the intracellular entry of the parasites must be considered to account for the difference observed between virulent and avirulent cells in their ability to parasitize macrophages. Whether this difference lies in the

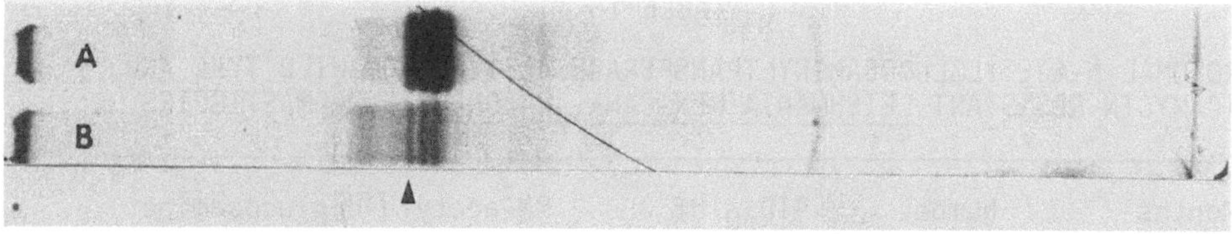

FIGURE 4. Quantitative difference of a major membrane glycoprotein (gp63) between virulent and avirulent L. m. amazonensis promastigotes determined by immunoprecipitation with a specific monoclonal antibody. [^{35}S]methionine labeling, immunoprecipitation, SDS-PAGE and autoradiography were done as described by Chang et al. (1986). A & B, immunoprecipitable gp63 from 10^8 promastigotes grown to late-log phase by weekly subculture *in vitro* for three (virulent) and 24 (avirulent) months, respectively. Arrow=66 kDa.

intracellular survival, differentiation or multiplication of the parasites, and how it may be related to the biochemical differences seen between virulent and avirulent cells (see below) remain to be determined. Nevertheless, these studies point to the intracellular events of promastigote-macrophage interactions as the cellular basis of leishmanial virulence to mice *in vivo*.

A Major Membrane Glycoprotein (gp63) as a Virulent Determinant of Leishmania

Avirulent cells appear to express much less gp63 than do the virulent cells, as assessed by using a monoclonal antibody specific to this molecule. Monoclonal binding of [^{35}S]methionine labeled and NP-40 extracted cell lysates gives several fold less radioactivity in the avirulent than in the virulent cells. This is demonstrable by SDS-polyacrylamide gel electrophoresis of gp63 by autoradiography (Fig. 4). Fluorometric assays using FITC-Fab fragment of the same monoclonal antibody (Chang et al. 1986) also indicate a 2-3 fold less gp63 expressed on the surface of avirulent than virulent cells (Chaudhuri & Chang 1987). These observations coupled with the finding of gp63 as an acid protease suggest that it is an important virulent determinant of leishmanias. Previously, we have proposed that gp63 may serve as ligands for binding of promastigotes to macrophage membrane receptors during intracellular entry of the parasites and also may serve the function to protect them from the lytic action of enzymes in the lysosomal compartment (Chang & Fong 1983; Chang & Chang 1986). Additional functions of leishmania gp63 as a protease may be proposed in the light of the present findings to include the following: (1) A nutritional role in breaking down proteins into peptides/amino acids for utilization by the parasites; (2) Protection of parasites against humoral immune factors, e.g. complement and antibodies, during their brief extracellular phase in the mammalian host; and (3) Modification of these and other host molecules to facilitate their intracellular parasitism of macrophages.

Association of Protein Glycosylation with Leishmanial Virulence

The effect of tunicamycin-treatment and induction of tunicamycin-resistance on leishmanial virulence strongly suggests that it may be related to a tunicamycin-sensitive enzyme, i.e. N-acetylglucosaminyltransferase in the dolichol pathway of protein glycosylation (Schwarz & Datema 1982). Thus, we have compared between virulent and avirulent cells the activity of this enzyme and 2-D- ^{3}H mannose incorporation into leishmanial macromolecules. Table 1 shows the results of our assays for the glycosyltransferase of leish-

TABLE 1

MICROSOMAL N-ACETYLGLUCOSAMINYLTRANSFERASE ACTIVITY OF WILD TYPE AND TUNICAMYCIN-RESISTANT *LEISHMANIA MEXICANA AMAZONENSIS* PROMASTIGOTES

Months in Culture	Number of Expt.	[a]ID_{50} of Tunicamycin (ug/ml)	[e]N-acetyl-[^{3}H]glucosamine incorporated (X 10^{-3} cpm/mg protein)
1-17	5	250[b]	71.7 ± 18.80
1-17	5	60[c]	23.6 ± 7.40
2- 4	3	3[d]	9.0 ± 0.85
9-12	3	3[d]	4.0 ± 0.51
18-21	3	3[d]	2.4 ± 0.62

[a]Determined on the basis of growth inhibition by tunicamycin
[b]Variants made resistant to and grown in the presence of tunicamycin at 80 ug/ml
[c]Variants made resistant to and grown in the presence of tunicamycin at 10 ug/ml.
[d]Wild type cells grown for different periods *in vitro* as promastigotes
[e]Assays done as described by Kink & Chang (1987) and, in the case of tunicamycin-resistant cells, after release of the drug pressure for 1-2 weeks.

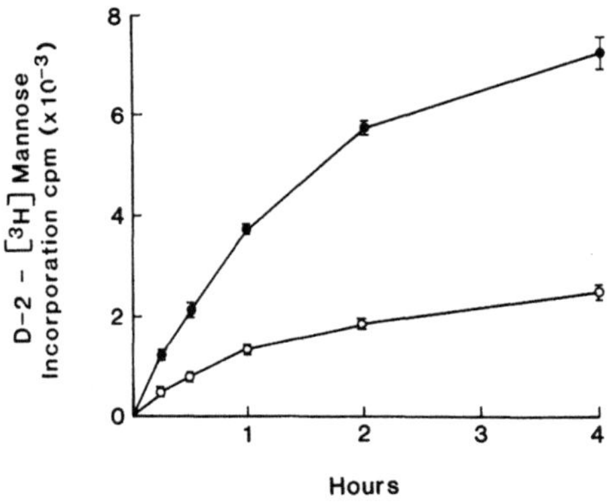

FIGURE 5. Incorporation of [^{3}H]mannose by virulent and avirulent *Leishmania mexicana amazonensis* promastigotes. Cells were incubated at 10^7/ml in a glucose-free medium at 27°C containing 2-D-[^{3}H]mannose (20.1 mCi/mmol) at 5 uCi/ml. Aliquots of 100 ul (=10^6 cells) in triplicate were taken at different time points as indicated and subjected to scintillation counting after precipitation with trichloroacetic acid. (•) & (o), virulent and avirulent cells obtained by growing promastigotes *in vitro* for 2 and 20 months, respectively.

manias with different virulent phenotypes. The activity of this enzyme in the wild type promastigotes diminishes gradually with periods of their *in vitro* cultivation as promastigotes, in parallel to their loss of virulence observed (Fig. 1). On the other hand, tunicamycin-resistant variants contain highly elevated activity of this enzyme, the degree of which is proportional to that of their drug-resistance. The enzyme activity remains elevated in tunicamycin-resistant cells, regardless of their *in vitro* culture periods, consistent with their maintenance of virulence (Fig. 2). Since N-acetylglucosaminyltransferase is the first enzyme in the dolichol pathway, its change in activity is expected to affect glycosylation of leishmania proteins with asparagine-linked oligosaccharides of the high mannose type (Parodi, Martin-Barrientos & Engel 1984). Indeed, we found that the overall glycosylation of proteins was 2-3 folds higher in the virulent than in the avirulent cells, as determined by 2-D-[^{3}H]mannose incorporation (Fig. 5). This difference is specific, as both cell types have a very comparable rate of growth and protein biosynthesis (not shown).

N-acetylglucosaminyltransferase Gene Amplification of Tunicamycin-resistant Leishmania

In the foregoing section, we described the biological effect of tunicamycin on leishmanial virulence. Of particular interest is the induction of tunicamycin-resistance, which confers on leishmanias their ability not only to maintain their virulent phenotype but also to recover it from avirulent cells. Our attention was drawn to the target enzyme of tunicamycin, N-acetylglucosaminyltransferase, which shows an increased activity in all virulent promastigotes, especially those made resistant to tunicamycin. Thus, regulation of leishmania virulence may be elucidated by understanding the molecular mechanism of tunicamycin-resistance.

Kink & Chang (1987) have shown that the mechanism of tunicamycin-resistance in leishmanias is based on the amplification of their gene(s) encoding N-acetylglucosaminyltransferase, resulting in its overproduction to overcome the inhibitory effect of the drug. The principal evidence for this conclusion are as follows: (1) Tunicamycin-resistant cells contain amplified DNA as discrete bands of restriction fragments on agarose gel electrophoresis and ethidium bromide staining; (2) The amplified DNA hybridizes with Alg 7, a plasmid with an insert of the yeast gene apparently coding for the glycosyltransferase; and (3) The degrees of this hybridization and DNA amplification are both proportional to the level of tunicamycin-resistance. The amplified DNA in tunicamycin-resistant cells were further studied by cloning their Bam HI fragments of 12, 14, 15 and 22 kbs in pBR 322 with *E. coli* and in genomic libraries (Detke, Kink & Chang In preparation). Southern and dot blot hybridization using cloned DNAs as probes against the total cellular DNAs showed their amplification to be up to 100-fold in the drug-resistant cells (Fig. 6), as found by hybridization with Alg 7. Restriction mapping of the amplified Bam HI fragments showed that they shared little or no sequence homology and were linked in the order of 12, 22, 14 and 15 kbs. Thus, the total length of the amplicon is about 63 kbs, which appear to exist as circular molecules, judging from their migration in pulse field gel electrophoresis and their isolation by the alkaline lysis method routinely used for plasmids (Unpublished). Northern analysis further revealed that the cloned amplified DNAs hybridized more strongly with a number of mRNA species from tunicamycin-resistant cells than those from the wild type (Unpublished). One or more species of these mRNAs must be the transcripts for the glycosyltransferase. Apparently, amplification of the gene(s) encoding this enzyme

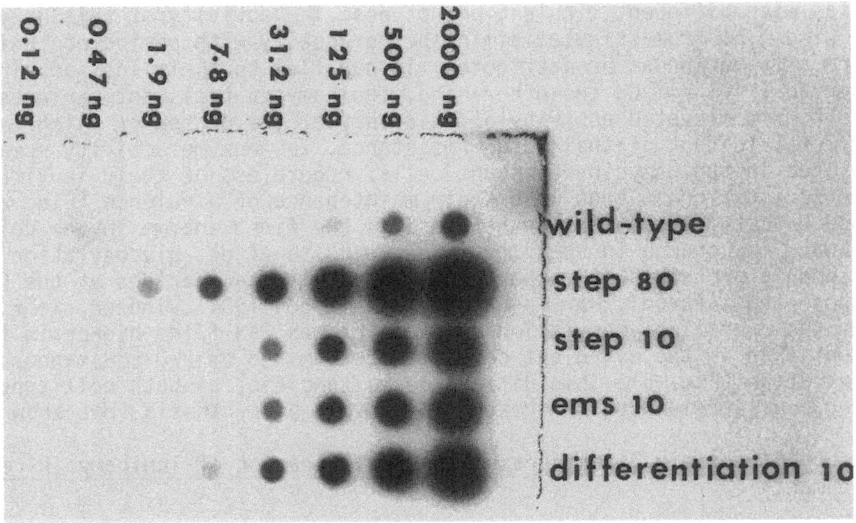

FIGURE 6. Amplification of presumptive Gene(s) encoding N-acetylglucosaminyltransferase in Tunicamycin-resistant Leishmania mexicana amazonensis. DNAs from the wild type, variants resistant to 10 ug/ml of tunicamycin (step 10, ems 10 & differentiation 10) and to 80 ug/ml of tunicamycin (step 80) were dot-bloted in descending concentrations on nitrocellulose paper. The blot was hybridized with amplified Bam HI fragments of 12, 14, 15 and 22 kbs, which were isolated from tunicamycin-resistant cells, inserted in pBR 322 and cloned in E. coli.

results in the accumulation of its mRNA in the tunicamycin-resistant cells, accounting for the increased level of the enzyme detected in them.

Molecular Mechanism of Leishmania Virulence and its Regulation

Evidence presented suggest that glycoproteins, such as gp63 are virulent determinants of leishmanias and that N-acetylglucosaminyltransferase in N-glycosylation of proteins is closely associated with their virulent phenotype. Thus, the molecular mechanism of leishmanial virulence may be based on the expression of leishmanial glycoproteins and its regulation by protein glycosylation.

We propose that gp63 is an important virulent factor of leishmanias because of its universal presence on the surface of all pathogenic species, its abundance (constituting up to 1% of the total cellular proteins) and its acidic protease activity. The possible role(s) of this molecule as a protease in intracellular parasitism of macrophages have been already discussed in the preceding section. Activity of the dolichol pathway for protein glycosylation appears to affect the expression of gp63 and possibly other glycoproteins as virulent determinants. Virulent phenotype of leishmanias, irrespective of the selective pressure applied, is always associated with a high activity of this pathway. How the N-glycosylation of proteins, such as gp63 via this pathway may affect its activity as a virulent determinant is at present unclear. Conceivably, inadequate glycosylation of proteins may result in their degradation or sequestration in inappropriate cellular compartment, as it has been suggested that the carbohydrate moiety of glycoproteins plays a role in stabilizing these molecules or in serving the function as signals for their appropriate intracellular traffics (Schwarz & Datema 1982). Alternatively, quantitative and qualitative changes in the glycosylation of proteins, e.g. gp63 may affect its molecular conformation and thus its enzymatic activity necessary for its expression as a virulent determinant.

N-glycosylation of proteins as an important regulator of leishmania virulence is suggested initially by our fortuitous finding of its association with tunicamycin-resistance. The induction of tunicamycin-resistance in leishmanias is made possible by their gradual acclimatization to increasing drug concentrations, which results in the selection of cells with amplification of gene(s) encoding N-acetylglucosaminyltransferase. It appears that transcriptions of the gene(s) lead to an over-production of this enzyme, rendering leishmanias not only resistant to tunicamycin but also efficient in protein glycosylation. The implication of these molecular events in the regulation of leishmania virulence is strongly suggested by its maintenance and recovery with the development of tunicamycin-resistance. It is possible that the experimental selection of promastigotes for tunicamycin-resistance represents an artificial pressure, which minics the one operational in natural infection to select for their virulent phenotype. The nature of this natural selective pressure is unknown. It must involve host factor(s), which are absent under the conditions used for the cultivation of promastigotes _in vitro_. Whatever the mechanism of the natural selection for leishmania virulence, it fosters the activity of the dolichol pathway for protein glycosylation, perhaps in a way similar to, but less dramatic than the induction of tunicamycin-resistance.The availability of tunicamycin-resistant leishmanias will at the very least facilitate our study on N-acetylglucosaminyltransferase, the structure of its gene(s) and its expression to elucidate their role in the regulation of leishmania virulence.

ACKNOWLEDGEMENTS

We thank A. E. Potthoff for technical assistance. This work was supported by Grant AI-20486 from the National Institute of Allergy and Infectious diseases of USA.

REFERENCES

Chang KP (1980) Human cutaneous _Leishmania_ in a mouse macrophage line: Propagation and isolation of intracellular parasites. Science 209: 1240-1242

Chang KP (1983) Cellular and molecular mechanisms of intracellular parasitism in leishmaniasis. Int Rev Cytol (Suppl) 14: 267-305

Chang CS, Chang KP (1986) Monoclonal antibody affinity purification of a leishmania membrane glycoprotein and its inhibition of leishmania-macrophage binding.Proc Natl Acad Sci USA 83: 100-104

Chang KP, Fong D (1983) Cell biology of host-parasite membrane interactions in leishmaniasis. In: Evered & Collins (eds) Cytopathology of parasitic diseases, CIBA Foundation Symposium vol. 99. Pitman Press, London, p 113

Chang KP, Fong D, Bray RS (1985) Biology of Leishmania and leishmaniasis. In: Chang & Bray (eds) Leishmaniasis, Human Parasitic Diseases vol. I. Elsevier, Amsterdam, Chapter 1, p 1

Chang KP, Hendricks LD (1985) Laboratory cultivation and maintenance of Leishmania. In: Chang & Bray (eds) Leishmaniasis, Human Parasitic Diseases vol. 1. Elsevier, Amsterdam, Chapter 13, p 214

Chang CS, Inserra TJ, Kink JA, Fong D, Chang KP (1986) Expression and size heterogeneity of a 63 kilodalton membrane glycoprotein during growth and transformation of Leishmania mexicana amazonensis. Mol Biochem Parasitol 18: 197-210

Chaudhuri G, Chang KP (1987) Acid protease activity of Leishmania gp63. In: Chang (ed) Host-parasite molecular recognition and interaction in protozoal infections, NATO ASI Series, This vol. Springer, Berlin Heidelberg New York

Etges RJ, Bouvier J, Bordier C (1986) The major surface protein of Leishmania is a protease. J Biol Chem 261: 9098-9101

Katakura K (1986) Studies on infectivity of Leishmania donovani promastigotes: Isolation, intracellular behavior and biological characteristics of virulent lines.Jikeikai Med J 33: 157-173

Kink JA, Chang KP (1987) Tunicamycin-resistant Leishmania mexicana amazonensis Expression of virulence associated with an increased activity of N-acetylglucosaminyltransferase and amplification of its presumptive gene. Proc Natl Acad Sci USA 84 (In press)

Schwarz RT, Datema R (1982) The lipid pathway of protein glycosylation and its inhibitors: The biological significance of protein-bound carbohydrates. Adv Carbohydr Chem Biochem 40: 287-374

INFECTIVITY OF *LEISHMANIA* PROMASTIGOTES IS DEPENDENT ON THE EXPRESSION OF A 65,000 DALTON SURFACE ANTIGEN

F. SANTORO, M. KWEIDER, J.L. LEMESRE, F. DARCY and F.S. RIZVI

Centre d'Immunologie et de Biologie Parasitaire, Unité Mixte INSERM 167 CNRS 624, Institut Pasteur, 1 Rue du Prof. Calmette, 59019 Lille Cédex, France

INTRODUCTION

Leishmania spp., a digenetic trypanosomatid protozoan, is an obligate intracellular parasite of mononuclear phagocytes transmitted by phlebotomine sandflies. The parasite cycles between a flagellated motile promastigote stage in the insect vector and a non-motile intracellular amastigote stage parasitizing the mammalian phagocytic cells. Differentiation into amastigote occurs after surface attachment and entry of promastigotes into mononuclear phagocytes.

A number of investigators have shown that promastigotes from the stationary phase culture were more infective than those from logarithmic phase both *in vivo* for experimental animals and *in vitro* for mouse peritoneal MØ (Giannini 1974, Keithly & Bienen 1981, Sacks & Perkins 1984, Rizvi *et al.* 1985). Also the late stage promastigotes obtained from infected sandflies were more infective *in vivo* & *in vitro* than the early stage forms. More recently, we and others (Rizvi *et al.* 1985, Sacks *et al.* 1985) have reported that the development of infective stage promastigotes is associated with changes in surface components.

To characterize the infective stage specific molecules and to evaluate their biological significance in promastigote-MØ interactions, several monoclonal antibodies (mAb) were generated against the surface components of the promastigotes of *Leishmania braziliensis* (MHOM/BR/00/LV65) and *L. chagasi* (MHOM/BR/81/IMPERATRIZ). Using these mAbs, we demonstrate that the development of the infective stage of *Leishmania* promastigotes is associated with modifications in antigenic expression on the promastigote surface.

NATO ASI Series, Vol. H11
Host-Parasite Cellular and Molecular Interactions in Protozoal Infections. Edited by K.-P. Chang and D. Snary

RESULTS AND DISCUSSION

For the present study the promastigotes of Leishmania spp. were cultivated in a monophasic medium to obtain a more or less uniform population of the parasite at different points of their growth curve. In our controlled cultures the promastigote of L. braziliensis remained in the active dividing phase (logarithmic) up to the 5th day and in the non-dividing phase (stationary) for the 6th and 7th day (Fig. 1). However, the stationary phase of L. chagasi started a day earlier.

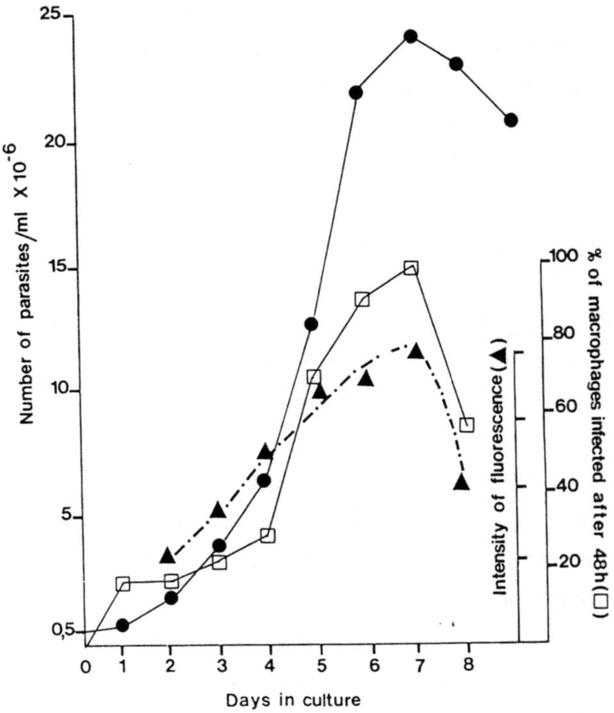

Figure 1. Comparison between the in vitro growth of L. braziliensis promastigotes (●——●), their infectivity towards mouse peritoneal MØ (□——□) and binding of mAb VD5/25 to the promastigote surface (▲-•-▲).

The sequential development of an infective promastigote population of L. braziliensis was also demonstrated in vitro (Fig. 1). As shown previously for other Leishmania spp. (Sacks & Perkins 1984, Rizvi et al. 1985), the generation of infective forms was related to the growth cycle and was restricted to the non-dividing promastigotes. The in vitro development of infective forms seems to be similar to the changes occuring during the growth of promastigotes in sandfly vector (Sacks & Perkins 1984). Moreover, the generation of an infective form in the sandfly suggests that the nutrient-depleted environment, resembling that of stationary cultures, exerts a selective pressure on the promastigotes, thereby giving rise to an infective form possibly resistant to enzyme degradation within the MØ. In fact, the labelled surface molecules of Leishmania parasites were shown to persist up to 36h within the MØ without being digested (Chang & Fong 1983).

In order to evaluate the surface antigenic expression of the promastigotes obtained at different time of growth, a battery of mAbs was generated against the promastigotes of L. braziliensis and L. chagasi. Fusion of SP2/O myeloma cells with spleen cells from mice immunized with two different promastigote populations generated several hybridomas secreting two different isotypes (IgM & IgG1k) of mAbs to the surface antigens. An IgM mAb against L. braziliensis (VD5/25) bound to the surface of promastigotes in IFA tests and precipitated two molecules of apparent Mr 65,000 and 50,000 (GP65-50) from the lysates of the radio—iodinated promastigotes. The IFA titer of this mAb increased progressively as the promastigotes developed in culture and reached at maximum with stationary phase promastigotes. Another IgM mAb against the log phase L. chagasi (IIIB1/26), which cross reacts with L. braziliensis, showed the same IFA titer throughout the developmental cycle. The identical results were obtained by flow cytometric analysis. The peak fluorescence intensity was in the increasing order as the parasite matured and was always maximum with the infective form (Fig. 1). This antigenic differentiation was not detected with anti L. braziliensis polyclonal rabbit antisera suggesting that only a few epitopes, including that recognized by VD5/25, have their expression effectively increased on the surface of infective promastigotes. The intracellular survival of L. braziliensis was significantly inhibited when the infective promastigotes were treated with VD5/25 (Fig. 2). It appears, therefore, that the increasing expression of GP65 on the promastigote surface represents an essential mechanism of Leishmania survival within the MØ.

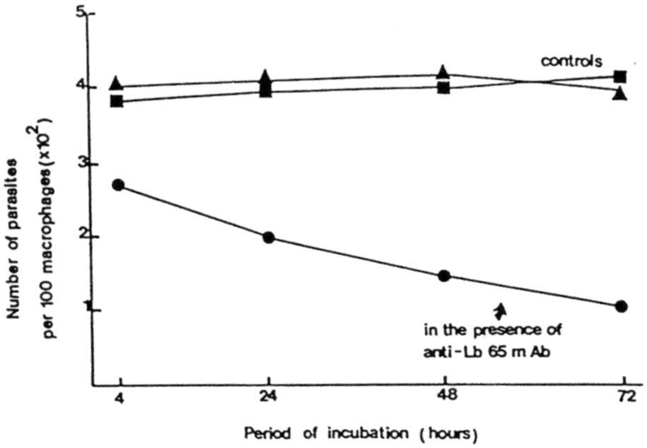

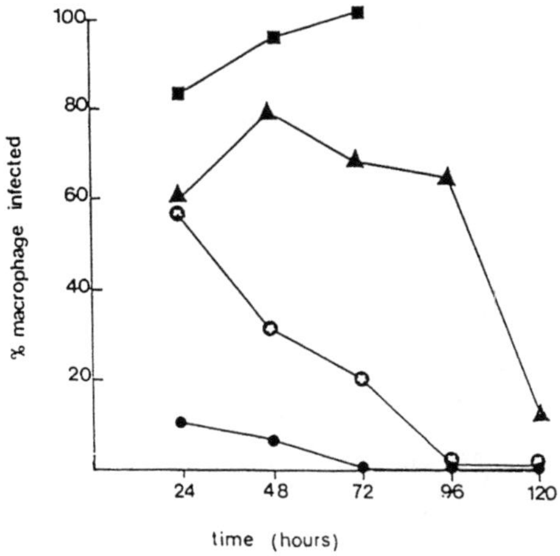

Figure 2. Survival of L. braziliensis treated with different mAbs in mouse peritoneal MØ in vitro. Controls represent the treatment of infective promastigotes with anti-L. chagasi IIIB1/26 (▲—▲) & an unrelated AA3/44 (■—■) mAbs.

Figure 3. Effect of different mAbs on the infectivity of L. chagasi towards mouse MØ in vitro. The promastigotes were treated with IIID4/9 (●——●), IIA4/3 (✪—✪), VH1/19 (▲——▲) and unrelated G5-54 (■——■) mAbs before co-culture with MO.

Among the four mAbs generated against the infective form of L. chagasi promastigotes, one mAb (IIID4/9 IgG1k) bound specifically to in vitro stationary phase and late stage sand-fly promastigotes, and recognised an integral membrane glyco-protein of Mr 63,000 (GP63, pI 7.2) specifically on the surface of the stationary phase and not the log phase promastigotes. Moreover, this mAb could inhibit the infectivity of Leishmania towards mouse MØ in vitro (Fig. 3) suggesting the critical role of the molecule recognised by IIID4/9 in L. chagasi MØ interactions. When lysates of surface-iodinated promastigotes from the log and stationary phase of culture were subjected to two-dimensional electrophoretic analysis, GP63 was revealed as two spots at apparent pI 6.6-7.0 on the surface of both log and stationary phase promastigotes while a third spot of pI 7.2 appeared specifically on the surface of stationary phase parasites (Fig. 4). Further, 2-D analysis of a mixture of immunoprecipitates obtained by IIID4/9 and VH1/19 (anti log phase mAb) confirmed that these mAbs recognize two mole-cules of different pI respectively (Fig. 4D).

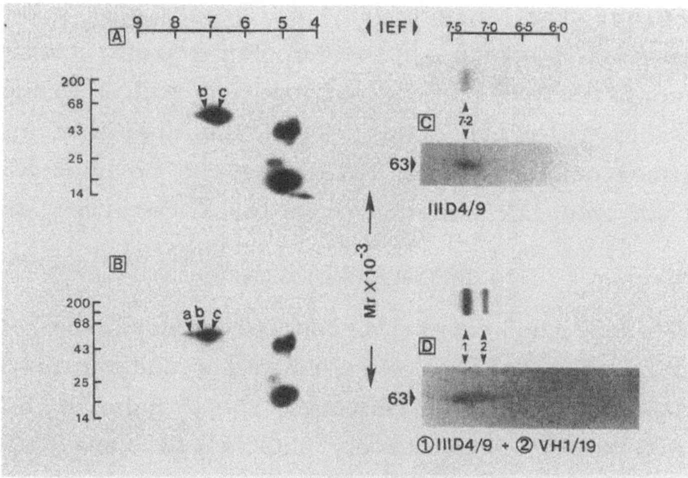

Figure 4. 2-D gel electrophoresis of surface-iodinated L. chagasi promastigotes. (A) log phase, (B) stationary phase, (C) immunoprecipitate using anti-stationary phase mAb and (D) a mixture of immunoprecipitates using anti-stationary and anti-log phase mAbs.

In view of the above findings one can not generalize the mechanism of the parasite MØ interactions to all Leishmania spp. It appears that at least two mechanisms of surface antigenic change are associated with the development of infective forms: the first suggested by Sacks et al. (1985) for L. major and Rizvi et al. (1985) for L. chagasi would be the appearance of new antigenic components and the second described here for L. braziliensis is the increasing expression of an epitope already existent on the surface of non-infective promastigotes. The mechanism controlling these expressions is unknown. Recent studies on the processing of GP65 have suggested that atleast three proteins of closely apposed Mr and with different levels of glycosylation represent the precursors of the mature forms (Chang et al. 1986). Also, GP65 is an amphiphilic membrane protein anchored by covalently bound myristic acid (Bouvier et al. 1985) and can also be found in a hydrophilic form. The reason why log phase parasites expressing a similar GP65 are not infective is still unknown. Perhaps the number of copies/cell of this antigen is important for their infectivity. Other mechanisms involving a lipid containing glycoconjugate and/or complement have also been suggested to play a role in the promastigote-MØ interactions.

In conclusion, the infectivity of Leishmania promastigote is associated with surface antigenic expression. Such modified expression contributes to the protection of leishmaniae in the hostile activities of host cells. The specific antibodies against isolated new antigenic molecules can be used as molecular markers to identify and characterize infective leishmanial forms.

ACKNOWLEDGEMENTS

We acknowledge the expert technical assistance of Didier Deslée, Marc Loyens, Marie-Pierre Fourmaux and Marie-Christine Liebart. This investigation received the financial assistance from The Rockefeller Foundation, UNDP/World Bank/WHO Special Programme for Research and Training in Tropical Diseases and EEC (grant no. TSD-M-010-F(RS)).

REFERENCES

Bouvier J, Etges RJ, Bordier C (1985) Identification and purification of membrane and soluble forms of the major surface protein of Leishmania promastigotes, J Biol Chem 260:15504 15509

Chang K P, Fong D (1983) Cell biology of host parasite membrane interactions in leishmaniasis In: Cytopathology of parasitic diseases Pitman Books, London, p 113–137

Chang SC, Inserra TJ, Kink JA, Fong D, Chang K P (1986) Expression and size heterogeneity of a 63 kDa membrane glycoprotein during growth and transformation of Leishmania mexicana amazonensis, Mol Biochem Parasitol 18:197–210

Giannini MS (1974) Effects of promastigote growth phase, frequency of subculture and host age on promastigote initiated infections with Leishmania donovani in golden hamster, J Protozool 21:521–527

Keithly JA, Bienen (1981) Infectivity of Leishmania donovani primary culture promastigotes for golden hamster, Acta Trop 38:85–89

Rizvi FS, Afchain D, Sherlock I, Sadigursky M, Capron A, Santoro F (1985) Infectivity of Leishmania promastigotes is associated with surface antigenic expression, Immunol Lett 11:317-323

Sacks DL, Perkins PV (1984) Identification of an infective stage of Leishmania promastigotes, Science 223:1417–1419

Sacks DL, Hieny S, Sher A (1985) Identification of cell surface carbohydrate and antigenic changes between noninfective and infective stages of Leishmania major promastigotes J Immunol 135:564-569

CHARACTERIZATION OF DEVELOPMENTALLY REGULATED MOLECULES OF LEISHMANIA

D. McMahon-Pratt,[1] C. L. Jaffe,[2] L. Kahl,[3] P. Langer,[4] K. Lohman,[1] A. Pan,[1] and L. Rivas[1]

[1]Department of Epidemiology & Public Health, Yale University School of Medicine, New Haven CT 06510, USA;[2]Weizmann Institute of Science, Rehovot, Israel;[3]Wellcome Research Laboratories, Beckenham, Kent, UK;[4]Department of Microbiology & Biochemistry, Laramie, WO 82071, USA

INTRODUCTION

Parasitic protozoa of the genus Leishmania have a relatively simple developmental life cycle. The flagellated promastigotes multiply within the alimentary tract of the sandfly vector. In the mammalian host, promastigotes enter macrophages and transform into aflagellated amastigote within the phagolyosome (1-3). Recent biochemical and immunologic analysis of Leishmania has focused on the surface membrane in attempt to elucidate the mechanisms of attachment, entry and maintenance of the parasite within macrophages (4-13). Immunologic analysis of cell surface of Leishmania has lead to the identification of species-specific antigens, promastigote- and amastigote-specific antigens and generation of monoclonal antibodies (14-20).

A number of monoclonal antibodies (M-2, M-4, M-6, M-8, T-4, T-8) (16,19,20,24) specific for distinct individual species of Leishmania by radioimmune binding assays recognize membrane components with similar M_r (50,000 daltons, non-reduced). These proteins represent a major surface radioiodinatable membrane component, which constitutes about 1% of the membrane protein; upon reduction a number of these protein molecules increase in M_r to (65,000 daltons). The epitopes recognized by antibodies M-2, M-4, M-6, M-8 and T-4 are promastigote stage specific. Although, the species-specificities of the monoclonal antibodies suggest that each molecule might be distinct, upon detergent solubilization (for immunoprecipitation) these determinants become exposed. The similar molecular properties suggested that these molecules might in fact be related. These possibilities were investigated. In addition data concerning the immunochemical and biochemical characterization of axenically cultured amastigotes are presented.

NATO ASI Series, Vol. H11
Host-Parasite Cellular and Molecular Interactions
in Protozoal Infections. Edited by K.-P. Chang and D. Snary

MATERIALS AND METHODS

Parasites

Leishmania stocks, WR309, L. major and WR352, L. donovani were obtained from Dr. L. Hendricks of the Walter Reed Army Institute of Research, Washington, D.C.; LTB0016, L. mexicana amazonensis was obtained from Dr. Phillip Marsden, Universidade de Brasilia, Brazil. Promastigotes were cultured in Schneider's Drosophilia media (28) supplemented with 20% heat inactivated fetal calf serum (Lot 26D0118, GIBCO, Grand Island, NY) containing gentamycin (50μg/ml). Axenic amastigotes were cultured as previously described (21).

Radiolabelling of Leishmania

Promastigotes were radiolabelled either using Na(^{125}I) and lactoperoxidase or metabolically with ^{35}S-methionine as previously described (18,20). Labelling with ^{125}I-TID (3-trifluormethyl)-3-(m-iodophenyl) diazarine (Amersham) was done as described (12,23) in the presence of glutathione.

Immunoprecipitation

Radiolabelled promastigotes ($5x10^8$) were solubilized in PBS containing 0.5% Nonidet P-40 (NP-40), 10mM ethylenediaminetetraacetic acid (EDTA) 1mM iodoacetamide (IAA), 1mM N-p-tosyl-L-lysine chloromethyl ketone (TPCK) and 1mM phenylmethylsulfonylfluoride (PMSF) for 1/2 hour at 0°C and the solutions clarified by centrifugation (one hour at 48,000 x g). The solubilized radiolabelled antigens were immunoprecipitated using the various monoclonal antibodies coupled (2mg/ml) to cyanogen bromide activated Sepharose 4BCL (Pharmacia, Uppsala, Sweden). Following a 30 minute incubation at 0°C, the resin was washed three times at 0°C in PBS containing 0.1% NP-40, followed by a wash in PBS containing 0.1% NP-40 and 0.5% sodium dodecylsulfate (SDS). The antigenic material was eluted from the resin using either sample loading buffer or 0.2% SDS depending upon whether the proteins were being treated with proteolytic enzymes or endoglycosidase H, respectively.

Enzymatic Analyses

Radioactivity labelled proteins eluted from affinity resins were used for peptide mapping studies. Trypsin (Sigma Chemical Co.) and S. aureus-V8 (Miles Pharmaceutical) were stored as stocks solutions (1 and 10 mg/ml) at -70°C. The proteins were digested at 37°C at variable concentration of enzyme (20-100μg) for a period between 2 to 20 hours. At the end of the digestion, the samples were analyzed using SDS-polyacrylamide gel electrophoresis (15%) (see below).

Samples treated by endoglycosidase H (New England Nuclear, Boston, MA) were eluted from the resin in 0.2% SDS to which was added an equal volume of 0.2M citric acid buffer, pH 5.5. The time course of deglycosylation was determined for each protein sample. Complete deglycosylation consistently occurred using lug of enzyme and a digestion time of four hours. Phospholipase C (Sigma Chemical Co.) digestion was carried out using non-radioactive M-2 (50μg) and 4U/ml of enzyme for 10-12 hours at 37°C. Phospholipase C activity was determined according to established procedures (25).

Gel Electrophoresis

Sodium dodecylsulfate polyacrylamide gel electrophoresis (SDS-PAGE) was carried out according to Laemmli (26). Autoradiography was done using Kodak X-Omat AR film (XAR-5). Charge-Shift Electrophoresis was performed according to established methods (27) using M-2 protein that had been isolated and purified using Brij-97h detergent according to methods previously described (12). Protease characterization was done using SDS-PAGE (26) to resolve the homogenate component of promastigotes or amastigotes followed by an overlay of 0.8% agarose containing casein (Carnation, 0.25%). The gel was incubated for several hours and then stained for protein using amido black (0.1% in 1:1::methanol:H_2O) to detect protease activity. For studies with protease inhibitors the homogenate was incubated at the given concentration of inhibitor prior to analysis as above.

Monoclonal Antibodies

The production and characterization of monoclonal antibodies M-2,M-4, M-6, M-8 and T-4 (16,19,20) have been described previously. Monoclonal antibody T-8 was produced against membrane preparations of Leishmania major (WR260B) according to established methods already described (16).

Radioimmune Binding Assays

Indirect radioimmune binding assays employing Leishmania membrane preparation or lysates of whole organisms were carried out essentially according to the methods previously described (16,20).

Southern Blot Analysis

Southern blot analysis (28) were performed using DNA isolated from L. mexicana amazonensis (LTB0016) (5μg/lane). Hybridizations were performed at 2XSSC, 0.1% SDS and 62°C. Controls employing 0.5μg of lambda phage DNA mixed with leishmanial DNA (2μg) were done to assure complete digestion occurred. Restriction endonucleases were used according to manufacture's specifications. The probe, a synthetic oligonucleotide representing a single sequence (62bp) (Yale University Oligonucleotide Lab) based on N-terminal sequence data, was N-labelled according to described procedures (29).

RESULTS AND DISCUSSIONS

Initial Observations

The molecule(s) associated with the antigenic determinants recognized by monoclonal antibodies T-4, M-6, M-8, T-8 and M-2 are surface membrane proteins with an estimated molecular weight of 50,000 daltons (11,12,19,20). Similar analyses of lactoperoxidase-catalyzed radioiodinated promastigotes show a major band in this molecular weight lane on nonreduced samples. Upon reduction, a portion of this radioiodinated material increases in molecular weight to approximately 60-65,000 daltons.

The specificities (Table 1) of monoclonal antibodies T-4, M-6, M-8 and M-2 for species of Leishmania have been reported previously (19,20). Antibodies M-6, M-8 are specific for Leishmania mexicana, antibody M-2 is specific for Leishmania mexicana amazonensis, while antibody T-4 is specific for Leishmania major. Monoclonal antibody T-8, however is cross reactive with all species of Leishmania. Antibody T-4 recognized a determinant that is accessible on L. major but that

Table 1

A Comparison of the Binding Reactivities[1] of Monoclonal Antibodies to Promastigote Membranes

Monoclonal antibody	Code	L. major	L. donovani	L. mexicana amazonensis	L. braziliensis braziliensis
LLXIX-1A5-G1	T-4	4.5	0.9	0.9	1.0
LXIX-3E12-F8	T-8	24.7	6.8	-	-
IX-2H7-E10	M-2	1.1	0.8	40.9	1.0
IX-1F9-D8	M-4	1.1	0.6	21.4	0.7
LXVII-4C7-B8	M-6	1.0	0.5	9.3	2.8
LXV11-4D8-B7	M-8	1.0	1.6	7.7	1.8

[1]The values indicate the ratio of cpm-monoclonal antibody containing culture supernatant bound to cpm-control culture supernatant based on an indirect radioimmune binding assay.

is cryptic until detergent solubilization on L. donovani (20). In immunoprecipitation experiments, M-2 was clearly specific for Leishmania mexicana amazonensis and did not bring proteins from L. braziliensis braziliensis, L. major or L. donovani. In immunoprecipitation, monoclonal antibodies T-4, M-6, M-8 and T-8 were clearly cross reactive with all species of Leishmania examined. The question arose as to whether the molecules identified by T-4, (M-6, M-8, T-8) and M-2 were in fact identical or represented a family of closely related molecules on the promastigote surface.

Competitive Absorption Experiments

To ascertain whether the determinants recognized by monoclonal antibodies M-2 and T-4 were in fact, on identical molecules competitive absorption experiments were done using the purified antibodies coupled (2mg protein/ml resin) to cyanogen bromide activated Sepharose.

Figure 1

Cross Absorption of $^{(125}I)$-Labelled **L. mexicana amazonensis** Promastigote

Absorption number	CPM $X10^{-5}$ Absorbed to 50ul of Affinity Resin: T-4 Sepharose 4B	M-2 Sepharose 4B
1	11.7	10.4
2	4.1	8.4
3	1.5	4.4
4	0.2	2.7
↓	↓ M-2 Sepharose 4B	↓ T-4 Sepharose 4B
5	8.6	8.2

Shown above are the results of sequential cross immunoabsorption experiments employing monoclonal antibodies T-4 and M-2 covalently coupled to Sepharose

Lactoperoxidase catalyzed radioiodinated promastigote of L. mexicana amazonensis strain, (LTB0016) were used for these experiments. The labelled organisms were solubilized in 1% Nonidet-40 (NP-40), clarified by centrifugation for 30 minutes (48,000 x g) and the supernatant used for repetitive absorptions. As seen in Figure 1, the amount of material left after four absorptions with T-4 and M-2 was less than approximately 1.7% and 26% of the initial absorption, respectively. Absorption of T-4 deleted material with M-2 Sepharose, however, resulted in a significant increase in number of counts per minutes bound to the M-2 affinity resin compared to that found for the final absorption T-4 (approximately 81% that found upon the initial absorption of the initial detergent solubilized lysate with M-2). Similarly, with the detergent solubilized lysate that had been relatively depleted in M-2, the T-4 Sepharose affinity removed 8.2 x 105 cpm or 70% that found in the primary absorption. These results suggest that the determinants recognized by monoclonal antibodies T-4 and M-2 are indeed on separate molecules.

Peptide Mapping Experiments

Although the determinants recognized by M-2 and T-4 appeared to exist on separate populations of similar M_r proteins, it was possible that these molecules were in fact members of a closely related family of proteins. Consequently, peptide mapping experiments were done to clarify this point. In preliminary experiments, affinity purified antigens derived from lactoperoxidase-catalyzed radioiodinated

Figure 2

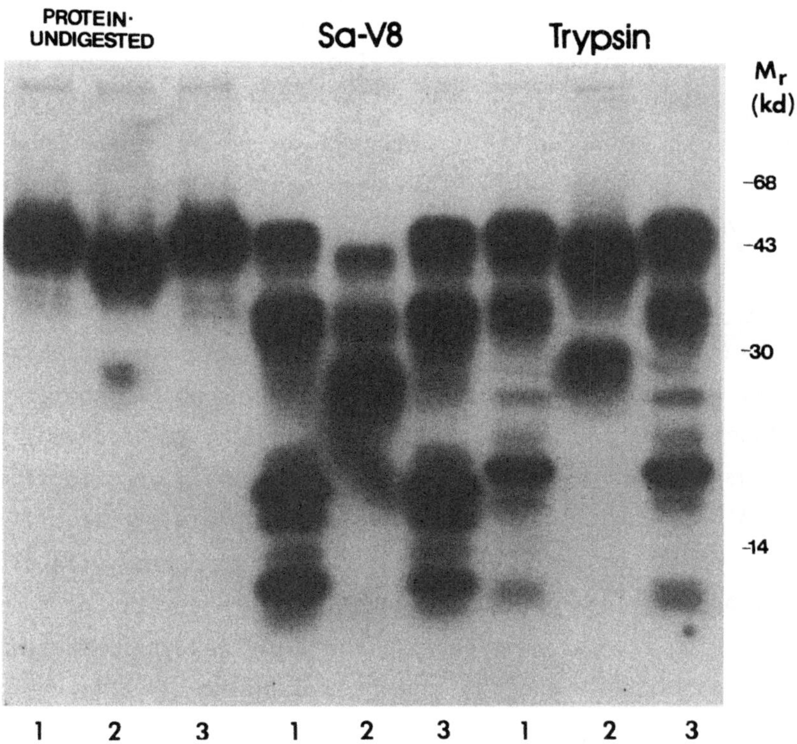

Shown are the autoradiographic analysis of partial proteolytic digestion experiments analysed on 15% SDS-PAGE and employing S. aureus-V8 protease and trypsin radioiodinated T-4 (lane 1), M-2 (lane 2) and M-6/M-8 (lane 3) were isolated from radioiodinated Leishmania mexicana amazonensis promastigotes.

promastigotes of L. mexicana amazonensis were used. As seen in Figure 2, partial enzymatic digestion with either trypsin or Staphylococcus aureus V-8 proteases resulted in identical polypeptide fragmentation patterns for the antigens T-4 and M-6/M-8. The antigen associated with monoclonal antibody M-2 appears to represent a distinct set of molecules.

Heterogeneity of Molecules Expressing the T-4 (M-6,M-8,T-8) Epitopes

As the molecules recognized by M-2 and T-4 (T-8) appeared to be possibly similar but not identical for L. mexicana, it was of interest to determine if, in fact, these molecules were identical/similar for other species of Leishmania. Unfortunately, as monoclonal antibody M-2 was specific for L. mexicana, these comparative studies were done only using T-4, M-6/M-8. The proteolytic fragmentation pattern observed with either trypsin of S. aureus-V8 were identical for the T-4 and M-6/M-8 molecules isolated from either L. mexicana or L. donovani. However, the molecules isolated from different species had distinct peptide mapping profiles which appear to depend upon variation in glycosylation. This heterogeneity has been observed by other investigators (5-7).

Molecular Characterization of M-2

The M-2 molecule has been characterized as membrane associated (11,12). The molecule is labelled by the hydrophobic photoactivatable radiolabelled probe, TID (3-trifluoromethyl)-3-(m-iodophenyl) diazarine. This hydrophobic probe is selectively incorporated into the 22Kd portion of the molecule that is completely refractile to protease digestion (using S. aureus V-8, trypsin, chymotrypsin, pepsin, or proteinase K). Consequently, the stability of this peptide may be due to its association with the lipid bilayer. In addition, recent experiments using Phospholipase-C and analysis by charge-shift electrophoresis suggest that the M-2 molecule, in contrast to GP-63 (8) is **not** linked to the membrane solely by a "phospholipid". Preliminary Southern blot analyses employing a DNA probe corresponding to a protein of the N-terminal sequence of M-2 show that in the case of certain restriction endonuclease (i.e. Hind III, Sal I, Bam HI) two distinct fragments hybridizing unequally. It is unclear at present whether this hybridization pattern represents an homologous (but not identical) gene or dispersed copies of the same gene (single/multiple copy). However, sequencing of genomic clones is currently in progress which should clarify this point.

Promastigote Stage-Specific of M-2

Immunoprecipitation experiments employing either mRNA translation production metabolically labelled promastigote and amastigotes of L. mexicana amazonensis and a polyvalent rabbit antiserum to M-2, suggest that the molecule (not just the M-2 epitope) is promastigote stage. No label was precipitated from amastigotes by either the monoclonal antibody or heteroserum in these experiments.

Table 2

Immunochemical[1] Characterization of Axenically Cultured Amastigotes of L. mexicana pifanoi

Monoclonal Antibody	Amastigotes Macrophase	Amastigotes Axenic	Promastigotes
CXIV-5C10-F1	17.8	18.4	1.7
CXIV-3G11-C10	7.8	7.1	2.1
CX111-2E6-D12	1.0	1.4	8.8
CXX11-2E1-G8	1.6	1.4	18.8
CX11-3B2-G2	1.9	2.6	27.2

[1]Shown are the results of indirect radioimmune binding assays (See Figure 1) employing monoclonal antibodies produced to membrane-enriched preparations L. m. pifanoi amastigotes (isolated from macrophages) and promastigotes.

Culture of Axenic Amastigotes

The extracellular axenic culture of amastigote of L. mexicana pifanoi (21) and the morphological characterization (electron microscopy) has been reported (30). Recent immunochemical studies employing specific monoclonal antibodies produced to amastigotes isolated from macrophages (Table 2) suggest that these axenically cultured forms are most closely related to the amastigote rather than the promastigotic stage. In addition, these cultured amastigotes have been found to possess the 30Kd thiol protease characteristic of L. mexicana amastigotes (31,32, Table 3). Consequently, these cultured parasites should be useful in further studies characterizing the developmental changes and regulation mechanisms of Leishmania. Preliminary studies indicate that certain strains of L. mexicana and L. braziliensis will transform into amastigotes under the described conditions. However, not all strains are capable of transformation. The reason for this variation is at present unclear, but may reflect the variation of either the requirements for metabolic nutrients or signals inducing transformation.

Table 3

Effect of Inhibitors on Amastigote Protease (30Kd)

Inhibitor	Concentration	Percentage of Initial Enzyme Activity
PMSF	2mM	97
Soybean Trypsin inhibitor	30μg/ml	95
EDTA	1mM	98
	5mM	103
	10mM	104
Pepstatin	3μg/ml	102
Iodoacetamide	2mM	8
Leupeptin	5μg/ml	7
Hg^{+2}	2mM	16
Dithiobisdipyridine	1mM	10

CONCLUSIONS

Peptide mapping experiments employing metabolic and surface radiolabelled proteins demonstrated that the molecules recognized by M-2 and M-4 were identical but distinct from the molecule(s) recognized by monoclonal antibodies T-4, T-8 and M-6/M-8. Sequential immunoprecipitation experiments confirmed the fact that the epitopes recognized by these two groups monoclonal antibodies reside on a distinct set of molecules. The M-2 glycoprotein has apparently unique 22Kd domain-associated with the membrane which is unsusceptible to either proteolytic digestion or phosphodipase-C treatment. Endoglycosidase H digestion data suggest that two/three carbohydrate chains are present on T-4 while M-2 has a single carbohydrate chain. Peptide mapping of the molecules recognized by T-4 and M-6,M-8 of *Leishmania donovani* are identical but distinct from those found in *Leishmania mexicana*. The apparent difference(s) found between species is most likely due to variation in glycosylation. Southern blot analysis using a probe corresponding to a region of the N-terminal sequence of M-2 suggests that a family of related molecules may exist.

Recent evidence (electron microscopy) suggests that amastigote of *L. mexicana pifanoi* can be cultured axenically. Subsequent immunochemical studies employing monoclonal antibodies confirm these results. In addition, the axenically cultured amastigotes possess the 30Kd thiol-protease characteristic of the *L. mexicana* amastigote stage.

REFERENCES

1. Zuckerman, A. and Lainson, R. (1977) Leishmania. **In:** Parasitic Protozoa (Kreier, J., Ed.), Vol. 1, pp. 58-133, Academic Press, New York.

2. Pearson, R.D., Wheeler, D.A., Harrison, L.H., and Kay, D.A. (1983) The immunobiology of leishmaniasis. Rev. Infect. Dis. 5, 907.

3. Killick-Kendrick, R. (1979) Biology of Leishmania in Phlebotomine sandfly. **In:** Biology of the Kinetoplastida (Lumsden, W.H.R. and Evans, D.A., eds.), Vol. 2, pp. 395-460, Academic Press, London.

4. Dwyer, D.M. and Gottlieb, M. (1983) The surface membrane chemistry of Leishmania: its possible role in parasite sequestration and survival. J. Cell Biochem. 23, 35.

5. Chang, K.P. (1983) Cellular and molecular mechanisms of intracellular symbiosis in leishmaniasis. Int. Rev. Cytol. (Suppl) 14, 267-305.

6. Chang, C.S., Inserra, T.J., Kink, J.A. Fong, D. & Chang, K.-P. (1986) Expression and size heterogeneity of a 63 kilodalton membrane glycoprotein during growth and transformation of Leishmania mexicana amazonensis. Mol Biochem. Parasitol. 18, 197.

7. Lepay, D.A., Nogueira, N. & Cohn, Z. (1983) Surface antigens of Leishmania donovani promastigotes. J. Exp. Med. 157, 1562.

8. Etges, R., Bouvier, J. & Bordier, C. (1986) The major surface protein of Leishmania promastigotes is anchored in the membrane by a myristic acid-labeled phospholipid. EMBO J. 5, 597.

9. Etges, R., Bouvier, J. & Bordier, C. (1986) The major surface protein of Leishmania promastigotes is a protease. J. Biol. Chem. 261, 9098.

10. Russell, D.G. & Wilhelm, H. (1986). The involvement of the major surface glycoprotein (gp63) of Leishmania promastigotes in attachment to macrophages. J. Immunol. 136, 2631.

11. McMahon-Pratt, D. & David, J.R. (1982) Monoclonal antibodies recognizing determinants specific for the promastigote stage of Leishmania mexicana. Mol. Biochem. Parasitol. 6, 317.

12. Kahl, L.P. & McMahon-Pratt, D. (1986) Structural and antigenic characterization of a species and promastigote specific Leishmania mexicana amazonensis membrane protein. (submitted).

13. Handman, E. and Goding, J.W. (1985) The Leishmania receptor for macrophages is a lipid-containing glycoconjugate. EMBO J. 4, 329.

14. Handman, E. and Hocking, R.E. (1982) Stage-specific, strain-specific and cross-reactive antigen of Leishmania species identified by monoclonal antibodies. Infect. Imm. 37, 28.

15. Delbarra, A.A.L., Howard, J.G., and Snary, D. (1982) Monoclonal antibodies to Leishmania tropica major: specificities and antigen location. Parasitology 85, 523.

16. McMahon-Pratt, D. and David, J.R. (1981) Monoclonal antibodies that distinguish between New World species of Leishmania. Nature 291, 581.

17. McMahon-Pratt, D., Bennett, E., and David, J.R. (1982) Monoclonal antibodies that distinguish subspecies of Leishmania braziliensis. J. Immunol. 129, 926.

18. Jaffe, C.L., Bennett, E., Grimaldi, Jr. G., and McMahon-Pratt, D. (1984) Production and characterization of species-specific monoclonal antibodies against Leishmania donovani for immunodiagnosis. J. Immunol. 133, 440.

19. McMahon-Pratt, D., Bennett, E., Grimaldi, G. & Jaffe, C.L. (1985) Subspecies-and Species-specific antigens of Leishmania mexivana characterized by monoclonal antibodies. J. Immunol. 134, 1935.

20. Jaffe, C.L. & McMahon-Pratt, D. (1983) Monoclonal antibodies specific for Leishmania tropica. I. Characterization of antigens associated with stage-and species-specific determinants. J. Immunol. 131, 1987.

21. Pan, A.A. (1984) Leishmania mexicana: Serial cultivation of intracellular stages in a cell-free medium. Exp. Parasitol. 58, 72.

22. Hendricks, L.D., Wood, D.E. and Hajduk, M.E. (1978) Hemoflagellates: commercially available liquid media for rapid cultivation. Parasitology 76, 209.

23. Brunner, J. and Semenza, G. (1981) Selective labeling of the hydrophobic core of membranes with 3-(trifluonamethyl)-3-(m-^{125}I-iodophenyl) diazirine, a carbene generating reagent. Biochem. 20, 7174.

24. Anderson, S., David, J.R., & McMahon-Pratt, D. (1983) In vivo protection against Leishmania mexicana mediated by monoclonal antibodies. J. Immunol. 131, 1616.

25. Kurioka, S. and Matsuda, M. (1976) Phospholipase C assay using p-nitrophenylphosphorylcholine together with sorbitol and its application to studying the metal and detergent requirement of the enzyme. Anal. Biochem. 75, 281.

26. Laemmli, U.K. (1970) Cleavage of structural proteins during the assembly of the head of bacteriophage T4. Nature 227, 680.

27. Helenius, A. and Simons, K. (1977) Charge shift electrophoresis. Simple method for distinguishing between amphiphilic and hydrophobic proteins in detergent solutions. Proc. Natl. Acad. Sci. USA 74, 529.

28. Southern, E (1975) Detection of specific sequences among DNA fragments separated by gel electrophoresis. J. Mol. Biol. 98, 503.

29. Maxam, A.M. & Gilbert, W. (1980) Sequencing end-labeled DNA with base-specific chemical cleavages. Method Enzymol. 65, 499.

30. Pan, A.A. & Pan, S.C. (1986) Leishmania mexicana pifanoi: Comparative fine structure of amastigotes and promastigotes in vitro and in vivo. Exp. Parasitol. 62, (in press).

31. North, M.J. & Coombs, G.H. (1981) Proteinases of Leishmania mexicana amastigotes and promastigotes: Analysis by gel electrophoresis. Mol. Biochem. Parasitol. 3, 293-300.

32. Pupkis, M.F. & Coombs, G.H. (1984) Purification and characterization of proteolytic enzymes of Leishmania mexicana mexicana amastigotes and promastigotes. J. Gen. Microbiol. 130, 2375-2383.

LEISHMANIA ANTIGENS PROBED BY MONOCLONAL ANTIBODIES TO LEISHMANIA MEMBRANE AND TO ACETYLCHOLINE RECEPTOR

K.Ph. Soteriadou, A.K. Tzinia, M. G. Hadziantoniou and S. J. Tzartos

Hellenic Pasteur Institute, 127 Vassilissis Sofias Avenue, Athens 11521, Greece.

INTRODUCTION

Leishmania donovani, the etiologic agent of kala-azar or visceral leishmaniasis, is the cause of significant mortality throughout the world. In its insect vector, the *phlebotomus* sandfly, the protozoan exists in the promastigote form. In the mammalian host, it is converted into the amastigote form and multiplies within the phagolysosomes of macrophages (1). The mechanism whereby the protozoan penetrates, survives and multiplies in macrophages is not yet well understood. It is likely that some factors of the parasite membrane play a crucial role in the binding and penetration of the parasite into the macrophage and its survival within the phagolysosomes (2).

A major *Leishmania* surface glycoprotein of 63.000 daltons, termed gp63, has been identified using kala-azar sera and human cutaneous leishmaniasis sera (3). The gp63 appears to be common among a great number of *Leishmania* species and is recognized by rabbit anti-*L. donovani* immune sera (4,5). Furthermore, a gp63 has been purified from *Leishmania mexicana amazonensis* (*L.m. amazonensis*) by monoclonal antibody affinity binding and this glycoprotein inhibits *Leishmania*-macrophage binding (6). The involvement of the *L.m. mexicana* gp63 in the attachement of *Leishmania* promastigotes to macrophages has been also demonstrated (7). It has been also reported that gp63 is an integral membrane protease (8,9,10).

In our investigation in order to study *Leishmania donovani infantum* (*L. infantum*) surface antigens, two kinds of monoclonal antibodies (mAbs) were used: a) mAbs produced against isolated *L. infantum* membranes and b) mAbs against the nicotinic

NATO ASI Series, Vol. H11
Host-Parasite Cellular and Molecular Interactions
in Protozoal Infections. Edited by K.-P. Chang and D. Snary

acetylcholine receptor (AChR), a muscle membrane glycoprotein. The anti-L. infantum mAbs recognized a 58.000 daltons polypeptide and cumulative results suggest the existence of a monomeric and an oligomeric form of the 58KD polypeptide. The observed inhibition of Leishmania promastigote-macrophage binding caused by these mAbs suggests that the 58KD polypeptide and the 58KD oligomer might be of pathophysiological significance. Furthermore, a sinificant number of the anti-AChR mAbs recognize a 58KD L. infantum polypeptide and one a 55KD polypeptide.

METHODS

Monoclonal Antibodies

Spleen cells of Balb/c mice immunized with isolated L. infantum membranes either intact or solubilized were fused with S194 mouse myeloma cells. Twenty-four stable hybridomas producing mAbs to promastigote membrane antigens were produced according to the "direct cloning" method (11) and detected by a solid-phase radioimmunoassay using intact isolated membranes.

The preparation and characterization of the anti-AChR mAbs used has been described (12,14). They derived from rats immunized with electric organ or mammalian muscle AChR or its SDS-denatured subunits.

Leishmania Preparations.

The L. infantum strain used was isolated in Greece from human visceral leishmaniasis. It was typed as Leishmania donovani infantum H HOM-Gr78-L_4 (L. infantum) (15). L. infantum promastigotes were cultured at 27°C in a monophasic medium and subcultured every 3 to 4 days (M. Hadziantoniou, PhD Thesis, unpublished). L. tropica (formerly L.t. minor) isolated in Greece, was kindly provided by Dr. A. Garifallou, Hellenic Pasteur Institute. L. mexicana amazonensis was kindly provided by Professor K.P. Chang, UHS/Chicago Medical School.

Membrane preparations of L. infantum, L. tropica or L. mexicana were obtained by disruption of the parasites followed by subfractionation by differential centrifugation and isolation on sucrose density gradients according to Dwyer (16).

Macrophage Preparations

Macrophages (murine line J774G8) were grown in RPMI-1640 with 15% fetal bovine serum. Log-phase cells were washed and then disrupted on ice in a Dounce-type homogenizer. Cell homogenates were centrifuged at 8.000 g for 30 min and the resulting membrane preparation was stored at -70°C until use.

Immunofluorescence.

Immunofluorescence experiments were carried out on intact promastigotes of L. infantum, L. tropica or L. mexicana (17). Leishmanias (10^6 to 10^7/ml) were placed on slides, air dried, and covered with PBS containing 5% FCS for 10 min. After two washings, the slides were incubated for 30 min with the hybridoma culture medium supernatant. Then they were rinsed and incubated with fluorescein-conjugated rabbit anti-mouse immunoglobulins. After washing, samples were examined with a fluorescence microscope.

Infection of Mouse Macrophages

Mouse macrophage cell line J774G8 were cultured on cover slips in Petri dishes. The cells were allowed to adhere at 37°C in an atmosphere of 10% CO_2 for 1 hr. They were then washed to remove non-adherent cells, and L. infantum promastigotes treated or not with mAbs were added at a ratio of 10 parasites per cell. After 1 hr, cover slips were washed vigorously with medium, fixed in methanol and stained with Giemsa. Promastigote attachement to macrophages was determined by counting 500 cells in a Giemsa-stained culture (18).

Solid-phase Radioimmunoassay

A solid phase radioimmunoassay (RIA) was used to screen mAbs reacting with L. infantum (19). Briefly, L. infantum membranes (or bovine serum albumin used as negative control) diluted in bicarbonate buffer pH 9.6 were plated in polyvinyl chloride flat-bottomed microtiter plates (Dynatech Laboratories, Inc., Alexandria, VA) and incubated overnight at 4°C. The microtiter plates were washed three times in PBS containing 0.05% Tween-20 and 0.02% NaN_3 and incubated for a minimum of 1 hr at room temperature with PBS containing 3% bovine serum albumin (BSA). mAbs were then added and incubated for 3 hr at room temperature or overnight at 4°C. Non-immune serum and cell culture medium were used as negative controls. The plates were washed 3 times with PBS, 0.05% Tween-20 and 0.02% NaN_3. Rabbit anti-mouse or rabbit anti-rat Ig serum was added to the wells and incubated for 2 hr at room temperature followed by three washings. Then ^{125}I-labelled Protein A was added and incubated for 1 hr at room temperature. Finally, the plates were washed 5 times and the radioactivity bound to each well was removed by 2% SDS and measured.

Electrophoresis and Western Blotting

Leishmania and macrophage membrane proteins were separated on 10% polyacrylamide gels in the presence of sodium dodecyl sulfate (SDS) according to Laemmli (20) and analyzed by Western blotting (21) using as probes the anti-*L. infantum* and anti-AChR mAbs. The *L. infantum* proteins were electrically transferred from SDS-polyacrylamide gels to nitrocellulose paper. The nitrocellulose sheets cut into strips were incubated for at least 2 hr at 40°C in 3% BSA in PBS and then overnight at 4°C in non-immune serum or mAbs. The strips were then washed 3 times and incubated with rabbit anti-mouse or rabbit anti-rat immunoglobulins for 3 hr at room temperature, washed again and finally incubated with ^{125}I-Protein A for 1 hr at room temperature. The strips were washed at least 4 times, air-dried and exposed for autoradiography with the use of Kodak X-Omat S film.

Proteolytic Digestion

Leishmania membrane preparations were incubated at room temperature for one hour in the presence of 2µg/well *S. aureus* V_8 protease. The resulting proteolytic products were analyzed by Western Blotting.

RESULTS

Identification and Characterization of Membrane Antigens Recognized by the anti-*L. infantum* mAbs.

The identification of the membrane antigens by ten anti-L. infantum mAbs was carried out by Western Blotting followed by autoradiography. Western blots of heat-denatured L. infantum membranes (under non reducing or reducing conditions of SDS-PAGE) showed that the ten mAbs tested recognize a 58.000 daltons (58KD) band. However, in the absence of heat denaturation (with or without reduction) the above mAbs gave distinct patterns: 8 out of 10 still recognize the 58KD band (e.g. LD16 and LD23) whereas two (LD20 and LD24) recognize also bands of higher mol. wt. (Figure 1).

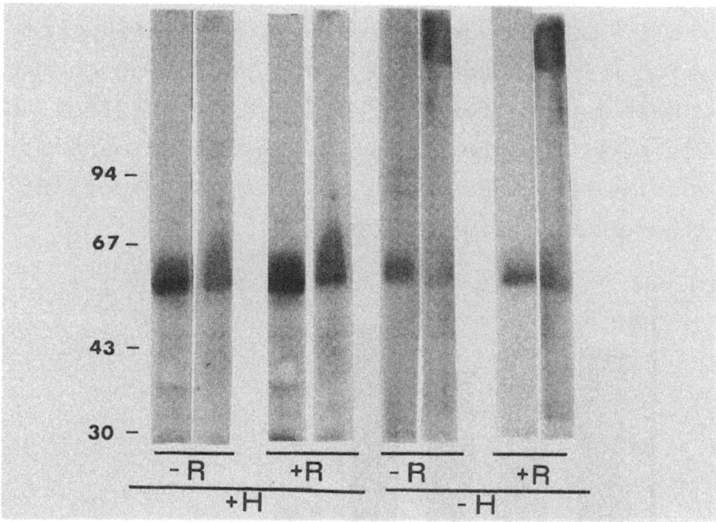

FIGURE 1. Autoradiographs of L. infantum membrane antigens identified by Western Blotting using as probes mAbs to L. infantum. Out of ten mAbs tested only two (one of each group) are presented here. SDS-PAGE (7.5%) performed under non reducing conditions (-R) or reducing conditions (+R). LD16 recognizes a 58KD band under all conditions (lanes 1,3,5,7). LD20 recognizes a 58KD band after heat denaturation (+H) (lanes 2,4). In the absence of heating (-H) LD20 recognizes high mol. wt. bands (lanes 6,8).

Limited proteolysis of L. infantum membranes with Staphylococcus aureus V_8 protease showed that mAbs LD16, LD20, LD23 and LD24 recognize the same digests (data not shown).

The localization of mAb binding to promastigotes was determined by indirect immunofluorescence using air-dried or methanol-fixed parasites. Both preparations gave similar results. LD16 and LD23 bind uniformly to the parasite surface whereas LD20 and LD24 label the anterior end and the flagellum (data not shown).

Inhibition of Parasite Attachement to Macrophages Caused by the anti L. infantum mAbs.

We examined the interference of the above mAbs with parasite penetration into mouse macrophages. Interestingly, mAbs LD16, LD23 and LD20, LD24 inhibit about 30% and 50% respectively parasite attachment to mouse macrophages *in vitro* whereas an anti-tubulin mAb and the anti-AChR mAbs did not affect promastigote binding to macrophages (Figure 2). The effect of the *anti-L. infantum* mAbs is specific since treatment of macrophages with these mAbs before infection had no effect and mAbs did not affect promastigote mobility. Therefore, the surface antigens recognized by these mAbs may play an important role in parasite-macrophage interaction

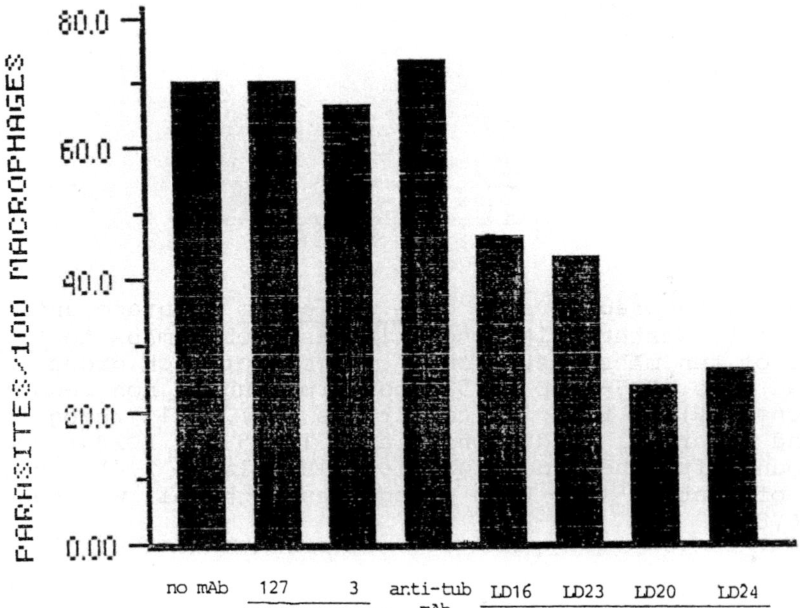

FIGURE 2. Effect of anti-L. infantum and anti-AChR mAbs on the attachment of promastigotes to macrophages.

Identification of *L. infantum* Membrane Antigens using as probes mAbs to AChR.

In an attempt to identify the nature of *Leishmania* membrane antigens that might be responsible for the appearance of antibodies against muscle elements in *Leishmania* infected patients (22) we have investigated the possible crossreactivity of one hundred and fifteen different AChR mAbs (12-14) with *Leishmania* membranes. These mAbs were screened for their crossreactivity with *L. infantum* intact membranes by solid-phase RIA. Out of 115, 34 mAbs exhibited strong (bound cpm > 5 times the background cpm) and 11 mAbs a weaker (bound cpm 2-5 times the background cpm) binding to *L. infantum* membranes. These mAbs are directed against all four AChR subunits. Western blots of *L. infantum* membranes under reducing (plus heating) or non reducing conditions (without heating) of SDS-PAGE showed that out of the 115 mAbs tested 23 recognize membrane proteins of *L. infantum*. These 23 mAbs were found to bind to intact membranes as well. 22 recognize a 58 KD band and one, mAb3, recognizes a band of 55KD. (Table 1, Figure 3).

TABLE 1

CROSS-REACTIVITY OF THE ANTI-ACETYLCHOLINE RECEPTOR MONOCLONAL ANTIBODIES WITH *L. INFANTUM* MEMBRANE ANTIGENS

No. of anti-receptor monoclonals	Crossreactivity with intact *L.I.* membranes	Antigen recognition (58KD or 55KD)
43 anti-α	16	10+1*
38 anti-β	18	9
5 anti-γ	2	1
14 anti-δ	6	2
15 ?	3	0
Total: 115	45 (39%) 34 5X backgr.	22+1*(20%)
(9 IgM, 106 IgG)	(7 IgM)	(5 IgM)

* mAb recognizing a band of 55KD.

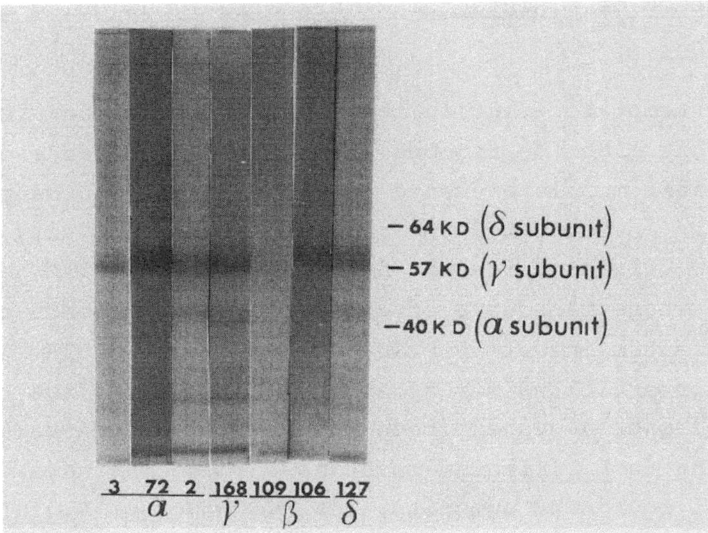

FIGURE 3. Binding of anti-AChR mAbs against all four subunits of AChR to Leishmania antigens. mAb 3 recognizes an antigen of L. infantum with approximate molecular weight of 55.000. Out of the 22 mAbs recognizing a protein of L. infantum with approximate molecular weight of 58.000 only mAbs 72,2,168,106,127 are shown here. mAb 109 does not crossreact with L. infantum.

Two of the above crossreactive mAbs were tested with other Leishmania species. They recognize on blots of L.m. amazonensis and L. tropica membranes bands of similar molecular weight as with L. infantum. They did not react however with membranes of the macrophage cell line J774G8 (data not shown). To assess their specificity of binding to Leishmania species they were preincubated with solubilized AChR-rich Torpedo, Leishmania or macrophage membranes (Figure 4). These inhibition experiments show clearly that binding of the above mAbs to L. infantum antigens was not mediated via the F_c portion but through the antigen binding site.

CONCLUSIONS

The similarities in antigen recognition of the anti-L. infantum mAbs tested which bind to a 58KD polypeptide after heat dena-

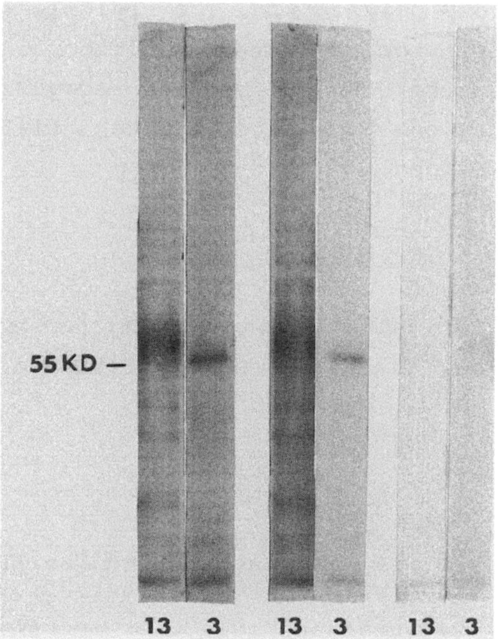

FIGURE 4. Specificity of the binding of mAbs 13 and 3 to L. infantum. mAbs 13 and 3 recognize proteins of L. infantum with approximate molecular weight of 58.000 and 55.000 respectively (lanes 1,2 from the left). Preincubation of the above mAbs with macrophage membranes had no effect on their binding to L. infantum proteins (lanes 3,4). Preincubation with solubilized Torpedo membranes resulted in inhibition of their binding to L. infantum (lanes 5,6) (antigen-antibody complexes formed not removed).

turation and the strong homology between their proteolytic digests on the one hand, and on the other hand the different pattern of antigen recognition by mAbs LD20 and LD24 in the absence of heating and the heterogeneity of their location revealed by immunofluorescence, may suggest the existence of a monomeric and an oligomeric form of the same polypeptide (58KD). The pathophysiological significance of the 58KD monomer and 58KD-oligomer may be crucial since the corresponding mAbs inhibit parasite-macrophage binding. It is likely that the 58KD L. infantum polypeptide identified in the present study might be the gp63 major surface antigen already identified by others (3-10). The differences between their findings and ours may be due simply to the strain and species studied.

Moreover, the anti-AChR mAbs recognize a 58KD L. infantum

polypeptide and a 55KD polypeptide. The biological significance of the cross-reactivity observed deserves further investigation. The possibility that the 58KD polypeptide recognized by the anti-AChR mAbs is the one identified by the anti-L. infantum mAbs needs to be examined.

ACKNOWLEDGMENTS

We thank Dr. A. Garifallou for providing Leishmania strains, A. Kokla for technical assistance and A. Fotiadou for typing the manuscript.

REFERENCES

1. Pearson RD, Wheeler DA, Harrison LH, and Kay HD (1983) The immunobiology of leishmaniasis. Rev. Infect. Dis. 5: 907-927.

2. Chang KP (1983) Cellular and molecular mechanisms of intracellular symbiosis in leishmaniasis. Int. Rev. Cytol. Suppl. 14:267-302.

3. Lepay DA, Nogueira N and Cohn Z (1983) Surface antigens of Leishmania donovani promastigotes. J. Exp. Med. 157:1562-1572.

4. Gardiner PR, Jaffe CL and Dwyer DM (1984) Subspecies-specific surface antigens of promastigotes of the Leishmania donovani complex. Infect. Immun. 43:637-643.

5. Colomer-Gould V, Quintao LG, Keithly J and Nogueira N (1985) A common major surface antigen on amastigotes and promastigotes of Leishmania species. J. Exp. Med. 162:902-916.

6. Chang CS and Chang KP (1986) Monoclonal antibody affinity purification of a leishmania membrane glycoprotein and its inhibition of Leishmania-macrophage binding. Proc. Natl. Acad. Sci. USA 83:100-104.

7. Russell DG and Wilhelm H (1986) The Involvement of the Major Surface glycoprotein (gp63) of Leishmania promastigotes in attachment to macrophage. J. Immunol. 136:2613-2620.

8. Bouvier J, Etges RJ and Bordier C (1985) Identification and Purification of Membrane and Soluble Forms of the Major Surface Protein of Leishmania promastigotes. J. Biol. Chem. 260:15504-15509.

9. Etges R, Bouvier J and Bordier C (1985) The major surface protein of Leishmania promastigotes is anchored in the membrane by a myristic acid-labeled phospholipid. EMBO J. 5:597-601.

10. Etges R, Bouvier J and Bordier C (1986) The Major Surface

protein of Leishmania Promastigotes Is a Protease. J. Biol. Chem. 261:9098-9101.

11. Tzartos SJ (1984) Monoclonal antibodies as probes of the acetylcholine receptor and myasthenia gravis. Trends Biochem. Sci. 9:63-67.

12. Tzartos SJ and Lindstrom JM (1980) Monoclonal antibodies to probe acetylcholine receptor structure: Localization of the main immunogenic region and detection of similarities between subunits.Proc. Natl. Acad. Sci. USA 77:755-759.

13. Tzartos SJ, Rand DE, Einarson BL and Lindstrom JM (1981) Mapping of surface structures of Electrophorus acetylcholine receptor using monoclonal antibodies. J. Biol. Chem. 256:8635-8645.

14. Tzartos S, Langeberg LK, Hochschwender S, Swanson LW and Lindstrom J (1986) Characteristics of monoclonal antibodies to denatured Torpedo and to native calf acetylcholine receptors: species, subunit and region specificity. J. Neuroimmunol. 10:235-253.

15. Tzamouranis N, Schnur LF, Garifallou A, Pateraki E and Sérié C (1984) Leishmaniasis in Greece. I. Isolation and identification of the parasite causing human and canine visceral leishmaniasis. Ann. Trop. Med. Parasitol. 78: 363-368.

16. Dwyer DM (1980) Isolation and partial characterization of surface membranes from Leishmania donovani promastigotes. J. Protozool. 27:176-182.

17. Handman E, and Remington JS (1980) Serological and Immunochemical characterization of monoclonal antibodies to Toxoplasma gondii. Immunol. 40:579-588.

18. Handman E and Hocking RE (1982) Stage specific, strain-specific and cross reactive antigens of Leishmania species identified by monoclonal antibodies. Infect.Immun. 37: 28-33.

19. Stocker JW and Heusser CH (1979) Methods for binding cells to plastic: application to a solid-phase radioimmunoassay for cell-surface antigens. J. Immunol. Methods 26: 87-95.

20. Laemmli UK (1970) Cleavage of structural proteins during the assembly of the head of bacteriophage T_4. Nature (London) 227:680-685.

21. Towbin H, Staehelin T and Gordon J (1979) Electrophoretic transfer of proteins from polyacrylamide gels to nitrocellulose sheets: procedure and some applications. Proc. Natl. Acad. Sci. USA 76:4350-4354.

22. Szarfman A, Khoury El, Cossio PM, Arana RM and Kagan IG (1975) Investigation of the EVI antibody in parasitic diseases other than american Trypanosomiasis. An anti-skeletal muscle antibody in leishmaniasis. Am. J. Trop. Med. Hyg. 24:19-24.

STRUCTURAL STUDIES ON SURFACE GLYCOPROTEINS OF LEISHMANIA PROMASTIGOTES: ISOLATION, AMINO ACID COMPOSITION AND AMINO TERMINAL SEQUENCE STUDIES

Robert W. Olafson,[1] Anne E. Wallis,[2] and Robert McMaster[2]

[1]Department of Bochemistry and Microbiology, University of Victoria, Victoria, British Columbia, Canada V8W 2Y2, and [2]Department of Medical Genetics, University of British Columbia, Vancouver, British Columbia, Canada V6T 1W5

INTRODUCTION

Successful initiation and maintenance of the parasitic state for *Leishmania* species of protozoan parasites, requires identification and specific adhesion to host cells followed by internalization and establishment of an intracellular existence. Clearly any strategy to block initial promastigote binding or subsequent spread of amastigotes will require knowledge about the molecular nature of host parasite cellular interactions. A number of laboratories have made good progress in this area demonstrating some of the characteristics of the macrophage receptors and promastigote ligands. At this time, evidence exists for a complicated interaction with a number of receptors including the iC3b complement receptor CR3 and the mannosyl/fucosyl receptor (Blackwell *et al*., 1985; Channon *et al*., 1984). Particular emphasis has been placed on those which bind a major surface glycolipid (Handman and Goding, 1985) and glycoproteins (Chang and Chang, 1986; Russell and Wilhelm, 1986) as well as covalently linked complement fragments (Blackwell *et al*., 1985). Very little attention has been placed on the other components of the parasite surface to assess whether the former ligands are singularly important or whether an array of glycosylated molecular species can interact with macrophage receptors via a redundant set of oligosaccharides. It is frequently observed for example, that monoclonal antibodies developed against cell surface antigens cross-react with many molecules of varying molecular weights - a phenomenon readily explained by a restricted set of oligosaccharides held in common by varying protein and lipid species. If the latter is found to be true, the utility of surface membrane ligands in application to prophylactic or therapeutic strategies could be seriously compromised, requiring identification and testing of all the major membrane species.

RESULTS

As part of our interest in the structure-function relationships of host lectin-parasite ligand complexes, we decided to investigate methods by which we could isolate all the major surface membrane glycoproteins of *L. donovani* for subsequent screening in a competitive promastigote macrophage-binding assay. In order to assess the complexity of the membrane

NATO ASI Series, Vol. H11
Host-Parasite Cellular and Molecular Interactions
in Protozoal Infections. Edited by K.-P. Chang and D. Snary

glycosylated species we carried out ^{3}H-mannose metabolic incorporation and tritiated sodium borohydride reductive labelling of whole cells. Mannose was labelled at C-2 and therefore the incoroporation results would be expected to reflect glycoconjugates of this monosaccharide. Figures 1 and 2 show the results of this study.

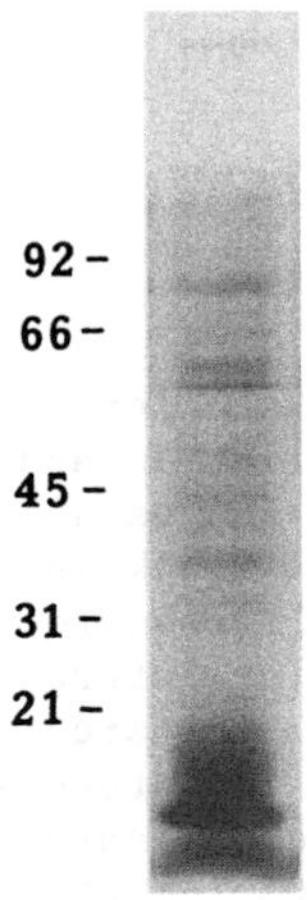

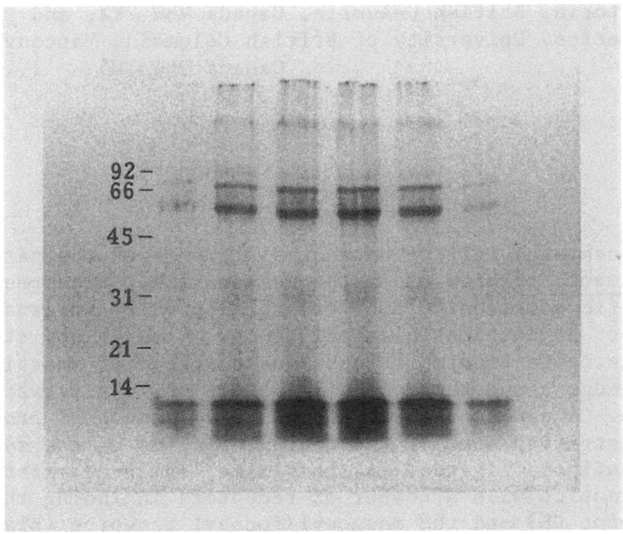

FIGURE 1. SDS-PAGE of glycosylated molecules from L. donovani membranes prepared from cells metabolically incorporated with 1 mCi of ^{3}H-mannose for 44 hours.

FIGURE 2. SDS-PAGE of glycosylated molecules from L. donovani labelled by the periodate/tritiated sodium borohydride procedure. Different loadings of the same material are shown.

The reductive labelling procedure indicated heaviest tritium incorporation into that region of the gel normally occupied by the major surface glycoprotein (MSG) (ca 55-63,000 D) as well as into a group of small molecular weight glycolipids (fronting band). Mannose incorporation was also heaviest in those molecular species with mobilities corresponding to approximately 80, 60, and less than 20,000 D. The low molecular weight glycolipid fraction (running ahead of the tracker dye) also incorporated large amounts of mannose, in agreement with the reductive labelling results and have been isolated in a separate study.

With the above information in hand, pellicular membranes from surface radioiodinated promastigotes were then extracted with Triton X-114 (Bordier, 1981) in the presence of EDTA, TLCK, TPCK, and E-60 protease inhibitors to obtain a preparation of ^{125}I-labelled integral membrane proteins and lipids. Figure 3 shows the complexity of the initial membrane preparation stained with Coomassie blue. Autoradiography of the SDS-PAGE separated, Triton extracted membranes, indicated the prominent 55,000 D MSG and at least 9 other species 30,000 D and larger. Many more smaller molecular weight molecules were poorly resolved at 20,000 D and lower.

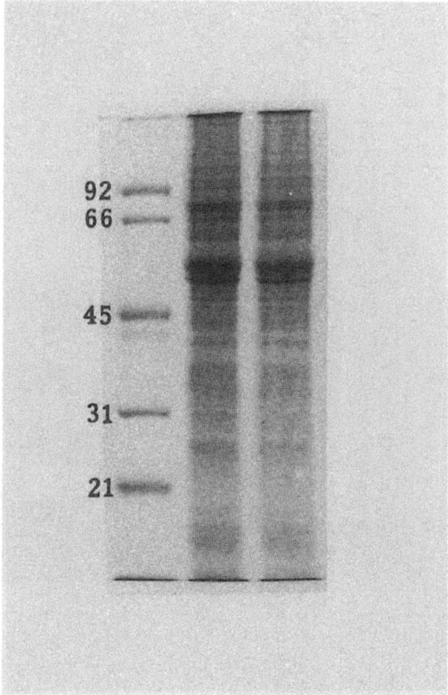

FIGURE 3. Coomassie blue stained gel of SDS-PAGE separated membrane proteins. Two loading of the same sample.

It should be emphasized that any peripheral membrane species were expected to be removed with the aqueous supernatant in this isolation methodology which was developed to partition integral membrane proteins as a function of their hydrophobicity. To the author's knowledge, the abundance and nature of the peripheral membrane species of the promastigote pellicular cell surface has as yet not been clearly delineated.

The detergent extracted membrane components from above were then acetone precipitated, dissolved in SDS-PAGE sample buffer and applied to an automated preparative SDS-PAGE apparatus (Fig. 4). The electrophoresis apparatus has been automated with an electronic system which actuates valves and controls the power and fraction collector.

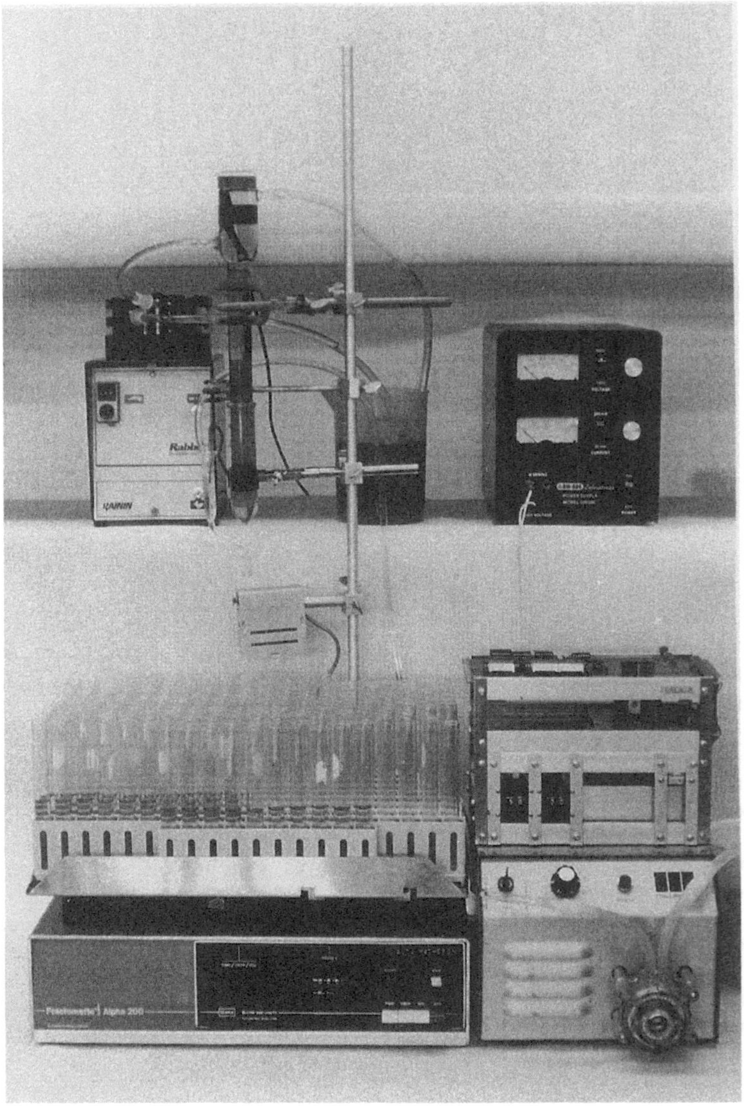

FIGURE 4. Automated preparative polyacrylamide gel apparatus.

Figure 5 indicates a typical elution profile of fractions monitored at 280 and 230 nm. Individual fractions were checked for purity by analytical SDS-PAGE which resulted in the expected molecular weight ladder of purified Coomassie and silver staining bands.

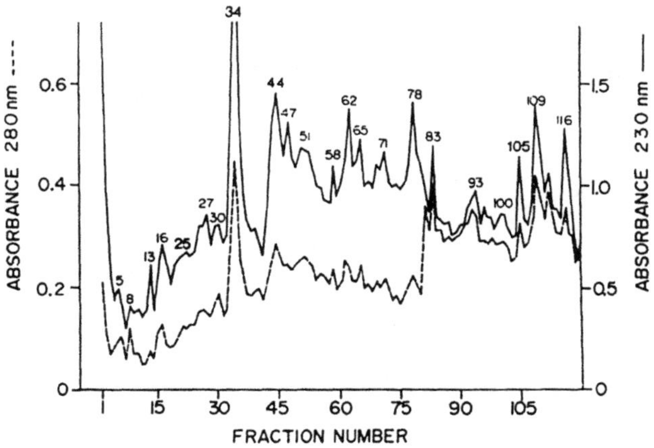

FIGURE 5. Electrophoretic profile of Triton X-114 extracted <u>L. donovani</u> membrane preparation solubilized in .1% SDS/8M urea.

Final purification of each fraction was performed by GP-HPLC using a 60 cm TSK-XL column eluted in 0.1% SDS, 80 mM ammonium acetate, pH 6.0. Small molecular weight contaminants were readily removed providing glycoproteins which on reduction and alkylation could be applied directly to the gas-phase sequencer for amino terminal sequence analysis (Fig. 6). Sample sizes of 20-30 picomoles were easily managed providing an open amino terminus was encountered.

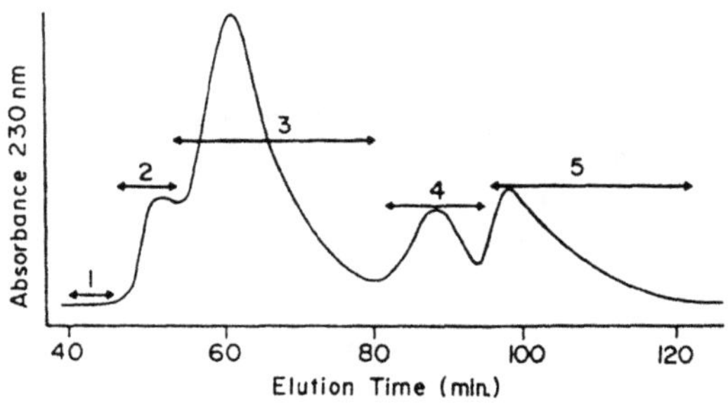

FIGURE 6. Gel permeation HPLC of typical preparative electrophoresis fraction eluted on a TSK-XL column as indicated above. Numbers refer to fractions cut for analysis.

Analytical SDS-PAGE (Fig. 7) indicated that even in the presence of SDS, the membrane proteins tended to aggregate with small molecular weight contaminants which could be removed on recycling those fractions through the HPLC. The leading chromatographic peak (Fig. 6, fraction 2) however, was invariably homogeneous.

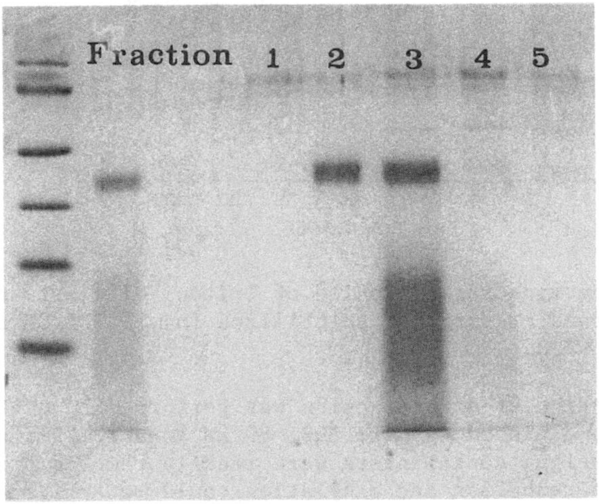

FIGURE 7. SDS-PAGE of typical GP-HPLC separated preparative electrophoresis fraction (peak fraction 44).

Using a similar approach, the MSG of L. major was isolated and the amino terminal sequence determined by gas-phase microsequence analysis. Oligonucleotide probes have been synthesized from the MSG sequence and have been used to probe a genomic library. In addition, this amino terminal sequence has now been synthesized using solid phase chemistry and the peptide is being studied in immunoprotection and promastigote-macrophage binding experiments, as well as being used to raise specific antisera. Further work in progress is directed towards isolation of the C-terminal peptide of this major surface glycoprotein.

The most surprising finding of this work however was obtained while attempting to verify that isolated glycoproteins were not glycosylation variants but were unique proteins. This data was collected by amino acid analysis of selected fractions across the electrophoretic profile. Table 1 shows that results of analysis of 30,000, 40,000, 45,000 and 55,000 molecular weight proteins.

TABLE 1
Amino Acid Compositions
For Selected Promastigote Membrane Proteins[a]

Amino Acid	Fraction 25 30,000 D (mol %)	Fraction 33 40,000 D (mol %)	Fraction 44 45,000 D (mol %)	Fraction 105 55,000 D (mol %)
ASX	12	9	13	11
THR	5	6	6	6
SER	17	8	13	10
GLX	14	12	13	14
PRO	-	3	-	5
GLY	20	10	16	12
ALA	6	9	8	10
VAL	7	6	5	5
MET	-	3	1	1
ILE	3	4	3	3
LEU	6	9	8	9
TYR	1	3	3	3
PHE	-	3	1	3
HIS	3	2	3	2
LYS	5	3	6	3
ARG	2	5	4	5
% hydrophobic residues	23	37	29	34

[a]All values rounded up to next highest integer.

While the data indicates that these molecules do not have identical compositions it also shows that they have much smaller hydrophobicities, as estimated from summing those aliphatic and aromatic residues analyzed, than would be expected for hydrophobic membrane molecules. These proteins were obtained from a membrane preparation, ^{125}I-labelled and partitioned with Triton X-114. The only explanations for these surprising results is that either these proteins have small highly localized areas of hydrophobicity for membrane interaction, or have lipids attached similar to the covalently attached myristic acid found on African trypanosome variable surface glycoproteins. These possibilities are now being actively researched.

ACKNOWLEDGEMENTS

This work was supported by the Natural Sciences and Engineering Research Council of Canada.

REFERENCES

Bordier C (1981) Phase separation of intetral membrane proteins in Triton X-114 solution J Biol Chem 256:1604-1607

Blackwell JM, Ezekowitz RAB, Roberts MB, Channon JY, Sim RB, Gordon S (1985) Macrophage complement and lectin-like receptors bind Leishmania in the absence of serum J Exp Med 162:324-331

Chang CS, Chang K-P (1986) Monoclonal antibody affinity purification of a Leishmania membrane glycoprotein and its inhibition of Leishmania-macrophage binding Proc Natl Acad Sci USA 83:100-104

Channon JY, Roberts MB, Blackwell JM (1984) A study of the differential respiratory burst activity elicited by promastigotes and amastigotes of Leishmania donovani in murine resident peritoneal macrophages Immunol 53:345-351

Handman E, Goding JW (1985) The Leishmania receptor for macrophages is a lipid-containing glycoconjugates EMBO 4:329-336

Russell DG, Wilhelm H (1986) The involvement of the major surface glycoprotein (gp63) of Leishmania promastigotes in attachment to macrophages J Immunol 136:2613-2610

MEMBRANE ATTACHMENT OF THE MAJOR SURFACE PROTEIN OF LEISHMANIA PROMASTIGOTES

C. Bordier[1], R. Etges[1], J. Bouvier[1], J. Ward[2], M. J. Turner[2], and M. L. Cardoso de Almeida[2]

[1]Institut de Biochimie, Universite de Lausanne, CH-1066 Epalinges, Switzerland.
[2]The Molteno Institute, MRC Biochemical Parasitology Unit, University of Cambridge, Cambridge CB2 3EE, England.

INTRODUCTION

A major suface glycoprotein with a subunit molecular mass of 63,000 Da has been identified on promastigotes of several species of **Leishmania** (Etges et al. 1987 for review). The amphiphilic glycoprotein possesses a neutral protease activity and has been designated promastigote surface protease, or PSP . The protease is active at the surface of the parasite or as soluble enzyme (Etges et al. 1986). In this paper, we shall describe succinctly the experiments which led to the elucidation of the anchoring mechanism of PSP to the pellicular membrane of the promastigote.

During purification by standard chromatographic procedures, the apparently homogenous protein was resolved into two fractions. The analysis by phase separation in the nonionic detergent Triton X-114 (Bordier 1981) demonstrated that the molecules found in one fraction bound detergent, as expected for an integral membrane protein, while the molecules found in the other fraction were not associated with detergent, and behaved like a water-soluble protein. The two forms of PSP were termed amphiphilic PSP (A-PSP, formerly Ap63) and hydrophilic PSP (H-PSP, formerly H-p63), respectively (Bouvier et al. 1985).

Analysis by Na.dodecylsulfate polyacrylamide gel electrophoresis (SDS-PAGE) after heat denaturation of the proteins showed that H-PSP migrated slightly slower than the original A-PSP extracted from the membrane. This behavior

NATO ASI Series, Vol. H11
Host-Parasite Cellular and Molecular Interactions
in Protozoal Infections. Edited by K.-P. Chang and D. Snary

immediately suggested a parallel between PSP of **Leishmania** and the variant surface glycoprotein (VSG) of **Trypanosoma brucei.**

The classical purification procedure for VSG yields a water-soluble protein, sVSG (Cross 1975). In contrast, the extraction of trypanosomes under strongly denaturing conditions like hot SDS solution, or in the presence of Zn^{++} or organomercury compounds, yields the original membrane form of VSG (mfVSG) which characteristically migrates slightly faster on SDS-PAGE than sVSG (Cardoso de Almeida and Turner 1983). The difference between the two forms of VSG resides in a dimyristylglycerol residue which is covalently attached to mfVSG, but which is absent from sVSG (Ferguson et al. 1985). Dimyristylglycerol functions as the sole hydrophobic moiety anchoring mfVSG to the trypanosome membrane. The following experiments were performed in order to compare the molecular nature of the membrane anchors of **Leishmania** PSP and **Trypanosoma brucei** VSG.

RESULTS AND DISCUSSION

Biosynthetic labeling of the promastigote surface protease with fatty acids.

L. major promastigotes were labeled for approximately one generation with ^{3}H-myristic acid and their proteins analyzed by SDS-PAGE. A major band labeled with the fatty acid comigrated with surface-iodinated PSP and was recovered with other amphiphilic molecules after phase separation in Triton X-114 solution (Bouvier et al. 1985, Etges et al. 1986). **L. major** promastigotes were similarly labeled with ^{3}H-palmitic acid (Figure 1). After 22 hours, the cells were washed and lysed in Triton X-114. Detergent-insoluble material was recovered as a pellet after centrifugation at 4°C (Figure 1, lane 1) and the supernatant was subjected to phase separation at 37°C. Two fractions, containing the hydrophilic molecules (lane 2) and the amphiphilic molecules (lane 3) were obtained and analysed by SDS-PAGE. Figure 1 represents a fluorogram of the gel. As expected, most of the radioactivity from the labeled fatty acid is recovered in the detergent phase after phase separation (lane 3). A-PSP is one of the few prominently labeled components of the detergent phase (arrow).

The labeling of PSP with both myristic and palmitic acids raises the question of the identity of the fatty acids incorporated in the enzyme. Myristic acid is the only fatty acid found in mfVSG from **Trypanosoma brucei** (Ferguson et al. 1985), suggesting that at least one enzyme in the biosynthetic pathway of the membrane anchor of VSG is specific for this fatty acid. Whether an identical specificity for myristic acid exists in the **Leishmania** pathway can only be hypothesized at the moment. During the long labeling period, **Leishmania** promastigotes could have transformed the labeled molecule into a different fatty acid before its incorporation into PSP. The limited amounts of protease available have for the moment precluded the direct chemical analysis of the fatty acid in A-PSP.

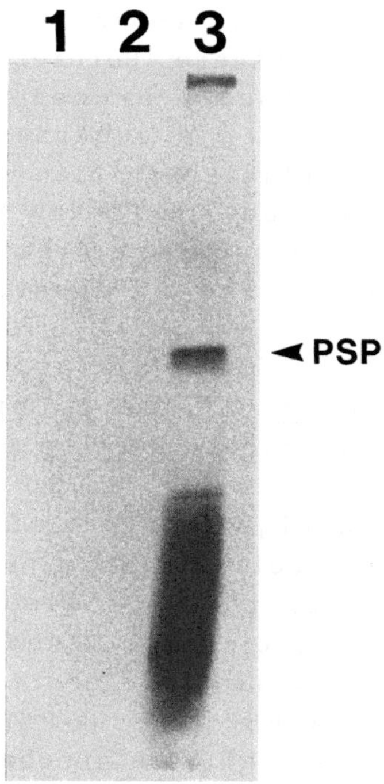

Figure 1. Palmitic acid labeling of L. **major**. L. **major** LEM 513 promastigotes (3 x 10^7 c/ml, 25 ml) were incubated in serum-free Schaefer's medium in the presence of ^{3}H-palmitate (Amersham). After 1 hr, 5% heat-inactivated foetal calf serum was added to the culture. After 22 hr, the cells were washed and lysed in 2 % Triton X-114. Detergent insoluble material was sedimented at 4° C (lane 1) and the supernatant was submitted to phase separation at 37° C. The aqueous and detergent enriched phases were analysed on lanes 2 and 3 respectively. The figure is a fluorogram of the samples after SDS-PAGE.

Digestion of amphiphilic PSP with phospholipases and generation of hydrophilic PSP in vitro.

The digestion of ^{3}H-myristic acid-labeled A-PSP with phospholipase C from **Bacillus cereus** eliminated the label from the protein, showing that the fatty acid was bound to the protein in the form of a phospholipid. Concommittant with the elimination of the labeled fatty acid by phospholipase C digestion, amphiphilic PSP was transformed into a water-soluble molecule which partitioned into the aqueous phase after separation in Triton X-114 solution (Etges et al. 1986). The membrane-bound phospholipase of **Trypanosoma brucei**, which is responsible for the release of sVSG, is also able to remove fatty acids from A-PSP and to generate a catalytically active hydrophilic form of the protease (Bordier et al. 1986, Etges et al. 1986). An assay for the T. **brucei** lipase was developed using radiolabeled A-PSP as a substrate. This assay takes advantage of the differential partition of A-PSP and of H-PSP, the product of the reaction, in Triton X-114 solution. On A-PSP, the lipase displays apparent first order kinetics with respect to the concentration of substrate (Ward et al. in press).

Lipase-dependent exposure of the **T. brucei** cross-reacting determinant on the promastigote surface protease.

Different variant antigenic types of T. **brucei** sVSG have a common antigenic crossreacting determinant (CRD) located on a carbohydrate moiety near the C-terminus of the polypeptide. This deteminant is cryptic in mfVSG and is exposed only upon the generation of sVSG by lipase digestion. In the experiment shown in Figure 2, A-PSP and H-PSP (generated during chromatographic purification) were incubated with or without B. **cereus** phospholipase C, analysed by SDS-PAGE, and electrophoretically transferred to a nitrocellulose sheet. Proteins were first revealed with Ponceau S (figure 2, Top). After complete destaining, the blot was incubated with affinity purified rabbit antibodies against the crossreacting determinant of T. **brucei** VSG (Cardoso de Almeida and Turner 1983). Bound antibodies were revealed with protein A-peroxidase. A-PSP does not react with the anti-CRD antibodies (Figure 2, lane 1). In contrast, A-PSP digested with phospholipase C (figure 2, lane 2) shows a strong reactivity with the antibodies, demonstrating the presence of the crossreacting determinant on **Leishmania** PSP. The same results were obtained with the phospholipase of T. brucei and with the PSP of **L. donovani** (Bordier et al. 1986). To our surprise, the H-PSP obtained during the purification of the enzyme did not display the crossreacting determinant, independently of lipase digestion (Figure 2, lanes 3 and 4). Possible explanations for the absence of CRD activity are that the enzyme responsible for the generation of H-PSP during extraction and purification of PSP is not a phospholipase C, or that additional enzymes rapidly modify the CRD and render it unreactive to the anti-CRD antibodies.

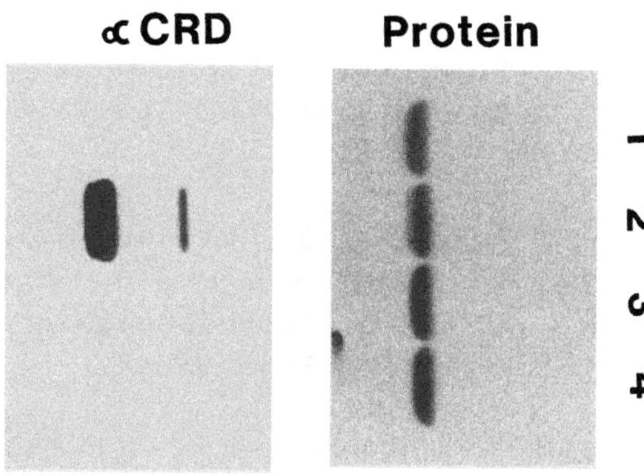

Figure 2. Detection of crossreacting determinant on **Leishmania** promastigote surface protease. 5 µg A-PSP and H-PSP from **L. major** LEM 513 were incubated without and with **B. cereus** phospholipase C for 30' at 30° C. After digestion the proteins were analysed by SDS-PAGE, transfered to nitrocellulose and stained with Ponceau S (top). After destaining, CRD was revealed with affinity purified antibodies raised against the CRD of sVSG (Cardoso de Almeida and Turner 1983). 1,2 : A-PSP. 3,4 : H-PSP. 1,3 : no lipase. 2,4 : 12.5 U/ml lipase.

In any case, the generation of H-PSP by endogenous **Leishmania** enzymes is extremely slow when compared to the same process taking place in trypanosomes on VSG. The existence of a phospholipase homologous to the T. brucei enzyme in **Leishmania** promastigotes remains to be demonstrated.

Whether a functional role is played by the glycophospholipid anchor of PSP during the life cycle of **Leishmania** also remains to be demonstrated. In this respect, it may be relevant to note that PSP appears to be synthesized by the amastigote form of the parasite, but that it cannot be detected at the surface of these cells. A possible explanation would be that the phospholipid anchor of PSP is cleaved by a macrophage phospholipase, allowing the protease to be released in a soluble form from the membrane of the parasite, free to exert its function at locations distant from the parasitophorous vacuole.

REFERENCES

Bordier, C. (1981) Phase separation of integral membrane proteins in Triton X-114 solution. J. Biol. Chem. 256, 1604-1607.

Bordier, C., Etges, R.J., Ward, J., Turner, M.J. and Cardoso de Almeida, M.-L. (1986) **Leishmania** and **Trypanosoma** surface glycoproteins have a common glycophospholipid membrane anchor. Proc. Natl. Acad. Sci. USA 83, 5988-5991.

Bouvier, J., Etges, R.J. and Bordier, C. (1985) Identification and purification of membrane and soluble forms of the major surface protein of **Leishmania** promastigotes. J. Biol. Chem. 260, 15504-15509.

Cardoso de Almeida, M.-L. and Turner, M.J. (1983) The membrane form of variant surface glycoprotein of **Trypanosoma brucei**. Nature 302, 349-352.

Cross, G.A.M. (1975) Identification, purification and properties of clone-specific glycoprotein antigens constituting the surface coat of **Trypanosoma brucei**. Parasitol. 71, 393-417.

Etges, R.J., Bouvier, J. and Bordier, C. (1986 a) The major surface protein of **Leishmania** promastigotes is anchored in the membrane by a myristic acid-labeled phospholipid. EMBO J. 5, 597-601.

Etges, R.J., Bouvier, J. and Bordier, C. (1986 b) The major surface protein of **Leishmania** promastigotes is a protease. J. Biol. Chem. 261, 9098-9101.

Etges, R.J., Bouvier, J. and Bordier, C. (1987) The promastigote surface proteases of **Leishmania.** in NATO ASI series .., p ..., Host-parasite cellular and molecular interactions in protozoal infections. K.P. Chang, Editor.

Ferguson, M.A.J., Haldar, K. and Cross, G.A.M. (1985) **Trypanosoma brucei** variant surface glycoprotein has a sn-1,2-dimyristyl glycerol membrane anchor at its COOH-terminus. J. Biol. Chem. 260, 4963-4968.

Ward, J., Cardoso de Almeida, M.-L., Turner, M.J., Etges, R. and Bordier, C. An assay of membrane-bound **Trypanosoma brucei** phospholipase using an integral membrane protein substrate and detergent phase separation. Molec. Biochem. Parasitol. (in press)

THE PROMASTIGOTE SURFACE PROTEASE OF LEISHMANIA

Robert Etges, Jacques Bouvier, and Clement Bordier

Institut de Biochimie, Universite de Lausanne,
CH-1066 Epalinges, Switzerland

INTRODUCTION

A neutral protease activity was observed to copurify with gp63, the major suface protein of **Leishmania** promastigotes. In **L. major** LEM 513, the protease is resistant to denaturation in sodium dodecylsulfate (SDS), and its activity can be detected after polyacrylamide gel electrophoresis in the presence of SDS with casein-agar overlay gels (Etges et al. 1986). Here we extend our identification of the promastigote surface protease (formerly gp63) to three strains of **Leishmania** previously shown by surface radioiodination, peptide mapping and antigenic crossreactivity to express gp63: **L. major** LRC-L137, **L. donovani infantum** LEM 75, and **L. tropica** LRC-L32 (Etges et al. 1985).

MATERIALS AND METHODS

Promastigotes of each strain were surface radioiodinated using IodoGen (Etges et al. 1985), lysed in nonionic detergent, and insoluble material was eliminated by centrifugation. Affinity-purified rabbit antibodies to the trifluoromethanesulfonic acid-deglycosylated gp63 of **L. major** LEM 513 (Bouvier et al. 1985) were added, and immune complexes were isolated by affinity to **Staphylococcus aureus** Protein A-conjugated Sepharose beads. Washed beads were resuspended in electrophoresis sample buffer containing 1 percent SDS, incubated at 37°C for 2 hr, and applied without additional heating or reduction to a 7.5-15 percent polyacrylamide gel containing 0.1 percent SDS and 240 ug/ml of copolymerized fibrinogen (Lonsdale-Eccles and Mpimbaza 1986). The gels were washed in 10 mM Tris HCl, pH 8.0, to eliminate the remaining detergent and incubated for 16 hr in the same buffer at 25°C to permit the promastigote proteases to digest the fibrinogen.

NATO ASI Series, Vol. H11
Host-Parasite Cellular and Molecular Interactions
in Protozoal Infections. Edited by K.-P. Chang and D. Snary

RESULTS

The immunoprecipitated protease activity of each strain is revealed as a clear band on a Coomassie blue-stained fibrinogen background. The dried gel was fluorographed to show the co-migration of the promastigote protease (Figure 1, left panel) with the major radioiodinated surface glycoprotein gp63 (Figure 1, right panel) of each strain. In each of the three strains, a precise comigration is observed, demonstrating that the immunoprecipitated proteases are surface proteins.

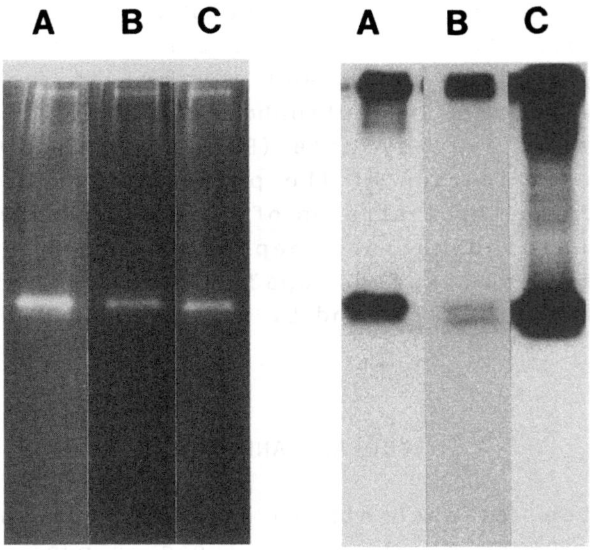

FIGURE 1. Enzymatic activity and surface radioiodination patterns of the promastigote surface proteases of **Leishmania major** (A), L. **donovani** (B), and L. **tropica** (C). The activity of the immunoprecipitated protease against fibrinogen of the three strains is shown in the left panel. The radioactive proteins revealed by fluorography of the dried gel are shown in the right panel.

DISCUSSION

The promastigote surface protease (formerly gp63) has been implicated in macrophage-promastigote interactions during infection of the mammalian host. Asparagine-linked glycans of the protease have been reported to bind to the macrophage mannose-fucose receptor (Russell and Wilhelm 1986). This mechanism would not reqire the activity of the protease. On the other hand, inactivated C3b (iC3b) has been suggested to function as a ligand to facilitate the phagocytosis of promastigotes by macrophages (Blackwell et al. 1985). The surface protease could catalyse the specific hydrolysis of C3b bound to the promastigote via the activation of the alternative pathway (Mosser and Edelson 1985), forming iC3b, thus enhancing the internalization of the promastigote via the complement receptor 3 (CR3) of the host macrophage, and diminishing the activation of the late, lytic components of the complement pathway.

The promastigote surface protease displays its maximum proteolytic activity in the neutral to alkaline pH range, suggesting that the enzyme may exert its function in other locations than in the parasitophorous vacuole (Etges et al. 1986). **Leishmania** promastigotes consume amino acids in preference to carbohydrate substrates during log-phase growth (Mukkada 1985), and all species of **Leishmania** require a source of preformed heme (Chang and Chang 1985). A possible role for the protease could be in the acquisition of these essential substrates from the hemoglobin contained in vector blood meals.

Neither the substrate specificity of the promastigote surface protease nor its enzymatic mechanism have been firmly established. It is unlikely to be a serine esterase, since its activity is unaffected by 1 mM diisopropyfluorophosphate (Etges et al. 1986). Nonspecific thiol-reactive compounds such as iodoacetamide, zinc, and mercury completely abrogate the activity of the protease, suggesting, but not proving, that it might be a cysteine protease.

A better understanding of the substrate specificity and enzymatic mechanism of the promastigote surface protease will allow the rational design of new, specifically targetted drugs that may prove to be useful in the treatment of the leishmaniases.

REFERENCES

Blackwell, J. M., A.B. Ezekowitz, M.B. Roberts, J.Y. Channon, R.B. Sim, and S. Gordon (1985) Macrophage complement and lectin- like receptors bind **Leishmania** in the absence of serum. J. Exp. Med. 162, 324-331.

Bouvier,J. R.J. Etges, and C. Bordier (1985) Identification and purification of membrane and soluble forms of the major surface protein of **Leishmania** promastigotes. J. Biol. Chem. 260: 15504-15509.

Chang, C.-S.and K.-P.Chang (1985) Heme requirements and acquisition by extracellular and intracellular stages of **Leishmania mexicana amazonenesis** Mol.Biochem.Parasitol.16:267-276.

Chang, K.-P. (1980) Endocytosis of **Leishmania**-infected macrophages. Fluorometry of pinocytic rate, lysosome-phagosome fusion and intralysosomal pH. In: H. van den Bossche (ed) The Host-Invader Interplay, Elsevier North Holland, Amsterdam, p. 231 - 234.

Chang, K.-P. (1983) Cellular and molecular mechanisms of intracellular symbiosis in leishmaniasis. Int.Rev.Cytol. Suppl. 14: 267-302.

Etges, R.J., J. Bouvier, R. Hoffman, and C. Bordier (1985) Evidence that the major surface protein of three **Leishmania** species are structurally related. Mol.Biochem.Parasitol.14: 141-149.

Etges R.J., J. Bouvier, and C. Bordier (1986) The major surface protein of **Leishmania** promastigotes is a protease. J. Biol. Chem. 261: 9098-9101.

Lonsdale-Eccles,J.D. and C.W.N. Mpimbaza (1986) Thiol-dependent proteases of african trypanosomes. Analysis by electrophoresis in sodium dodecyl sulfate/polyacrylamide gels copolymerized with fibrinogen. Eur. J. Biochem. 155, 469-473.

Mosser,D.M. and P.J.Edelson (1985) Activation of the alternative complement pathway by **Leishmania** promastigotes: parasite lysis and attachment to macrophages. J. Immunol.132: 1501-1505.

Mukkada, A.J. (1985) Energy metabolism in **Leishmania**. in: K.-P.Chang and R.S. Bray, Human Parasitic Diseases, Vol. 1, Elsevier, Amsterdam, New York, London. p.49-64.

Russell D.G. and H.Wilhelm (1986) The involvement of gp63, the major surface glycoprotein, in the attachment of **Leishmania** promastigotes to macrophages. J.Immunol.136:2613-2620.

ACID PROTEASE ACTIVITY OF LEISHMANIA GP63 AND ITS POSSIBLE ROLE IN VIRULENCE

Gautam Chaudhuri and K. -P. Chang

Department of Microbiology/Immunology, UHS/Chicago Medical School, N. Chicago IL 60064, USA

INTRODUCTION

Virulent determinants of most parasitic protozoa remain largely unknown. Of particular interest are leishmanias, which live in the digestive tract of the sandfly vector as extracellular promastigotes and, when delivered into the mammalian hosts, in the phagolysosomes of their macrophages as intracellular amastigotes (Chang, Fong & Bray 1985). Leishmanial virulence thus must hinge on their ability not only to multiply in these varied environments but also to evade the extracellular and intracellular digestive enzymes and immune lytic factors therein. Here, we present evidence that a major surface glycoprotein identified and purified previously as gp63 (Chang & Chang 1986) is an acid protease. Its broad spectrum of activity and increased expression on the cell surface of virulent phenotype suggest that this novel molecule is an important virulent factor in leishmanial promastigotes.

MATERIALS AND METHODS

Organism and its Maintenance

Leishmania mexicana amazonensis and other species were grown at 27°C as promastigotes in Medium 199 with heat-inactivated fetal bovine serum. Virulent cells of L. m. amazonensis were those passaged weekly in vitro for less than 25 times (1 passage=5-6 generations) and produced lesions in mice within 20-25 days. Avirulent cells were those subcultured in vitro for more than a hundred weekly passages and produced no palpable lesions in mice after infection for 90 days or longer.

Protease Assays

Proteolytic activity of purified gp63 (Chang & Chang 1986) was assayed using [^{125}I]Na labeled bovine serum albumin (BSA) as substrate. Immunoglobulin G, human complement (C3), hemoglobulin and rat liver lysosomal proteins (Chang & Fong 1983) labeled with [^{125}I], and [^{3}H]leucine-labeled hemoglobulin and azocasein were also used. Radioiodination of proteins were done with [^{125}I] Na by the iodine monochloride catalyzed method (Golstein, Basu & Brown 1983). The protease assays with radioiodinated proteins measured the formation of trichloroacetic acid soluble [^{125}I] tyrosines or their derivatives, but not deiodination, as extraction of inorganic iodide from the acid soluble reaction products did not reduce the net increase in radioactivity as a result of the protease action (Goldstein & Brown 1974).

NATO ASI Series, Vol. H11
Host-Parasite Cellular and Molecular Interactions
in Protozoal Infections. Edited by K.-P. Chang and D. Snary

RESULTS AND DISCUSSION

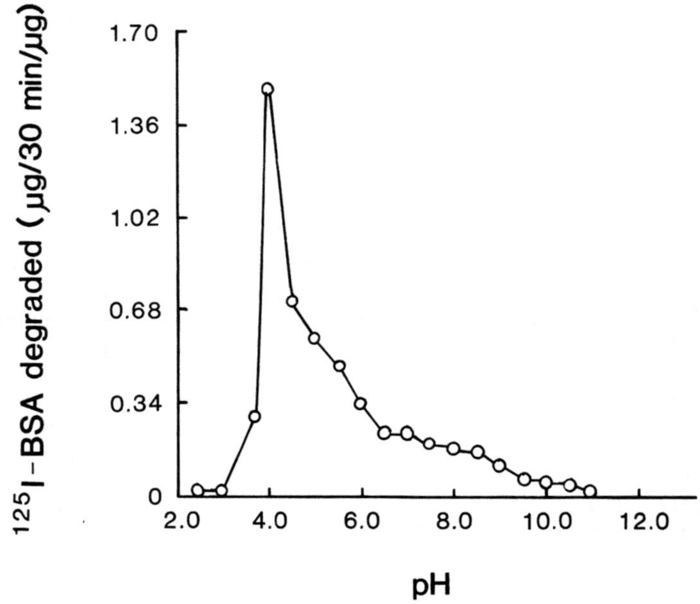

FIGURE 1. Effect of hydrogen ion concentration on the protease activity of purified gp63 at 37°C. The proteolytic activity of purified gp63 on[^{125}I] BSA was measured at 0.5 pH unit intervals between pH 2.6 and 11.0 using Britton-Robinson universal buffer (Britton & Welford 1937) containing 29 uM each of citric acid, potasium dibasic phosphate, sodium barbital and boric acid.

The association of the protease activity with purified gp63 was indicated by several lines of experimental evidence: (1) Samples of gp63 from 12 independent isolations gave the same specific activity of 1.73 ± 0.17 ug BSA degraded/30 min/ug gp63 protein; (2) Immunoprecipitation of gp63 by specific monoclonal antibody from the enzyme preparation removed the protease activity; and (3) The proteolytic activity comigrated with the only protein band of gp 63 in 5% polyacrylamide gel after electrophoresis under non-reducing, non-dissociating conditions.

The gp63 protease from L. m. amazonensis was charaterized and found to differ from other proteases of Leishmania reported (Steiger et al. 1979; Fong & Chang 1981; Coombs 1982; Simon & Mukkada 1983; Etges, Bouvier & Bordier 1986). This is the only leishmanial protease with a broad pH spectrum, although it is most active at pH 4.0 (Fig. 1). It also has a wide temperature spectrum of 4 to 80°C: enzyme activity increases with increasing temperatures, reaching an optimum at 60°C. There is little or no loss of enzymatic activity of gp63 after incubation at either pH 4.0 or pH 8.5 for 24 hours at 37°C, suggesting that the protease is stable at this temperature without inactivation or autodegradation.

The leishmanial gp63 degraded non-glycosylated proteins, e.g. albumin and hemoglobulin more readily than glycoproteins, e.g. immunoglobulin G, complement (C3) and lysosomal proteins (Table 1). Degradation of the lysosomal proteins by gp63 protease was about 5-fold higher than that by a representative

TABLE 1
PROTEOLYSIS OF DIFFERENT SUBSTRATES BY GP63 OF LEISHMANIA MEXICANA AMAZONENSIS

Substrate[a]	Specific activity (ng/30 min/ug gp63) at pH: 4.5	7.5
Bovine serum albumin	750	78
hemoglobulin	730	73
Immunoglobulin G	263	34
Complement C3	378	42
Lysosomal proteins	218	27

[a]Protease activity was assayed at 37°C using [^{125}I]-labeled substrates.

lysosomal acid protease, cathepsin D, at pH 4.5, 37°C, although both were equally active in degrading BSA under identical conditions. About 50-80% of the protease activity of gp63 was inhibited by 1 mM 1,10-phenanthroline or 100 ng/ml bestatin or 1 mM each of the following heavy metal ions: Pb^{2+}, Fe^{2+}, Fe^{3+}, Cu^{2+}, Cr^{6+}, Ni^{2+}, Hg^{2+}, Cd^{2+}, Zn^{2+}, Ag^{+}. About 25% of the enzymatic activity was inhibited by 1 mM phenylmethylsulphonyl fluoride or 100 ug leupeptin per ml. Other compounds known to inhibit protease from different sources (North 1982) had no effect on the proteolytic activity of leishmania gp63. Its activity was unaffected by non-ionic detergents, e.g. Triton X-100 or Nonidet P-40 (1%), but was sensitive to 0.1% sodium dodecyl sulphate (SDS). Thus, gp63 seems to be a unique protease, which cannot be grouped with known enzyme classes (North 1982). The proteolytic activity of gp63 from *L. m. amazonensis* reported here differs from that reported from *L. major* in that the latter is resistant to 0.1% SDS and has an alkaline pH optimum (Etges, Bouvier & Bordier 1986). It is uncertain if these differences are due to the assay conditions used in the respective studies or are the inherent properties of gp63 from different *Leishmania* species.

The N-linked oligosaccharides of gp63 (Chang et al. 1986) appear to be important for its protease activity. Pre-treatment of gp63 with protease-free N-glycanase (Genezyme Co.) at pH 8.6 (60 units of the enzyme/ml in 0.2 M sodium phosphate buffer) for 24 hours at 37°C reduced the protease activity to 2% of the control simultaneously prepared under the same conditions without N-glycanase. As has been suggested for other glycoproteins (Olden et al. 1982), the N-linked oligosaccharides may be important in stabilizing the gp63 molecules to prevent their autodegradation. Alternatively, they may be necessary for maintaining certain configuration of the molecules for enzymatic activity.

The amount of this surface gp63 and its protease activity were both higher in the virulent than in the avirulent cells of *L. m. amazonensis* (Table 2). The gp63 molecules per cell were determined fluorometrically (McKinney, Spillane & Pearce 1966) using FITC-Fab fragments of a monoclonal antibody specific to this molecule. The mean flagellar and body surface areas were determined to be 52.99 $\pm$ 12.93 um^2 and 185.83 $\pm$ 2.79 um^2, respectively. Based on these determinations, the number of gp63 molecules/um^2 was calculated and found to be 2.5-fold higher on the virulent than on the avirulent cells. This amount of difference was verified by determining the amount of gp63 relative to the total cellular proteins. The protease activity of gp63 *in situ* was compared between the virulent and avirulent promastigotes using cells fixed with 0.5% glutaraldehyde for 30 min at 4°C in Hanks' balanced salt

TABLE 2
PROTEASE ACTIVITY OF GP63 AND ITS MOLECULAR DENSITY ON THE SURFACE OF VIRULENT AND AVIRULENT PROMASTIGOTES OF LEISHMANIA MEXICANA AMAZONENSIS

Cell Type	Surface Density of gp63 (Molecules/um^2)	% Total Protein of gp63	Surface Protease Activity at pH 4.0 (ng/hour/10^7 cells)
Virulent	2,500	1.23	338
Avirulent	1,000	0.54	146

solution. Virulent cells gave 2.3 fold more proteolytic activity per cell as compared to avirulent cells (Table 2). Expression of *Leishmania* surface protease is thus related to the parasite virulence.

The universal presence of gp63 on the surface of *Leishmania* promastigotes signifies its functional importance. This is further underscored by finding protease activity associated with this molecule in *L. major* (Etges, Bouvier & Bordier 1986) and in *L. m. amazonensis* reported here.

ACKNOWLEDGMENTS

We thank A.E. Potthoff for technical assistance. This work was supported by Grants AI-20486 from US National Institute of Allergy & Infectious Diseases, PCM-8309895 from US National Science Foundation and No. 350305 from the World Health Organization/ Special Programme for Research and Training in Tropical Diseases/World Bank.

REFERENCES

Britton B, Welford M (1937) Universal buffer. J Chem Soc 1937: 1848

Chang CS, Chang KP (1986) Monoclonal antibody affinity purification of a leishmania membrane glycoprotein and its inhibition of leishmania-macrophage binding. Proc Natl Acad Sci USA 83: 100-104

Chang KP, Fong D (1983) Cell biology of host-parasite membrane interactions in leishmaniasis. In: Evered & Collins (eds) Cytopathology of Parasitic Diseases, CIBA Foundation Symposium, vol. 99. Pitman, London, p 113

Chang KP, Fong D, Bray RS (1985) Biology of leishmania and leishmaniasis. In: Chang & Bray (eds) Leishmaniasis, Human Parasitic Diseases vol. 1. Elsevier, Amsterdam, p 1

Chang CS, Inserra TJ, Kink JA, Fong D, Chang KP (1986) Expression and size heterogeneity of a 63 kilodalton membrane glycoprotein during growth and transformation of *Leishmania mexicana amazonensis*. Mol Biochem Parasitol 18: 197-210

Coombs GH (1982) Proteinases of *Leishmania mexicana* and other flagellate protozoa. Parasitology 84: 149-155

Etges R, Bouvier J, Bordier C (1986) The major surface protein of Leishmania is a protease. J Biol Chem 261: 9098-9111

Fong D, Chang KP (1981) Protease activity of a parasitic protzoan, Leishmania mexicana. J Cell Biol 91: 43A

Goldstein JL, Brown MS (1974) Binding and degradation of low density lipoproteins by cultured human fibroblasts. J Biol Chem 249: 5153-5162

Goldstein JL, Basu SK, Brown MS (1983) Receptor-mediated endocytosis of low-density lipoprotein in cultured cells. Method Enzymol 98: 241-260

McKinney RM, Spillane JT, Pearce GW (1966) A simple method for determining the labeling efficiency of fluorescein isothiocyanate products. Anal Biochem 14: 421-428

North MJ (1982) Comparative biochemistry of proteinases of eukaryotic microorganisms. Microbiol Rev 46: 308-340

Olden K, Bernard BA, White SL, Parent JB (1982) Functions of carbohydrate moieties of glycoproteins. J Cell Biochem 18: 313-335

Simon MW, Mukkada AJ (1983) Intracellular protein degradation in Leishmania tropica promastigotes. Mol Biochem Parasitol 7: 19-27

Steiger RF, Van Hoof F, Bontemps J, Nyssens-Jardin M, Druetz JE (1979) Acid Hydrolases of trypanosomatid flagellates. Acta Trop 36: 335-341

THE ROLES OF SURFACE MEMBRANE ENZYMES AND TRANSPORTERS IN THE SURVIVAL OF LEISHMANIA

Dennis M. Dwyer

Cell Biology and Immunology Section, Laboratory of Parasitic Diseases, National Institute of Allergy and Infectious Diseases, National Institutes of Health, Bethesda, MD 20892, USA

INTRODUCTION

The means by which Leishmania survive within their several hosts remain unknown. However, as their surface membrane interfaces with host environments, it must fulfill numerous vital functions requisite for maintaining the parasite. Thus, knowledge of the biochemistry and physiology of these membranes is relevant to defining the mechanisms of parasite survival. This topic has been reviewed recently (Dwyer & Gottlieb, 1983, 1985) and the current report concerns some recent advances in this area.

RESULTS AND DISCUSSION

Three externally oriented surface membrane (SM) phosphomonoesterases, viz. a nonspecific acid phosphatase (MAcP), a 5'- and a 3'-nucleotidase (5'- and 3'-NT) have been identified in L. donovani (see: Dwyer & Gottlieb, 1983, 1985). Acid phosphatases are generally considered as constituents of, and markers for lysosomes. However, the unusual SM orientation of the Leishmania MAcP suggests that it might be involved in targeting, sequestration and protection of the parasite in macrophage phagolysosomes. Recently, an L. donovani MAcP was isolated and shown to be both resistant to toxic oxygen metabolites (Saha et al., 1985), and to inhibit their production by human neutrophils (Remaley et al., 1985). Further, this MAcP was also reported to dephosphorylate both inositol di- and triphosphates as well as some phosphoproteins (Das et al., 1986) suggesting, that this enzyme could potentially inactivate several important host regulatory mechanisms. Thus, this SM enzyme could play important roles in both establishing and maintaining the parasite within its hosts.

NATO ASI Series, Vol. H11
Host-Parasite Cellular and Molecular Interactions in Protozoal Infections. Edited by K.-P. Chang and D. Snary

Surface membrane 5'-nucleotidases have been reported from a wide variety of cell types and organisms including mammalian systems. In contrast, SM 3'-nucleotidases appear more restricted in their distribution (i.e. found in some plants and bacteria but not mammalian cells). Recently, the 3'-NT of L. donovani was isolated and shown to be a mannose-containing glycoprotein with a M_r=43,000 (Gottlieb & Zlotnick, 1986). This was confirmed via immunoprecipitation of this activity and from Western blots using an antiserum made against the purified 3'-NT (Pastakia, Gottlieb & Dwyer, unpublished). This enzyme activity appears to be inducible and is stimulated under conditions of purine deprivation (Gottlieb, 1985; Pastakia, Dwyer & Gottlieb, unpublished). Further, the purified 3'-NT has also been shown to possess exoribonuclease activity i.e. liberating 5'-nucleotides from RNA (Gottlieb & Zlotnick, 1986). Preliminary results have shown that the solubilized L. donovani 5'-NT is also a mannose-containing glycoprotein (Odera, Bates & Dwyer, unpublished). In immunochemical studies using an antiserum made against purified rat liver 5'-NT (supplied by P. Luzio, Cambridge University), a single band of M_r=72,000 was observed in SDS-PAGE autoradiograms of immunoprecipitates of ^{125}-I surface labeled L. donovani (Bates, Luzio & Dwyer, unpublished). Similar results were also obtained in Western blots with this antiserum and isolated L. donovani SM. These results suggest that 5'-nucleotidase might be an evolutionarily conserved functional surface membrane protein. Trypanosomatids can not synthesize purines de novo and hence rely totally upon their hosts to supply these. Thus, the Leishmania SM 5'-NT, which hydrolyzes both ribo- and deoxyribonucleotides, and the 3'-NT which hydrolyzes ribonucleotides and has direct 3'-exoribonuclease activity, undoubtedly function as a salvage pathway to provide the parasite with essential purine nucleosides generated from the products of host endogenous DNase and RNase activities. It is conceivable that the leishmanial SM 5'- and 3'-NT could produce levels of extracellular adenosine beyond the parasite's own need and ability to transport. As adenosine is involved in numerous regulatory functions, the parasite could influence host regulatory processes via excess production of this nucleoside. Obviously, these two SM enzymes must play important roles in the survival of the parasite within its insect and mammalian hosts.

In addition to the foregoing, three distinct lipolytic enzyme activities: phospholipase (PLase)-A1, -A2 and -C have been identified and characterized in isolated L. donovani SM (Wassef, Fioretti & Dwyer, manuscript submitted). These were differentiated on the bases of substrate specificity/preference, nature and yield of hydrolysis products, selective stability, metal requirements, pH optima and response to bile salts. Phosphatidyl-choline (PC), -ethanolamine (PE) and -inositol (PI) were used as lipase substrates as they consitute the major phospholipids of the isolated L. donovani SM (Wassef et al., 1985). These were radiolabeled biosynthetically: PI with myo-[^{14}C (U)] inositol; PC, PE and PI in the 2-acyl position with [1-^{14}C]arachadonate and PE alone at the 2-acyl position with [1-^{14}C]oleic acid and purified to homogeneity. The 2-acyl-[^{14}C]-labeled substrates contained > 98% of the radiolabel in the 2-position. Analyses of the SM hydrolysis products of 1-acyl-2-[1-^{14}C]oleoyl-PE showed: [^{14}C]labeled lyso-PE (i.e. deacylated in position 1), free-oleic acid and diacylglycerol indicative of Plase-A1, -A2 and -C activities, respectively. No labeled phosphatidic acid product was found in these assays suggesting that the SM apparently lack any PLase-D-like activity. The isolated SM also hydrolyzed phosphatidyl-myo-[^{14}C]inositol into:[^{14}C]labeled lyso-PI, glycerophosphoryl-inositol and free inositol-1-phosphate further demonstrating the three PLase activities. The PLase-A1 had a pH optimum of 3.5 hydrolyzing ≧3.3 nmol [^{14}C]lyso-PE/mg SM protein/min. from 1-acyl-2-[1-^{14}C]oleoyl-PE. This activity declined to 50% at pH 7 and was undetectable at pH 8. At its optimum pH, PLase-A1 had a substrate preference for 2-oleoyl-PE≧ 2-arachidonoyl-PE> 2-arachidonoyl-PC> PI analogs. The PLase-A2 demonstrated two peaks of activity with 2-[1-^{14}C]oleoyl-PE: a major peak at pH 9 and a smaller one at pH 5 liberating 3.5 and 1.2 nmol/[^{14}C]fatty acid/mg SM protein/min., respectively. With 2-[^{14}C]arachidonoyl-PE, the PLase-A2 showed activity over a broad pH range:peaking at pH 7 but maintaining two-thirds of its maximal activity at >pH 9 and <pH 6. This enzyme hydrolyzed 2-arachidonoyl -PE≧ 2-oleoyl-PE> 2-arachidonoyl-PI>> 2-arachidonoyl-PC. The PLase-A2 activity was stable to both heat (70°C/30' and 100C/5') and five freeze-thaw cycles. It was stimulated by Ca^{++} (1 mM) and Mg^{++} (2 mM) and unaffected by K^{+} (≦100 mM) or deoxycholate (1.5 mM). However, at 5 mM DOC this enzyme was inhibited by ∿60%. In contrast, the PLase-C was labile to both heat (50°C/5') and freeze-thawing. The PLase-C showed two peaks of activity with 2-[^{14}C]oleoyl-PE: one at pH 9.5 and another at pH 4.5

liberating 2.3 and 0.8 nmol diacyglycerol/mg SM protein/min., respectively. This enzyme hydrolyzed 2-arachidonoyl-PE more efficiently over a broad pH range between 6.5 - 7.5. The PLase-C was stimulated 2-fold by CA^{++} (1 mM), unaffected by Mg^{++} (≦15 mM) or DOC (≦5 mM) and showed a substrate preference: 2-arachidonoyl-PE≧ 2-oleoyl-PE> 2-arachidonoyl-PI>2-arachidonoyl-PC. The three PLase activities all had preference for PE. As, this is the major phospholipid in L. donovani SM, it seems likely that these enzymes play roles in membrane restructuring/turnover in this organism. In other systems, PLase-C has been implicated as playing key roles in trans-membrane signaling events and in exocytosis. Conceivably, the endogenous PLase-C could be involved in the turnover/release of p63, the major amphiphillic lipid-containing SM antigen/protease of Leishmania (Etges et al., 1986). PLase-A1 and -A2 have been shown in other systems to be important in deacylation-reacylation cycling events, in the remodeling of fatty acid components esterified into membrane phospholipids, and in affecting the fluidity of membranes. In addition to their endogenous SM roles, the parasite PLases could have roles in locally altering or restructuring the phospholipids of host cell phagolysosomal membranes, imparting changes in the fluidity and recycling of these membranes and thus influencing the physiology of this cellular compartment. These parasite enzymes undoubtedly also have nutritive functions in the metabolism of phospholipids obtained from their hosts. Further, as free-fatty acids and lyso-phospholipids have cytotoxic and cytolytic detergent-like properties, it seems possible that the parasite PLase-A1 and -A2 might contribute in part to the histopathology associated with leishmaniasis.

Surface membrane transport systems play significant roles in both cellular nutrition and regulation. Such systems are of obvious critical import to parasite survival. Recently, a pentose transport system was identified and characterized in L. donovani promastigotes (Pastakia & Dwyer, manuscript submitted). Radiolabeled ribose was rapidly accumulated and reached steady-stage levels within 4 min. at an intracellular concentration of 26 nmol/mg cell protein. Of the intracellular ribose accumulated, 30% was free sugar and 70% was phosphorylated, reaching steady state within 0.5 min. The intracellular free-ribose pool demonstrated rapid efflux in exchange with 2-deoxyribose. Characteristic of carrier-mediated uptake, the promastigotes showed saturation kinetics for ribose with an apparent K_m=2 mM and a V_{max} = 6 nmol/mg cell protein/min. Ribose uptake was inhibited by 2-deoxyribose with a K_i = 2 mM. Inhibition of ribose uptake by a variety of sugars, adenosine and proline was tested. None of the hexoses (e.g. mannose, galactose, etc.) inhibited ribose transport. Neither adenosine, having a ribose moiety, nor proline, an amino acid, both of which are actively transported by Leishmania on specific carriers, affected ribose uptake. Further, ribose uptake was not affected by certain pentoses (e.g. 3-deoxyribose, L-ribose, L-xylose) or pentitols (e.g. L-arabitol). However, D-arabinose and D-xylose significantly inhibited ribose uptake. These results demonstrated that the ribose carrier had specificity for certain pentoses and was distinct from other SM transporters for nucleosides, amino acids and hexoses. Ribose uptake was unaffected by metabolic inhibitors (i.e. NaN_3, KCN, NaF or iodoacetate), ATPase inhibitors (arsenate or oubain), uncoupling agents (2,4-DNP or FCCP) or the ionophores nigericin and valinomycin. The ribose uptake was dramatically inhibited by the sulfhydryl group inhibitors NEM and pHMB. The NEM effect was reversed by dithiothreitol indicating the sulfhydryl nature of the carrier. These results indicated that ribose uptake does not occur via an active transport mechanism but rather by a facilitated diffusion carrier system. Ribose transport is not prevalent among eukaryotic microorganisms; however, in Leishmania the pentose carrier would confer an advantage. As pentoses would be present in both the macrophage phagolysosome and the insect gut as products of nucleic acid digestion, the parasite could utilize these via its SM transporter. The ribose carrier operating in conjunction with transporters for preformed purines would provide the parasite with essential precursors for nucleoside and nucleotide biosynthesis. In addition, the parasite could use ribose as an alternative energy source via a pentose shunt pathway.

L. donovani promastigotes also possess a high affinity, carrier-mediated, stereo-specific SM transport system for D-glucose (Zilberstein & Dwyer, 1984a). It is unusual for a eukaryotic organism to possess such a high affinity transporter for glucose. However, among the hexoses tested, glucose is the primary sugar actively accumulated by L. donovani (Pastakia & Dwyer, manuscript submitted). This transport system would allow the parasite to maintain high levels of intracellular glucose and thus normal glucose metabolism throughout its development in both the insect and mammalian hosts. Recently, the L. donovani SM glucose transporter was specifically labeled, identified and characterized (Zilberstein et al., 1986). Cytochalasin B has been shown as a potent inhibitor of D-glucose transport in mammalian cells and used to specifically label this carrier via photochemical cross-linking methods. Cytochalasin B, but not cytochalasin E, effectively inhibited D-glucose transport in L. donovani promastigotes. Its IC_{50} for inhibition of 2-DOG transport was 55 μM. Hydroxysuccinimidyl-4-azidobenzoate was used to photochemically cross-link [^{3}H]cytochalasin B (CB) to the glucose transporter in isolated surface membranes and plasma membrane vesicles of L. donovani. In SDS-PAGE, a broad band of [^{3}H]CB labeled activity was present between M_r=20-30,000. D-, but not L-glucose inhibited the amount of radioactive CB incorporated into this band by ~60%. Isoelectric focusing results demonstrated two isoforms of the CB laebled molecule having apparent pI's of 6.6 and 6.8. These results suggested that the molecule contained some microheterogeneity possibly due to glycosylation. Lectin binding studies demonstrated that the CB labeled transporter was specifically bound and eluted from Con-A beads indicative of a mannose-containing glycoprotein. In SDS-PAGE, the mannose eluted material showed one major band of CB labeled activity of M_r = 19,000 and a minor one of M_r = 27,000 both of which were inhibitable with D-glucose. The Leishmania glucose transporter shares some features in common with those of mammalian cells (e.g. substrate specificity and inhibition by CB) suggesting some evolutionary conservation of the binding site in these carriers. However, the parasite carrier is an active transporter which has ~ a 300 fold higher affinity for 2-DOG than the facilitated diffusion-mediated carriers of mammalian cells. These properties of the Leishmania carrier might reflect an adaptation of this transporter to accommodate fluctuations in the host environment (e.g. those within the sand fly gut).

Recently, it was demonstrated that the active transport of both D-glucose and L-proline in L. donovani promastigotes is accomplished via a proton motive force-driven mechanism (Zilberstein & Dwyer, 1985). Thus, a proton electrochemical gradient is created across the parasite SM which is coupled to these transport systems by maintaining symport translocation of substrates with protons. Proton pumps play important roles in regulating both the magnitude of the proton motive force and maintaining intracellular pH homeostasis. As Leishmania encounters extreme environmental changes during development, its surface membrane proton pump is of critical survival value (i.e., for maintaining nutrient transport and intracellular pH). Membrane bound proton-translocating ATPases (H^+-ATPase) have been shown in a variety of systems to be energy transducers i.e. coupling the energy released from ATP hydrolysis to generate a proton electrochemical gradient. A SM ATPase has been identified, localized cytochemically and characterized in L. donovani which might satisfy the criteria of a primary proton pump (Zilberstein & Dwyer, manuscript in preparation; Zilberstein et al., in this volume).

From the foregoing, it is obvious that some progress has been made toward defining the biochemical composition and physiological functions of the Leishmania surface membrane. Identification of such functional membrane proteins has allowed speculation concerning their potential roles in parasite maintenance, development and survival. Isolation and characterization of these and other functional surface membrane constituents will permit testing of such hypotheses. Further, in a practical sense, the identification of unique parasite surface membrane constituents (e.g. MAcP, 3'-NT and H^+-ATPase) provides potential candidates for vaccine development and targets for the design of new or more effective drug therapies. In the latter regard, recent results with both L. donovani and L. major (Zilberstein & Dwyer, 1984b) have demonstrated the feasibility of this approach.

ACKNOWLEDGEMENT

I gratefully thank Mrs. Wilma Davis for her skillful and expeditious preparation of this manuscript.

REFERENCES

Das S, Saha AK, Remaley AT, Glew RH, Dowling JN, Kajiyoshi M, Gottlieb M (1986) Hydrolysis of phosphoproteins and inositol phosphates by cell surface phosphatase of Leishmania donovani. Mol Biochem Parasitol 20:143-153

Dwyer DM, Gottlieb M (1983) The surface membrane chemistry of Leishmania: its possible role in parasite sequestration and survival. J Cell Biochem 23:35-45

Dwyer DM, Gottlieb M (1985) The biochemistry of _Lesihmania_ surface membranes. In: Chang K-P, Bray RS (eds) Leishmaniasis. Elsevier, Amsterdam New York Oxford, p 31

Etges R, Bouvier J, Bordier C (1986) The major surface protein of _Leishmania_ promastigotes is a protease. J Biol Chem 261:9098-9101

Gottlieb M (1985) Enzyme regulation in a trypanosomatid: effect of purine starvation on levels of 3'-nucleotidase activity. Science 227:72-74

Gottlieb M, Zlotnick GW (1986) 3'-nucleotidase of _Leishmania donovani_ --evidence for exonuclease activity. In:Agabian N, Goodman H, Noguiera, N (eds) Molecular strategies of parasitic invasion. UCLA Symposia on Molecular and Cellular Biology, New Series. Vol. 42, Alan R. Liss Inc., New York

Remaley AT, Glew RH, Kuhns DB, Basford RE, Waggoner AS, Ernst LA, Pope MB (1985) _Leishmania donovani_: surface membrane acid phosphatase blocks neutrophil oxidative metabolite production. Exp Parasitol 60:331-341

Saha AK,, Das S, Glew RH, Gottlieb M (1985) Resistance of leishmanial phosphatases to inactivation by oxygen metabolites. J Clin Microbiol 22:329-332

Wassef MK, Fioretti TB, Dwyer DM (1985) Lipid analyses of isolated surface membranes of _Leishmania donovani_ promastigotes. Lipids 20:108-115

Zilberstein D, Dwyer DM (1984a) Glucose transport in _Leishmania donovani_ promastigotes. Mol Biochem Parasitol 12:327-336

Zilberstein D, Dwyer DM (1984b) Antidepressants cause lethal disruption of membrane function in the human protozoan parasite _Leishmania_. Science 226:977-979

Zilberstein D, Dwyer DM (1985) Proton motive force-driven active transport of D-glucose and L-proline in the protozoan parasite _Leishmania donovani_. Proc Natl Acad Sci USA 82:1716-1720

Zilberstein D, Dwyer DM, Matthaei S, Horuk R (to be published) Identification and biochemical characterization of the plasma membrane glucose transporter of _Leishmania donovani_. J Biol Chem

THE PLASMA MEMBRANE H^+-ATPase OF Leishmania donovani PROMASTIGOTES

Dan Zilberstein[1], Haynes W. Sheppard[2] and Dennis M. Dwyer[3]

[1]Department of Biology, Technion-Israel Institute of Technology, Haifa 32000, Israel, [2]Naval Biosciences Laboratory, Naval Supply Center, Oakland, CA 94625, U.S.A. and [3]Laboratory of Parasitic Diseases, NIAID, National Institutes of Health, Bethesda, MD 20892, U.S.A.

INTRODUCTION

The trypanosomatid protozoan Leishmania has a digenetic life cycle in which both developmental stages are continuously exposed to harsh environments, first as extracellular flagellated promastigote forms in the alimantary tract of their sandfly vectors and subsequently as obligate intracellular amastigote forms within phagolysosomal compartments of mammalian macrophages (Chang and Dwyer 1978, Dwyer 1979). The adaptive mechanisms by which these organisms survive in the widely different environmental conditions are still largely unknown but obviously involve surface interactions since all physiologic and biochemical interactions occur across such membranes. Previously, we have demonstrated that L. donovani possess active transport systems which are proton motive force driven (pmf), i.e. a proton electrochemical gradient ($\Delta\mu_{H^+}$) is created across the parasite surface membrane which is coupled to transport systems by maintaining symport translocation of the specific substrate with protons (Zilberstein and Dwyer 1985). Such a $\Delta\mu_{H^+}$ is composed of a chemical gradient (ΔpH) and of membrane potential ($\Delta\psi$). The size of the electrical membrane potential across L. donovani promastigote plasma membranes was recently determined to be 90 mV. Furthermore, a direct correlation between the size of the $\Delta\mu_{H^+}$ and the transport of L-proline have been demonstrated (Zilberstein, Langley and Dwyer, Submitted for publication). As pmf drives energy transduction processes across the Leishmania plasma membrane, it was anticipated that a proton translocating ATPase (H^+-ATPase) acting as a primary proton pump should be present in the plasma membrane of this organism. In the current report we sought to verify this hypothesis.

MATERIALS AND METHODS

Parasite maintenance and growth conditions. A cloned line of L. donovani promastigotes strain 1-S was maintained and grown as previously described (Dwyer 1977).

Enzyme assays. ATPase activity was determined as described by Zilberstein et al. (1986). Acid phosphatase activity was determined as described by Gottlieb and Dwyer (1981).

Isolation of plasma membranes. Plasma membrane ghosts were isolated according Gottlieb and Dwyer (1981). Plasma membrane vesicles were isolated as recently described by Zilberstein et al. (1986a).

Determination of internal pH. Plasma membrane vesicles (200 μg protein) were incubated in Tris/(2-N morpholino)ethane sulfonate at pH 6.7 containing 2 mM Mg-ATP and 5 μM acridine orange. Acidification was detected as previously described (Cidon et al. 1983).

Protein assays. Protein was determined according to Lowry et al. (1951).

NATO ASI Series, Vol. H11
Host-Parasite Cellular and Molecular Interactions
in Protozoal Infections. Edited by K.-P. Chang and D. Snary

RESULTS AND DISCUSSION

Previously, we have characterized Mg-ATPase activities in both isolated plasma membranes and mitochondria of L. donovani promastigotes. The mitochondria of this organism possesses an F_0,F_1-ATPase resembling the corresponding enzyme in bacteria and mitochondria of eukaryotic organisms (Zilberstein et al. 1986): the enzyme has an optimal activity at pH 8.0 and it is sensitive to oligomycin and azide. Moreover, antibodies generated against the highly conserved β subunit of F_1-ATPase from both E. coli, spinach chloroplast and yeast mitochondria (Rott and Nelson 1981) cross react with a 56kDa protein of L. donovani mitochondria, indicating that the β subunit in this organism is also conserved (Zilberstein and Dwyer, submitted for publication).

The plasma membrane ATPase differs from that of the mitochondrion i.e. it has optimal activity at pH 6.5, it does not react with anti β antiserum and it is insensitive to both oligomycin and azide (Fig. 1). Further, the plasma membrane ATPase is highly sensitive to orthovanadate demonstrating an IC_{50} of 7.5 μM, Fig 1. Moreover, N'N' dicyclohexylcarbodiimide (DCCD) also inhibits the plasma membrane ATPase (IC_{50}=33 μM). On the basis of its ion requirement and specific inhibition, the plasma membrane ATPase did not possess Na^+,K^+-, Ca^{2+}- or dynein-ATPases (Zilberstein and Dwyer, submitted for publication).

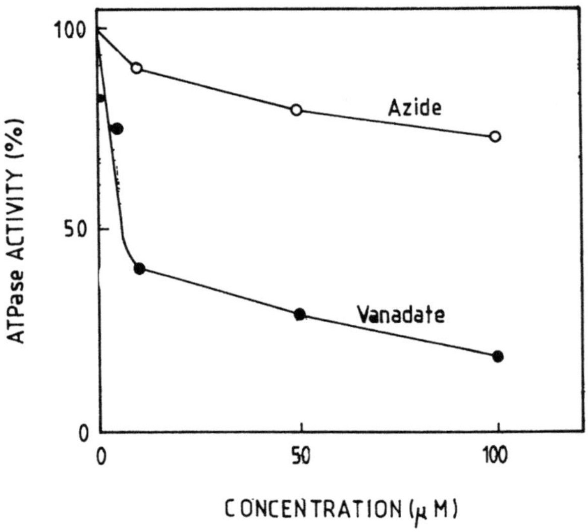

Fig. 1. Effects of azide and orthovanadate on ATPase activity in isolated plasma membranes of L. donovani promastigotes. 100% activity = 125 nmoles P released/min/mg protein.

The foregoing were carried out using isolated plasma membrane ghosts which possessed attached subpellicular microtubules (Dwyer 1980). However, in order to measure proton pumping activity sealed vesicles are required. We therefore modified the procedure for isolation plasma membranes and were able to isolate plasma membrane vesicles which lacked any demonstrable attached cytoskeletal elements (Fig. 2). These plasma membrane vesicles were shown to be derived from the parasite surface since they contained all of the plasma membrane functions outlined above (Sheppard Zilberstein and Dwyer, in preparation). Using antibodies against the β subunit, the plasma membrane vesicle preparation demonstrated no mitochondrial contamination (Fig. 3B). We used the unique L. donovani plasma membrane bound acid phosphatase as an enzyme marker of the plasma membrane (Zilberstein and Dwyer, 1986a).

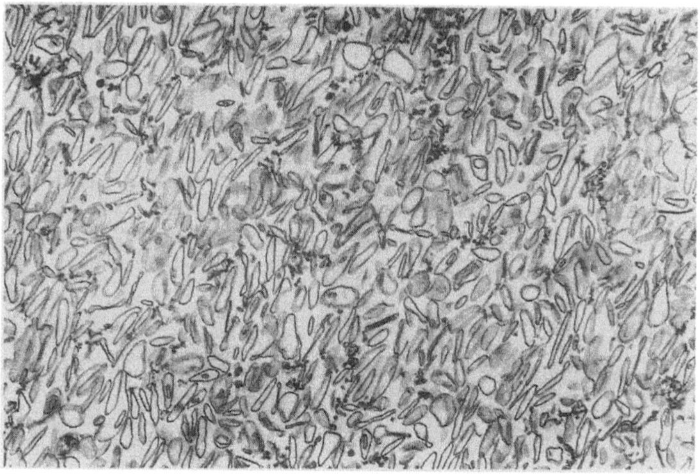

Fig. 2. Plasma membrane vesicles of L. donovani promastigotes.

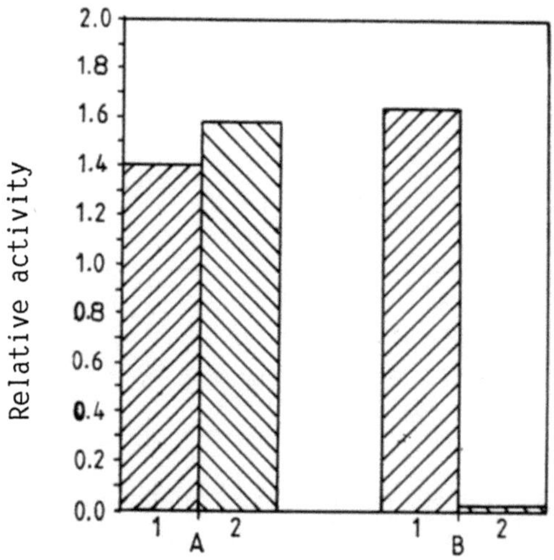

Fig. 3. Immunological detection of the β subunit in isolated mitochondria (A) and plasma membrane vesicles (B) of L. donovani promastigotes. Membrane proteins were separated on 9% SDS PAGE (Laemmli 1970) and transferred to nitrocellulose paper (Towbin et al. 1979) and further analysed for reactivity with anti E. coli β subunit antiserum.

Fig. 4. Purification of Plasma membrane ATPase in the isolated vesicles. Relative activity = ATPase/acid phosphatase activities. A. promastigotes lysate, B. Vesicle band. Vanadate sensitive ATPase (1) and azide sensitive ATPase (2) activities were measured.

The vanadate sensitive ATPase was copurified with the plasma membrane vesicles whereas the mitochondrial F_0,F_1-ATPase (azide sensitive) was absent from this fraction (Fig. 4). It was also demonstrated that at least 50% of the sealed plasma membrane vesicles had an inside out orientation and thus they should acidify their internal space upon addition of exogenous ATP. Plasma membrane vesicles were incubated in the presence of acridine orange, to detect acidification within the vesicles (Cidon et al. 1983). As shown in Fig.5, addition of ATP resulted in the acidification of the vesicle internal space. The pH gradient formed was reversed upon the addition of carbonylcyanide p-(trifluoromethoxy)phenylhydrazone (FCCP). However, no acidification was observed when ATP was added to the vesicles in the presence of either FCCP or DCCD. The foregoing results, although preliminary, indicate that the plasma membrane ATPase is a proton pump. Further investigations of the proton pumping activity of the leishmanial plasma membrane ATPase are being carried out and will be published elsewhere.

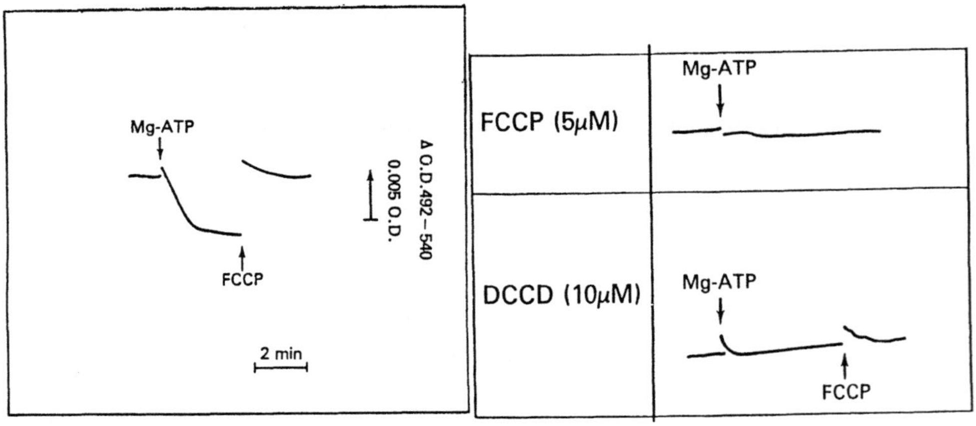

Fig. 5. ATP driven proton uptake in *L. donovani* promastigote plasma membrane vesicles

The foregoing results and those of previous studies, describing proton movement across the *L. donovani* promastigote plasma membrane are illustrated by the model in Fig. 6. Intracellular ATP is utilized by the plasma membrane H^+-ATPase to generate a $\Delta\mu_{H^+}$. The proton electrochemical gradient which is created across the promastigote plasma membrane is coupled to transport systems of nutrients and drives them by maintaining symport translocation with protons. Proton pumps were previously demonstrated to be involved in maintaining intracellular pH homeostasis (Padan et al. 1981). The cumulative data suggest that the *L. donovani* plasma membrane H^+-ATPase is responsible for maintaining both intracellular pH homeostasis and providing the chemiosmotic energy to drive transport of nutrients in these cells. Therefore, this enzyme is of critical import to and is essential for parasite survival (Zilberstein and Dwyer 1984, 1985a and Zilberstein Langley and Dwyer, submitted for publication).

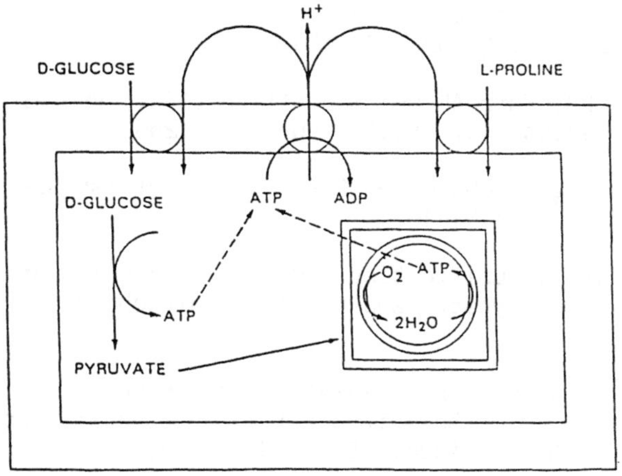

Fig. 6. Proton traffic across the L. donovani promastigote plasma membrane.

ACKNOWLEDGMENT

We would like to thank Dr. Nathan Nelson for providing us with the antibodies. This work was partially supported by a grant from Bat-Sheva De Rothchild Foundation.

REFERENCES

Chang KP and Dwyer DM (1978) Leishmania donovani-hamster macrophage interaction in vitro: cell entry, intracellular survival and multiplication of amastigotes. J. Exp. Med. 147:515-530.

Cidon S, Ben-David H and Nelson N (1983) ATP-driven proton fluxes across membranes of secretory organelles. J. Biol. Chem. 258:11684-11688.

Dwyer DM (1977) Leishmania donovani: Surface membrane carbohydrates of promastigotes. Exp. Parasitol. 41:341-358.

Dwyer DM (1979) Membrane interactions between Leishmania and host cells. In: Schlessinger D (ed), Microbiology 1979, pp 130-134, American Society of Microbiology, Washington, DC.

Dwyer DM (1980) Isolation and partial characterization of surface membranes from Leishmania donovani promastigotes. J. Protozool. 27:176-181.

Gottlieb M and Dwyer DM (1981) Leishmania donovani: Surface membrane acid phosphatase activity of promastigotes. Exp. Parasitol. 52:117-128.

Laemmli UK (1970) Cleavage of structural proteins during the assembly of th head of bacteriophage T4. Nature 277:680-685.

Lowry OH, Rosebrough NS, Farr AL and Randall RJ (1951) Protein measurement wit the folin phenol reagent. J. Biol. Chem. 193:265-275.

Padan E, Zilberstein D and Schuldiner S (1981) pH homeostasis in bacteria. Biochim Biophys. Acta 650:151-166.

Rott R and Nelson N (1981) Purification and immunological properties of proton ATPase complexes from yeast and rat liver mitochondria. J. Biol. Chem 256:9224-9228.

Towbin H, Staehelin T and Gordon J (1979) Electrophoretic transfer of protein from polyacrylamide gels to nitrocellulose sheets: Procedure and som applications. Proc. Natl. Acad. Sci. USA 76:4350-4354.

Zilberstein D and Dwyer DM (1984) Antidepressants cause lethal disruption of membrane function in the human protozoan parasite Leishmania. Science 226:977-979.

Zilberstein D and Dwyer DM (1985) Proton motive force-driven active transport of D-glucose and L-proline in the protozoan parasite Leishmania donovani. Proc. Natl. Acad. Sci. USA 82:1716-1720.

Zilberstein D and Dwyer DM (1985a) Antiprotozoal activity of tricyclic compounds. Science 230: 1063-1064.

Zilberstein D, Shirvan MH, Barile MF and Rottem S (1986) The β subunit of the F_0,F_1-ATPase is conserved in Mycoplasma. J. Biol. Chem. 261:7109-7111.

Zilberstein D, Dwyer DM, Matthaei S and Horuk R (1986a) Identification and biochemical characterization of the glucose transporter from the human protozoan parasite Leishmania. J. Biol. Chem. 261, in press.

PHOSPHOMONOESTERASES AND PROTEINASES OF LEISHMANIAS

G.H. Coombs[1], H.F. Hassan[1], B.C. Lockwood[2], D.J. Mallinson[1], M.J. North[2], J.M. Alunda[1] and L. Tetley[1]

[1]Department of Zoology, University of Glasgow, Glasgow G12 8QQ, Scotland, U.K. and [2]Department of Biological Science, University of Stirling, Stirling FK9 4LA, Scotland, U.K.

INTRODUCTION

Our knowledge of the chemistry of the surface of leishmanias has increased considerably over recent years and notable amongst the advances have been the discoveries of the surface location of three phosphomonoesterase activities (3'-nucleotidase, 5'-nucleotidase and acid phosphatase) (Dwyer & Gottlieb 1984, Gottlieb & Dwyer 1981) and a proteinase (Etges, Bouvier & Bordier 1986). The functional significance of these enzymes is not entirely clear, although several interesting possibilities have been proposed. The 3'-nucleotidase may operate as a nuclease and so act together with the 5'-nucleotidase in providing nucleosides for the parasite (Zlotnick, Wyman & Gottlieb 1985). A surface acid phosphatase of *Leishmania donovani*, purified from promastigotes and characterised by its resistance to tartrate, was shown to inhibit the production of oxygen metabolites by neutrophils (Remaley, Kuhns, Basford, Glew & Kaplan 1984); a similar action by the enzyme *in situ* could be central to the survival of the parasite upon entry into macrophages. There is little information concerning the role of the surface proteinase, whereas it has been postulated (Pupkis & Coombs 1984) that an amastigote-specific cysteine proteinase activity of *L. mexicana mexicana*, which apparently also occurs in *L. m. amazonensis* but not *L. major* or *L. donovani* (Pupkis, Tetley & Coombs 1986), may aid parasite survival by modifying the conditions of the parasitophorous vacuole. These studies have provided a fascinating insight into the way in which leishmanias may be adapted for survival in macrophages and how this apparently varies with species. These hypotheses are yet to be confirmed, however, and other roles of the enzymes can be envisaged (see, for example, Dwyer & Gottlieb 1985). Some important function for the parasite in the vector cannot be discounted, especially as most of the data collected so far have come from studies on cultured promastigotes which are thought to be similar to the insect forms. In an attempt to provide more information on these enzymes and the parts they play, we have investigated their occurrence in a range of trypanosomatids and various forms of leishmania.

MATERIALS AND METHODS

Leishmania donovani (LV79), *L. mexicana mexicana* (M379), *L. m. amazonensis* (LV78), *L. major* (LV39), *L. tarentolae*, *Crithidia fasciculata* (ATCC 11745), *Herpetomonas muscarum muscarum* (ATCC 30260), *H. m. ingenoplastis* (ATCC 30269) and *Trypanosoma brucei brucei* (EATRO

NATO ASI Series, Vol. H11
Host-Parasite Cellular and Molecular Interactions
in Protozoal Infections. Edited by K.-P. Chang and D. Snary

1125) were grown, harvested and lysed as described previously (Hassan & Coombs 1986a, Mallinson & Coombs 1986). Phosphomonoesterase activities were measured by the methods described by Hassan and Coombs (1986b) and surface activity was determined by electron microscopic studies using the procedures described previously (Gottlieb & Dwyer 1981, Dwyer & Gottlieb 1984, Gottlieb, Gardiner & Dwyer 1986). Proteinase activities were analysed using 7.5% (w/v) polyacrylamide slab gels which contained copolymerised gelatin at a final concentration of 0.2% (w/v) and a SDS-discontinuous buffer system (Lockwood, North, Scott, Bremner & Coombs 1986).

RESULTS AND DISCUSSION

Phosphomonoesterases

Each of the trypanosomatids investigated, listed in Table 1, with the exception of T. b. brucei, were found to contain relatively high levels of the three phosphomonoesterase activities; T. b. brucei, however, was devoid of detectable 5'-nucleotidase. All activities except for H. m. ingenoplastis and L. major acid phosphatase were sedimentable > 50% by centrifugation of cell homogenates at 105,000 x g for 1 hour (Hassan & Coombs 1986c). In contrast, only about 25% of the acid phosphatase activity of log-phase and PNA^- (metacyclic) promastigotes of L. major (see Mallinson & Coombs 1986) was sedimentable by centrifugation.

Greater variation was found when the occurrence of phosphomonoesterase activities on the surface of the organisms was investigated. The results are summarised in Table 1. We confirmed previous reports that all three activities occur on L. donovani promastigotes; the situation with amastigotes is apparently similar (see Dwyer & Gottlieb 1985). We also found all three activities on the surface of promastigotes (log- and stationary-phase) and isolated amastigotes of L. m. mexicana (acid phosphatase is shown in Figure 1) and two on L. m. amazonensis promastigotes. Recently Pimenta and De Souza (1986) reported the occurrence of acid phosphatase on the surface of both infective and non-infective promastigotes of this latter species, but its absence from amastigotes. In contrast, however, L. major promastigotes (log-phase and metacyclics) lacked surface acid phosphatase (Figure 1c) under the conditions employed in these studies. Similarly, none of the activities were found with L. tarentolae or C. fasciculata despite previous mention of their presence in the latter organism (Gottlieb 1985). Herpetomonas muscarum muscarum proved most interesting, a very heavy deposit of acid phosphatase reaction produce being detected on the cell's surface (Figure 1d); this was due to a 10 mM tartrate-resistant activity.

An acid phosphatase is released into the growth medium from L. donovani promastigotes (see Lovelace, Dwyer & Gottlieb 1986) and could play an important part in parasite survival. We have shown that L. m. mexicana behaves similarly, whereas there was no detectable release of acid phosphatase from L. major promastigotes. Enzyme release was not found with the lower trypanosomatids except for C. fasciculata, with which some tartrate-sensitive enzyme activity was recovered in the growth medium.

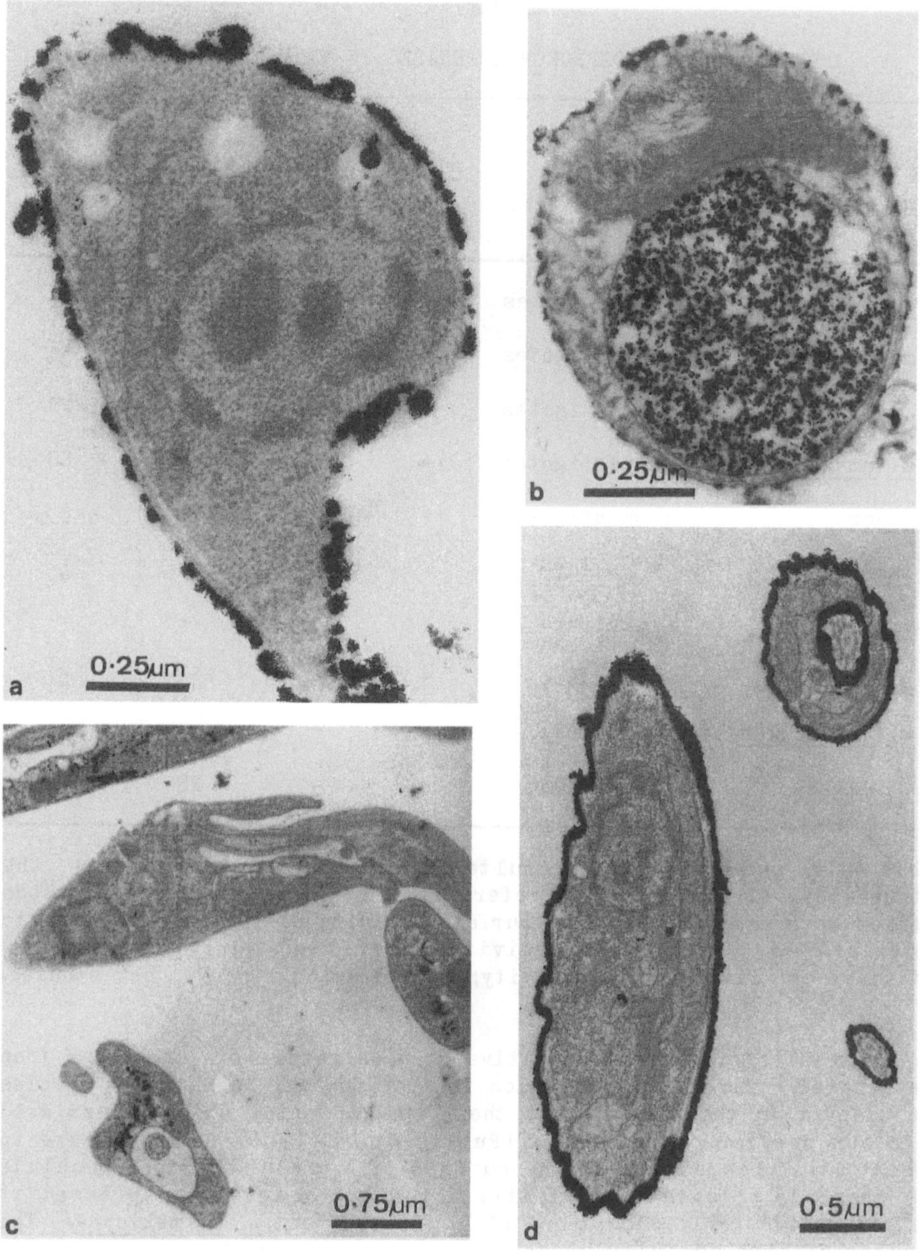

FIGURE 1. Cytochemical staining for acid phosphatase using glycerol-2-phosphate as substrate. (a) L. m. mexicana promastigotes. (b) L. m. mexicana amastigotes. (c) L. major promastigotes, a stationary phase population containing about 50% metacyclics. Note the lack of reaction product on the cell surface but its presence within the cell. (d) H. muscarum muscarum, with a thick deposit of reaction product on the surface (Hassan, Mallinson, Tetley and Coombs).

TABLE 1

OCCURRENCE OF PHOSPHOMONOESTERASES IN TRYPANOSOMATIDS

Species	Form	surface nucleotidases		acid phosphatase	
		3'-	5'-	surface	excreted
L. donovani	promastigotes	++	++	++	++
L. m. mexicana	promastigotes	++	++	++	++
	amastigotes	+	+	+	NS
L. m. amazonensis	promastigotes	++	NS	++	NS
L. major	promastigotes	++	++	-	±
L. tarentolae	culture	-	-	-	±
C. fasciculata	culture	-	-	-	+
H. m. muscarum	culture	-	+	+++	±
H. m. ingenoplastis	culture	-	+	-	±
T. b. brucei	procyclic	+	-	NS	NS

[a]This table summarises the results obtained in our laboratory; the cytochemical staining methods referred to in the Materials and Methods section were used to study the surface activity of the enzymes. Key to symbols: +++, very high activity; ++, high activity; +, low activity; ±, little or no activity; -, no activity; NS, not studied.

The results of these comparative studies raise more questions than they answer. The major conclusion is that not all leishmanial species are adapted in the same way to their environments, which agrees well with our previous conclusions (Pupkis *et al*. 1986). Our failure to detect acid phosphatase on the surface of *L. major* does not preclude the presence of such an activity; clearly, however, if present it differs considerably from those of *L. donovani* and *L. m. mexicana*. The finding of high activity present on the surface of *H. m. muscarum* is especially intriguing and provides strong evidence for a role of this enzyme in the invertebrate; we have insufficient information at present to understand why only some insect trypanosomatids possess the enzyme or its role when present. These findings, although providing no support for the proposal, do not rule out the possibility that the surface acid phosphatase of *L. donovani* and *L. mexicana* plays some crucial part in survival of the parasites in macrophages. They suggest, however, that other factors must also be involved and that the presence of a similar enzyme on the ancestral insect forms could have been a useful preadaptation for parasitism of mammals. The finding that C. fasciculata releases acid phosphatase into the medium can be interpreted in a similar fashion. We need more information on the environmental conditions of the various trypanosomatids in their insect hosts to enable us to understand the parts the enzymes play.

Proteinases

We have previously shown that there is considerable variation in the proteolytic activities of different forms (North & Coombs 1981, Coombs 1982) and different species (Pupkis et al. 1986) of Leishmania. For example with L. m. mexicana, the log-phase promastigotes contain relatively low levels of activity which are predominantly due to particulate proteinases whereas the amastigotes have high activity due mainly to soluble enzymes (Coombs 1982). We have now analysed the total proteinase activities of three species of Leishmania with a view to establishing whether they have enzymes in common and the extent to which there are developmentally-regulated activities in these species. The highly sensitive electrophoretic method used in this study has previously proved extremely useful in analysing trichomonad

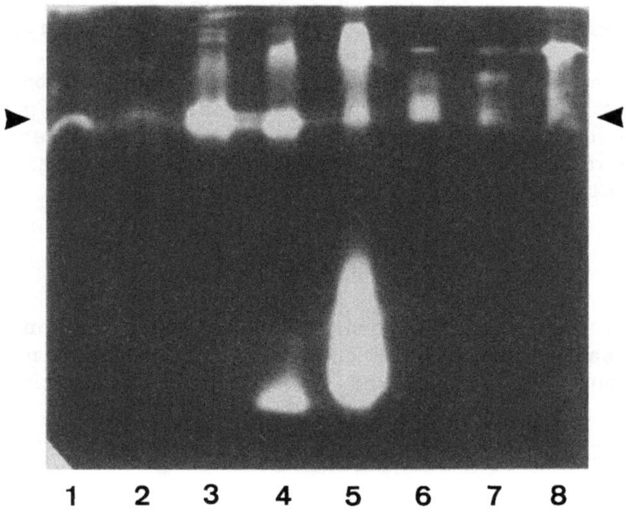

FIGURE 2. Proteinase activities in Leishmania lysates detected on gelatin-containing gels. Samples were run in the direction cathode (top) to anode (bottom). After electrophoresis the gel was incubated at pH 5.5 in 0.1 M acetic acid/sodium acetate and 1 mM dithiothreitol for 4 h before staining. The samples (with protein loaded) were: (1) L. major, log-phase promastigotes (100 µg); (2) L. major, PNA⁻ (73.5 µg); (3) L. m. mexicana, log-phase promastigotes (50 µg); (4) L. m. mexicana, stationary-phase promastigotes (50 µg); (5) L. m. mexicana, amastigotes (25 µg); (6) L. donovani, log-phase promastigotes (100 µg); (7) L. donovani, stationary-phase promastigotes (100 µg); (8) L. donovani, amastigotes (100 µg).

proteinases (Lockwood et al. 1986), particularly since it provides data on molecular weights. Proteinase activity appears as clear zones in the stained gel (Figure 2). The gel shows enzymes active at pH 5.5. All of them with the exception of a high M_r proteinase in L. major were also observed at pH 8. One proteinase (indicated by arrowhead in Figure 2) was common to all forms of the three species. It has an apparent M_r in the 60000 to 70000 dalton range, which is similar to that of the promastigote surface proteinase (p63) described by Etges et al. (1986). Since it is the only proteinase detected by this method at pH 8 in L. major, the pH and species used for the studies on p63, it is possible that they are the same enzyme. If this is so, we should be able to use different extraction procedures to demonstrate two bands corresponding to the amphiphilic and hydrophilic forms described by Etges et al. (1986).

Significant differences between species with respect to low molecular weight proteinases were detected (Figure 2). Leishmania mexicana mexicana amastigotes were shown to contain multiple high activity forms which correspond to those previously described (Pupkis & Coombs 1986). High mobility isoenzymes were previously detected using haemoglobin-containing gels (North & Coombs, 1981). There were no equivalent activities in amastigotes of L. donovani, confirming previous results (Pupkis et al. 1986), nor in log-phase promastigotes of any species. Most interesting was the finding that, in contrast to the log-phase promastigotes (containing few if any metacyclics), stationary-phase cells (comprising > 50% putative metacylics) of L. m. mexicana possessed a high activity enzyme with an apparent molecular weight slightly less than those of the amastigotes. A possible interpretation is that a proteinase is synthesised by metacyclic promastigotes of L. m. mexicana as a preadaptation for survival in mammalian macrophages. The variation in the high molecular weight enzymes is less dramatic, although there is an indication that some are developmentally regulated.

ACKNOWLEDGEMENTS

J.M. Alunda thanks the Spanish Ministry of Education and Science and H.F. Hassan thanks the University of Salahaddin, Iraq, for financial support. This work was supported in part by the MRC

REFERENCES

Coombs GH (1982) Proteinases of Leishmania mexicana and other flagellate protozoa. Parasitology 84, 149-155.

Dwyer DM, Gottlieb M (1984) Surface membrane localisation of 3'- and 5'-nucleotidase activities in Leishmania donovani promastigotes. Mol Biochem Parasit 10, 139-150.

Dwyer DM, Gottlieb M (1985) The biochemistry of Leishmania surface membrane. In: Chang K-P, Bray RS (eds) Leishmaniasis, Elsevier Science Publications, Amsterdam, New York, London, pp 32-47.

Etges R, Bouvier J, Bordier C (1986) The major surface protein of Leishmania promastigotes is a protease. J biol Chem 261, 9098-9101.

Gottlieb M (1985) Enzyme regulation in a trypanosomatid: effect of purine starvation on levels of 3'-nucleotidase activity. Science 227, 72-74.

Gottlieb M, Dwyer DM (1981) Leishmania donovani: surface membrane acid phosphatase activity of promastigotes. Exp Parasit 52, 117-128.

Gottlieb M, Gardiner PR, Dwyer D (1986) 3'-Nucleotidase activity in procyclic and bloodstream stages of Trypanosoma rhodesiense. Comp Biochem Physiol 83B, 63-69.

Hassan HF, Coombs GH (1986a) A comparative study of the purine- and pyrimidine-metabolising enzymes of a range of trypanosomatids. Comp Biochem Physiol 84B, 217-223.

Hassan HF, Coombs GH (1986b) Purine metabolising enzymes in Entamoeba histolytica. Mol Biochem Parasit 19, 19-25.

Hassan HF, Coombs GH (1986c) Phosphomonoesterase activities of Leishmania mexicana mexicana and other flagellates. Submitted to Molec Biochem Parasit

Lockwood BC, North MJ, Scott KI, Bremner AF, Coombs GH (1986) A highly sensitive, electrophoretic method for the analysis of trichomonad proteinases. Submitted to Mol Biochem Parasit

Lovelace JK, Dwyer DM, Gottlieb M (1986) Purification and characterisation of the extracellular acid phosphatase of Leishmania donovani. Mol Biochem Parasit 20, 243-251.

Mallinson DJ, Coombs GH (1986) Molecular characterisation of the metacyclic forms of Leishmania. IRCS Med Sci 14, 357-358.

North MJ, Coombs GH (1981) Proteinases of Leishmania mexicana amastigotes and promastigotes: Analysis by gel electrophoresis. Mol Biochem Parasit 3, 293-300.

Pimenta PF, De Souza W (1986) Cytochemical localization of acid phosphatase in Leishmania mexicana amazonensis. J Submicrosc Cytol 18, 127-132.

Pupkis MF, Coombs GH (1984) Purification and characterisation of proteolytic enzymes of Leishmania mexicana mexicana amastigotes and promastigotes. J Gen Microbiol 130, 2375-2383.

Pupkis MF, Tetley L, Coombs GH (1986) Leishmania mexicana: amastigote hydrolases in unusual lysosomes. Exp Parasit 62, 29-39.

Zlotnick GW, Wyman C, Gottlieb M (1985) Identification of the active subunits of the 3'-nucleotidase of Leishmania donovani. Fed Proc 44, 1332.

THE STRUCTURE, LOCALIZATION, AND FUNCTION OF THE LIPOPHOSPHOGLYCAN OF *LEISHMANIA DONOVANI*

Salvatore J. Turco, Carole L. Johnson, Diane L. King, Palmer A. Orlandi, and Briana L. Wright

Department of Biochemistry
University of Kentucky Medical Center
Lexington, Kentucky 40536 USA

INTRODUCTION

The protozoan parasite *Leishmania donovani* is able to successfully live in two harsh hydrolytic environments in its digenetic life cycle: as an extracellular promastigote in the alimentary tract of its sandfly vector and as an intracellular amastigote in lysosomes of macrophages of its mammalian host. That *L. donovani* has adapted to survive in such hostile environments is most likely due to protection conferred by specialized molecules on the parasite's cell surface.

RESULTS AND DISCUSSION

General characteristics of lipophosphoglycan

In addition to glycoproteins and glycolipids, *Leishmania donovani* promastigotes synthesize and express on their cell surface a novel glycoconjugate called lipophosphoglycan. This complex carbohydrate is a major glycoconjugate since it contains over half of the total carbohydrate bound to macromolecules in *L. donovani*. Based on the results described below and elsewhere (Turco, et.al. 1986), a tentative structure of lipophosphoglycan is proposed:

Gal-->[P-6Gal(β1,4)Man-]$_{16}$-->[Gal$_4$,Man$_2$]>GlcNH$_2$-->Lyso-alkyl PI

Lipophosphoglycan is a heterogeneous glycoconjugate (molecular weight range of 10-20-Kd) containing as its salient feature a repeating disaccharide unit of mannose, galactose, and phosphate in a ratio of 1:1:1. The repeating unit is attached to an unusual lyso-alkyl phosphatidylinositol via an hexasaccharide carbohydrate core.

Structural characterization of lipophosphoglycan

Purification of lipophosphoglycan. A rapid and simple method for purifying the *L. donovani* lipophosphoglycan to homogeneity has been

NATO ASI Series, Vol. H11
Host-Parasite Cellular and Molecular Interactions
in Protozoal Infections. Edited by K.-P. Chang and D. Snary

developed. Lipophosphoglycan can be extracted from the parasites with the solvent water:ethanol:diethylether:pyridine:NH_4OH (15:15:5:1:0.017) after removal of lipids and metabolites as described elsewhere (Turco, et.al., 1984). Subsequently, lipophosphoglycan can be precipitated from contaminating proteins by adding an equal volume of methanol and freezing the sample at -20°C. After resuspending in the original solvent, the lipophosphoglycan preparation can be shown to be pure by a number of chromatographic and electrophoretic protocols.

Characterization of the repeating disaccharide unit. A surprising finding was the observation that lipophosphoglycan is extremely labile to mild acid hydrolysis (0.02 N HCl, 15 min, 60°C) yielding the phosphorylated disaccharide P->6Gal(β1,4)Man as the main carbohydrate product. The presence of 4-substituted mannose in lipophosphoglycan was determined by methylation linkage analysis (GLC-mass spectrometric analysis of the partially methylated alditol acetates). This was a significant discovery considering that 4-substituted mannose is highly unusual in a glycoconjugate in a eukaryotic cell and, thus, may prove to be highly antigenic. A repeating unit of P->6Gal(β1,4)Man is also unique in a eukaryotic glycoconjugate. The location of the phosphate group on the 6-hydroxyl of the galactose residue was determined with ^1H-NMR by Dr. Steven Homans, Department of Biochemistry, University of Oxford, Oxford, England. Currently, the configuration of the phosphate group attached to the anomeric carbon of the mannose residue is being determined. The method that is being used is to perform alkaline degradation of lipophosphoglycan along with appropriate standards and analyzing the products chromatographically.

Characterization of the lipid moiety of lipophosphoglycan. The lipid core of lipophosphoglycan, thought to be the anchoring unit for this cell surface macromolecule, has been isolated, purified and characterized. Analysis has shown the glycan portion of lipophosphoglycan to be covalently linked to a novel lyso-phosphatidyl inositol core in which the glycerol backbone is in ether linkage with a alkyl side chain. Support for this conclusion is based on results with treatment of lipophosphoglycan with 10% NH_4OH or a phosphatidylinositol-specific phospholipase C (PI-PLC) isolated from *Staphylococcus aureus* released the same lipid species. Structural characterization of the PI-PLC released lipid via gas-liquid chromatography-mass spectrometry indicated the presence of two products: one with a saturated, unbranched C_{24} alkyl chain comprising approximately 78% of the lipid with the remaining 22% as a saturated, unbranched C_{26} alkyl chain. Periodate sensitivity demonstrated that the alkyl chain is linked to the C-1 position of the glycerol backbone. Ammonolysis of the intact lipophosphoglycan also revealed the presence of a octasaccharide structure linked to the inositol of the lipid moiety. Results from further digestion with nitrous acid indicated the presence of an unacetylated glucosamine attached to the lyso-PI core. The glycolipid core structure of the cell surface lipophosphoglycan suggests that it may play an important role in the ability of the lipophosphoglycan to protect the parasite from degradation under the extreme hydrolytic conditions in both its insect vector and mammalian host. Identification of the lipid moiety was highly important. Since lipophosphoglycan is known to be shed into the medium, it is possible that a lipid cleavage enables it to be released from the cells. This would be analogous to the phenomenon with the related trypanosomatid *Trypanosoma brucei* in which an endogenous PI-specific phospholipase C liberates the variable surface glycoprotein from the cell surface into the culture medium (Ferguson, et.al., 1985).

Determination of inositol as a substituent. Demonstration that lipophosphoglycan possesses inositol was obtained by two separate methods. L. donovani cells were incubated with [^{3}H]inositol and incorporation of label into isolated lipophosphoglycan was quantitated and identified. Secondly, the presence of inositol in lipophosphoglycan was demonstrated by gas-liquid chromatography. Along with other information from compositional analysis, it was determined that there are an average of sixteen phosphorylated disaccharide units attached to a carbohydrate core containing four galactose, two mannose, and one glucosamine bound to an inositol residue.

Cell localization of lipophosphoglycan

In another significant finding, it was established that lipophosphoglycan is located on the cell surface of Leishmania donovani. This finding was determined by a galactose oxidase/NaB[^{3}H]$_4$ technique of radiolabeling cell surface macromolecules. Incorporated radioactivity was analyzed by gel filtration on Sephadex G-100, chromatography on ricin agglutinin-coupled agarose, and lability to mild acid hydrolysis. The rational for the use of this procedure was based on the knowledge that the lipophosphoglycan possesses a terminal galactose residue. It has been reported that several glycoconjugates of Leishmania tropica major were able to be radiolabeled by the galactose oxidase/NaB[^{3}H]$_4$ method (Slutzky and Greenblatt, 1982; Handman, et.al. 1984), one of which may be similar in structure to the L. donovani lipophosphoglycan. Interestingly, metabolically-labeled lipophosphoglycan was observed to be shed from the cell surface into the culture medium. Partial characterization of the lipophosphoglycan released into the culture medium revealed no obvious structural difference compared to the cellular lipophosphoglycan. However, a detailed characterization of the lipophosphoglycan found in the medium is necessary to determine whether a minor but significant modification in the structure occurred during release from the cell surface. Nevertheless, a cell surface topology of this unusual glycoconjugate of L. donovani suggests that it may play a major role in host cell-parasite interactions.

Growth phase expression of lipophosphoglycan

With the use of a monoclonal antibody, Handman and Goding (1985) have shown that a lipid-containing glycoconjugate on the surface of L. tropica major parasites is necessary for the initial infection of macrophages. Sacks and coworkers (1984,1985) have reported that the same organism maintained in stationary phase of growth is more infectious than logarithmically-growing cells. These two observations would indicate that perhaps the lipid-containing glycoconjugate of L. tropica major is present in highest amounts on the surfaces of cells maintained in stationary phase. However, our results have shown that even though lipophosphoglycan is expressed by L. donovani maintained in stationary phase, phosphoglycan is present preferentially during logarithmic phase of growth. This conclusion is based on the results of three experiments: 1) incorporation of [3H]mannose into lipophosphoglycan during 1 h of metabolic labeling of the cells; 2) cell surface labeling of lipophosphoglycan by the galactose/ NaB[^{3}H]$_4$ method; and 3) total carbohydrate determination of the material extracted by water/ethanol/diethylether/pyridine/NH_4OH (15:15:5:1:0.017). Moreover, another observation by Sacks and coworkers (1985) suggests that L. tropica major maintained in logarithmic phase possess more ricin binding sites. If the L. donovani lipophosphoglycan (which is known to be a site for ricin binding) is the same molecule as the lipid-containing glycoconjugate of L. tropica major, then it would appear that lipophosphoglycan is necessary but not sufficient for entry into the macrophage. In support of this conclusion is the recent observation by Chang and workers (1986) that a 63 kDa N-linked glycoprotein of Leishmania mexicana amazonensis may play a role in the process of infection of macrophages.

Selection and characterization of glycosylation mutants of L. donovani

The availability of mutant Leishmania donovani with altered surface carbohydrates should provide an opportunity to perform a variety of studies on the structure, biosynthesis, and function of these substances. It became apparent that glycosylation mutants of L. donovani would be extremely beneficial for determining the biochemistry not only of lipophosphoglycan, but also for glycoproteins and glycolipids that may be important in the life cycle. Isolating such mutants by procedures employing plant lectins as the selective agents is being employed. Two lectins used for the selection of glycosylation mutants of L. donovani are concanavalin A (specific for mannose) and ricin (specific for galactose).

Selection and characterization of a ricin-resistant clone. A variant cell line of Leishmania donovani was recently selected for resistance to the cytotoxic lectin ricin agglutinin and shown to be defective in the synthesis and expression of lipophosphoglycan. Compared to the parental line, the variant line continuously passaged in culture medium containing 0 µg/ml, 80 µg/ml, and 200 µg/ml of ricin exhibited, respectively, 45-, 90-, and 1,180-fold resistance to the toxic effects of the lectin. The synthesis of lipophosphoglycan in the variant line was judged to be <1% relative to wildtype as determined by incorporation of radioactive mannose and galactose and by analyses on SDS polyacrylamide gels, Sephadex G-100, ricin agglutinin-bound agarose, and DEAE cellulose. These results suggest that the variant line cannot assemble and express lipophosphoglycan and indicate that this variant cell line may prove useful in the study of the structure, biosynthesis, and function of this glycoconjugate.

Use of the ricin-resistant clone. This variant clone already has been used to obtain useful information. In contrast to the resistance to the cytotoxic effects of ricin, the mutant clone was found to be approximately 7-fold more sensitive to the lectin concanavalin A than wildtype cells. The reason for the increased sensitivity of the mutant line to concanavalin A remains to be established. In unpublished observations, the glycoproteins of L. donovani possess only high mannose-type carbohydrate chains. Since there is an absence of expression of lipophosphoglycan in the mutant cells, there may be an unmasking of some cell surface glycoproteins normally covered by lipophosphoglycan in wildtype cells and a consequent increase in the number of concanavalin A binding sites. In addition, the mutant line served as an important control in the experiments establishing that lipophosphoglycan is located on the cell surface in wildtype cells.

Clones with altered surface lipophosphoglycans will play a major role in future investigations in determining the structure, biosynthesis, and function of lipophosphoglycan. We are currently applying techniques for growing macrophages in culture, infecting them with wildtype L. donovani promastigates, and harvesting and quantitating amastigotes. Once these techniques are mastered, similar experiments will be performed with the lectin-resistant clones, which will be examined for their ability to enter and survive in macrophages. Failure of the mutant parasites to thrive in these expected environments would be a major step in determining the function of lipophosphoglycan.

ACKNOWLEDGEMENTS

The authors are indebted to Drs. Robert Lester and Buddy Ullman for many helpful discussions during the course of this work. This investigation was supported by the UNDP/World Bank/WHO Special Programme for Research and Training in Tropical Diseases, by N.I.H. Grant AI02941, and by a N.I.H. Research Career Development Award AM01087 to S.J.T.

REFERENCES

Chang CS and Chang K-P (1986) Monoclonal Antibody affinity purification of a Leishmania membrane glycoprotein and its inhibition of Leishmania-macrophage binding. Proc Natl Acad Sci USA 83: 100-104

Chang CS, Inserra TJ, Kink JA, Fong D, and Chang K-P (1986) Expression and Size Heterogeneity of a 63-Kd Membrane Glycoprotein During Growth and Transformation of Leishmania Mexicana Amazonensis. Mol Biochem Parasitol 18: 197-210

Ferguson MAJ, Low MG, and Cross GAM (1985) Glycosyl-sn-1,2-dimyristyl phosphatidylinositol is Covalently Linked to Trypanosoma brucei Variant Surface Glycoprotein. J Biol Chem 260: 14547-14555

Handman E, Greenblatt CL, and Goding JW (1984) An amphipathic sulphated glycoconjugate of Leishmania: characterization with monoclonal antibodies. EMBO J 3: 2301-2306

Handman E, and Goding JW (1985) The Leishmania receptor for Macrophages is a lipid-containing glycoconjugate. EMBO J 4: 329-336

Sacks DL, and Perkins PV (1984) Identification of an Infective Stage of Leishmania Promastigotes. Science 223: 1417-1419

Sacks DL, Heiny S, and Sher A (1985) Identification of Cell Surface Carbohydrate and Antigenic Changes Between Noninfective and Infective Developmental Stages of L. major Promastigotes. J Immunol 135: 564-569

Turco SJ, Wilkerson MA, and Clawson DR (1984) Expression of an Unusual Acidic Glycoconjugate in L. donovani. J Biol Chem 259: 3883-3889

LEISHMANIA HOST-PARASITE INTERACTIONS : THE DEVELOPMENT OF CHEMOTHERAPEUTIC TARGETS AND SPECIFIC DRUG DELIVERY SYSTEMS I. LIPOPROTEIN MEDIATED ANTILEISHMANIAL CHEMOTHERAPY.

David T. Hart

International Institute for Cellular and Molecular Pathology, B-1200 Brussels, BELGIUM

INTRODUCTION

The amastigote stages of *Leishmania* are intracellular parasites of the reticuloendothelial system. Amastigote adaptation for intracellular survival and the molecular interaction of the parasite with the host cell are poorly understood. The urgent need to develop new and effective antileishmanial drugs, together with a specific drug delivery system, has initiated investigation of this poorly understood area of *Leishmania* biochemistry and cellular biology. The ultimate aim of these non-empirical investigations is therefore to identify unique and essential aspects of the host-parasite relationship, which have potential for exploitation in both the development of chemotherapeutic agents and novel methods for drug delivery.

The subcellular organisation of *Leishmania* energy metabolism has been found to be at total variance to the mammalian host (Hart & Opperdoes, 1984). The occurrence of glycosomes, unique subcellular organelles which contain essential anabolic, catabolic and anaplerotic pathways, constitutes an exquisite target for chemotherapeutic attack. Two of the pathways localised in these organelles, glycolysis and β-oxidation, have been shown to be important for amastigote viability (e.g. Hart *et al.*, 1981).

While little is known about the relative abundance of carbohydrates and fatty acids during the chronic stages of infection, β-oxidation has been found to be 10-fold higher in

NATO ASI Series, Vol. H11
Host-Parasite Cellular and Molecular Interactions
in Protozoal Infections. Edited by K.-P. Chang and D. Snary

the amastigote stage, which presumably reflects an essential adaptation for intracellular survival (Hart & Coombs, 1982). It was further postulated that the receptor-mediated uptake and degradation of lipoproteins by the host macrophage could constitute the source of fatty acids and sterols for amastigote metabolism. As part of a non-empirical approach to antileishmanial drug design and delivery, the aforementioned postulate was investigated.

MATERIALS AND METHODS

Leishmania

The promastigote stages of Leishmania major and Leishmania mexicana mexicana were maintained as described in Hart and Opperdoes (1984).

J774G8 Macrophages

The macrophage-like cell line of Ralph et al. (1975) was maintained in RPMI 1640 growth medium, supplemented with Hepes 10 mM, glutamine 10mM and FCS-HI 20%. A modified method of that described by Chang (1980) was used to obtain infected macrophages and to quantify amastigote viability (manuscript in preparation).

Lipoproteins

Low-density lipoproteins (LDL) were isolated and labelled from both rabbit and human serum by the method of Shepherd et al. (1976), and Acetylated-LDL was prepared as Basu et al. (1976). Accumulation and degradation of lipoproteins was assayed as described by Goldstein et al. (1979). Colloidal gold labelling and morphological uptake studies were performed as described by De Roe et al. (1982). The fatty acid derivative of Adriamycin was complexed with LDL as in Masquelier et al. (1986) and was a generous gift from Dr M. Masquelier.

RESULTS

From Figure 1a and b it can clearly be seen that low-density lipoproteins, in their native form as well as their acetylated form, are both taken up and degraded by J774G8 macrophages, via a receptor-mediated process. These results were also demonstrable using peritoneal macrophages isolated from Balbc mice (data not shown). Acetylated-LDL was accumulated and degraded at higher rates when compared to normal LDL, and degradation of both normal and acetylated forms of LDL was somewhat reduced in infected macrophages. Leishmania infected J774G8 macrophages therefore demonstrate elevated levels of lipoprotein uptake, and somewhat reduced capacity for degradation. Similar results were obtained using peritoneal macrophages (data not shown), and it is therefore presumed that elevated lipoprotein uptake and decreased degradation capacity is associated with amastigote survival.

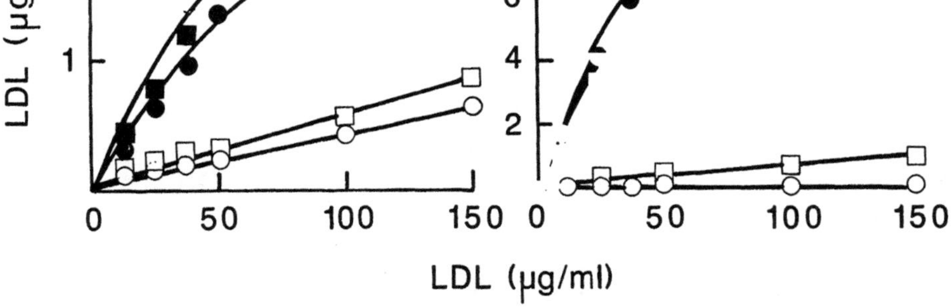

FIGURE 1a-b. Lipoprotein metabolism in J774G8 macrophages. O, J774G8 and □, L. m. mexicana infected J774G8 macrophages. Native LDL unshaded symbols and acetylated LDL shaded symbols.

Lipoprotein drug complexes were found to be excellent vehicles for drug delivery (Figure 3a,b & c). The antileishmanial activity of the fatty acid derivative of Adriamycin could be greatly enhanced via lipoprotein delivery, and 10 to 100-fold increases in anti-leishmanial action could be demonstrated. Acetylated-LDL proved to be the most effective drug targeting method, which correlates well with the higher uptake and degradation shown in Figure 1a and b. In contrast, the free drug was shown to possess only moderate antileishmanial activity, even at the highest concentration tested (10^{-6} M). Indeed, the LD_{50} for the free drug after 24 hrs was shown to be $>10^{-6}$ M, while for the normal and acetylated LDL-drug complex, it was $<10^{-7}$ M and $<10^{-8}$ M respectively.

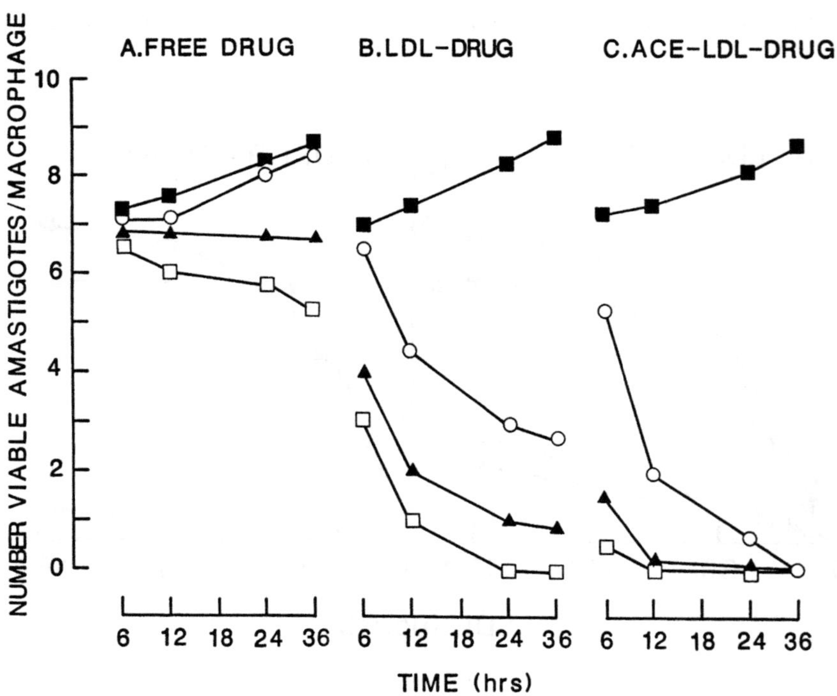

FIGURE 3a-c. Lipoprotein mediated drug targeting. ■, control; ○, 10^{-6} M; ▲, 10^{-7} M and □ 10^{-8} M drug.

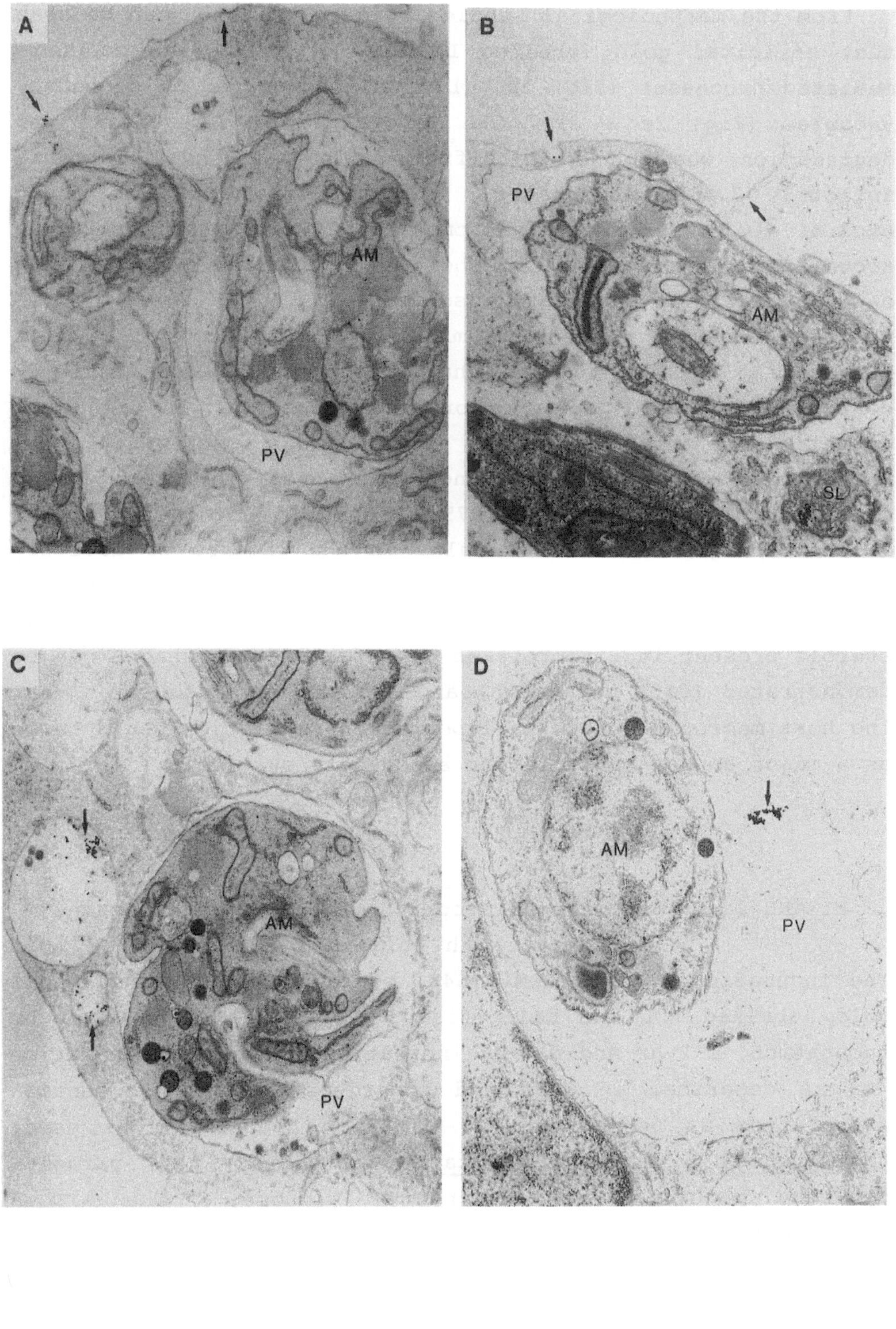
A
AM
PV
B
PV
AM
SL
C
AM
PV
D
AM
PV

From the morphological studies (Figure 2a-d) it can be seen that colloidal gold labelled LDL is taken up via receptor-mediated processes (Fig. 2a & b) and accumulated in secondary vacuoles (Fig. 2c & d). The topography of LDL uptake and degradation was found to differ in normal and Leishmania infected J774G8 macrophages. In the former LDL was found exclusively in secondary lysosomes, while in the latter it was accumlated predominantly in the parasitophorous vacuole, and only partially in secondary lysosomes.

Taken together, the biochemical and morphological results (Figures 1 & 2) suggest that lipoproteins are actively degraded in the parasitophorous vacuole. It is well established that secondary lysosomes fuse with the parasitophorous vacuole during Leishmania infection (e.g. Alexander & Vickerman, 1975; Chang, 1983) and that lysosome-phagosome fusion results in the release of actively hydrolytic lysosomal contents into the parasitophorous vacuole. The hydrolytic release of the fatty acids and sterols present in lipoproteins by infected macrophages can be demonstrated (data not shown) and support the suggestion that the host macrophage role in lipoprotein regulation could serve as a major source of lipid for amastigote survival.

FIGURE 2a-d. Electron micrographs of J774G8 macrophages incubated with colloidal gold labelled LDL. The incubation medium (RPM1 1640) contained 10µg/ml colloidal gold labelled LDL and delipified FCS (20%). a and b, 1 hr incubation, c 2 hr and d 4 hr incubation. Sample preparation was as described in Hart and Opperdoes (1984) and section preparation as De Roe et al. (1982). Abbreviations used: AM, Leishmania mexicana mexicana amastigotes; PV, parasitophorous vacuole; SL, secondary lysosome. The arrows indicate the colloidal gold particles.

DISCUSSION

The postulated occurrence of receptor-mediated uptake and degradation of lipoproteins by infected macrophages (Hart & Coombs, 1982) has been unequivocally confirmed by both biochemical and morphological data. Receptor-mediated uptake and degradation of both native LDL and acetylated LDL has been demonstrated, with the latter modified lipoprotein having the highest rates of uptake and degradation. In addition, *Leishmania*-infected macrophages demonstrate an enhanced capacity for uptake and possibly reduced capacity for degradation of lipoproteins, suggesting that lipoprotein may play an important role in amastigote survival.

Mammalian macrophages have been shown to maintain cellular sterol levels via both *de novo* synthesis and receptor-mediated processes (e.g. Goldstein *et al.*, 1986). The occurrence of the latter process could therefore result in increased host compatability for amastigote survival. In a parallel sterol metabolism study we have shown that sterol synthesis in infected macrophages is almost completely (>90%) inhibited (Hart *et al.*, manuscript in preparation) while parasite sterol biosynthesis continues unabated. The repression of host sterol biosynthesis is therefore compensated by the increase in receptor-mediated capture and concomitant decrease in degradation, at least in the early stages of infection. It is therefore possible that *Leishmania* may manipulate and survive the host-macrophage antimicrobial activity via preturbation of sterol homeostasis.

The occurrence of a specific receptor for LDL and acetylated LDL on *Leishmania* infected macrophages has been exploited *in vitro* to selectively deliver antileishmanial drugs. The fatty acid derivation of Adriamycin can easily be complexed with native LDL and the acetylated form without changes in the lipoprotein stability (Hart *et al.*, unpublished observations). The antileishmanial action of the native or acetylated LDL-complexed drug is considerably higher than the free drug, and the latter modified form of LDL was found to be the most efficient targeting system, presumably due to the higher rates of uptake and degradation.

Lipoproteins such as LDL constitute ideal drug transport systems due to their hydrophobic interior, which contains 1000-2000 molecules of nonesterified cholesterol, and which could be substituted to contain a similar quantity of drug. Similarly the hydrophobic polar shell could be used to transport hydrophobic and polar drugs, while the apoprotein moeity could be used (with or without the lipid phases) to transport covalently linked pro-drug conjugates. The pro-drug conjugate would then be degraded by either the phago-lysosomal and/or leishmanial proteases, to locally release the active drug molecule. The intrinsic advantages of multiple drug administration could be efficiently and conveniently accomodated by lipoprotein drug delivery, even if the drugs of choice had vastly differing physical properties. Lipoproteins are therefore proposed as a new, novel and extremely efficient, drug delivery system for antileishmanial chemotherapy.

ACKNOWLEDGEMENTS

The interest of Drs Michèle Masquelier, Jim Shepherd, Colette De Roe and Fred Opperdoes are most gratefully acknowledged.

REFERENCES

Alexander J and Vickerman K (1975) Fusion of host cell secondary lysosomes with the parasitophorous vavuoles of Leishmania mexicana-infected macrophages. J Protozool 22:502-508.

Basu SK, Goldstein JL, Anderson GW & Brown MS (1976) Degradation of catanioized low density lipoprotein and regulation of cholesterol metabolism in homozygous familial hypercholesterolemia fibroblasts. Proc Natl Acad Sci 73:3178-3182.

Chang K-P (1980) Human cutaneous leishmaniasis in a mouse macrophage line: Propagation and isolation of intracellular parasites. Science 209:1240-1242.

Chang K-P (1983) Cellular and molecular mechanisms of intracellular symbiosis in leishmaniasis. Int Rev Cytol 14:267-305.

De Roe C, Courtoy PJ, Quintart J & Baudhuin P (1982) Molecular aspects of the interactions between protein colloidal gold and cultured cells: Application to galactomylated serum albumin and rat hepatocytes. Arch Int Physiol Biochem 90:186-187.

Goldstein JL, Ho YK, Basu SK & Brown MS (1979) Binding site on macrophages that mediate uptake and degradation of acetylated low density lipoprotein, producing massive cholesterol deposition. Proc Natl Acad Sci 76:333-337.

Hart DT and Coombs GH (1982) Leishmania mexicana: Energy metabolism of amastigotes and promastigotes. Exp Parasitol 54:397-409.

Hart DT and Opperdoes FR (1984) The occurrence of glycosomes (microbodies) in the promastigote stage of four major Leishmania species. Mol Biochem Parasit 13:159-172.

Hart DT, Vickerman K & Coombs GH (1981) Transformation in vitro of Leishmania mexicana amastigotes to promastigotes: Nutritional requirements and the effects of drugs. Parasitology 83:529-541.

Masquelier M, Vitols S & Peterson C (1986) Low density lipoprotein (LDL) as a carrier of anti-tumoral drugs: in vivo fate of drug-LDL complexes in mice. Cancer Res 46:3842-3847.

Ralph P, Prichard J & Cohn M (1975) Reticulum cell sarcoma: An effecter cell in antibody-dependent cell-mediated immunity. J Immun 114:898-905.

Shepherd J, Bedford DK & Morgan HG (1976) Radioiodination of human low density lipoprotein: A comparison of four methods. Clinica Chimica Acta 97:97-109.

Recognition processes involved in binding and ingestion of L. donovani by host macrophages

*A. O. Wozencraft and J. M. Blackwell

Department of Tropical Hygiene, London School of Hygiene and Tropical Medicine, Keppel Street, London WC1E 7HT, U.K.

INTRODUCTION

The causal organism of visceral leishmaniasis, the protozoan parasite *Leishmania donovani*, displays a specific tropism for mononuclear phagocyte (MP) populations within the vertebrate host. Infection is initiated upon invasion of host MPs by the promastigote form of the parasite, following inoculation by the sandfly vector. A number of studies have been carried out to determine which receptors on the MP surface are involved in this interaction (Blackwell, 1985). Our laboratory has previously established that the complement type 3 receptor (CR3), which recognizes complement component iC3b, plays a major role in the invasion of resident murine peritoneal macrophages (RPM) *in vitro* by *L. donovani* LV9 promastigotes (Blackwell *et al*., 1985). These studies provided compelling evidence to suggest that opsonization of the promatigote by locally secreted, RPM-derived complement facilitates uptake via CR3 in the absence of an exogenous serum source. The presence of RPM-derived complement on the parasite surface has since been visualized by immunoelectron microscopy (Wozencraft *et al*., 1986). In other studies, it has also been established that complement (Mosser and Edelson, 1984) and CR3 (Blackwell *et al*., 1985; Mosser and Edelson, 1985) play a major role in promastigote invasion of MPs *in vitro* in the presence of serum. However, it remains to be determined how complement opsonization affects the fate of the parasite *in vivo*: as well as facilitating uptake, it may increase exposure to complement-mediated killing. In the present study, we have addressed this question by measuring the complement binding capacity of log and stationary phase promastigotes and have related these findings to differences in their infectivity (Sacks and Perkins, 1984) both *in vitro* and *in vivo*.

Current address: Immunology Section, Department of Biophysics, Cell and Molecular Biology, King's College, University of London, Manresa Road, London SW3 6LX, U.K.

NATO ASI Series, Vol. H11
Host-Parasite Cellular and Molecular Interactions
in Protozoal Infections. Edited by K.-P. Chang and D. Snary

MATERIALS AND METHODS

Parasite maintenance

L. donovani (LV9) was maintained by continuous amastigote passage: amastigotes were isolated from hamster spleens and promastigotes were obtained by in vitro transformation and maintenance by weekly passage. Promastigote numbers were assessed daily using a Helber bacteriological counting chamber and parasite viability was tested by fluorescent dye uptake/exclusion techniques (see Channon et al., 1984).

Immunoelectron microscopy

Following 20 min. parasite incubation with cytochalasin B-treated RPM, C3 was detected on parasite surface membranes using γ-globulin fraction rabbit anti human C3 (which also recognises murine C3) and gold-labelled Protein A (see Wozencraft et al., 1986).

Assessment of promastigote infectivity

Parasites were washed, counted and resuspended in RPMI to a final concentration of 1 x 10^{8} ml^{-1}. A 0.2 ml volume of this suspension was injected into 6 - 8 week old female BALB/c mice. These were sacrificed after 14 days and parasite load were assessed by the LDU method outlined by Bradley and Kirkley (1977). The in vitro binding capacity of these promastigote populations to RPM was measured using the method described by Blackwell et al., (1985).

Parasite triggering of the Alternative Complement Pathway (ACP)

ACP activity of a stock human serum source was assessed by measuring the non-specific lysis of rabbit erythrocytes (Riches and Stanworth, 1980). The total ACP-fixing capacity of parasite populations was measured by comparing CH values for erythrocyte lysis after parasite incubation with this serum source (Wozencraft and Blackwell, 1986).

ABBREVIATIONS:

ACP, alternative complement pathway; CP, classical complement pathway; C3, complement component 3; CR3, complement type 3 receptor; MAC, membrane attack complex; MP, mononuclear phagocyte; NMS, normal mouse serum; RPM, resident murine peritoneal macrophages

Solid phase ELISA measurement of C3 on the promastigote membrane

C3 fixation on the promastigote surface was assessed by a modification of the method outlined by Engels et al. (1985). Log and stationary phase promastigotes were incubated in human serum depleted of classical pathway (CP) activity (by Mg^{++}, EGTA-treatment), and C3 deposition was detected by sequential incubation with γ-globulin fraction rabbit anti-human C3 and swine anti-rabbit Ig (peroxidase linked) before the addition of O.P.D. substrate (see Wozencraft and Blackwell, 1986).

RESULTS

Immunoelectron microscopy

The detection of gold particles on the surface of promastigotes (Plate 1A) and amastigotes (Plate 1B) in the absence of an exogenous serum source demonstrated parasite opsonization by locally-secreted, RPM-derived C3. The addition of 10% normal mouse serum (NMS) slightly increased the extent of gold deposition (Plate 1C) and led to promastigote damage. The nucleophile, SSH, which prevents covalent binding of C3, completely abrogated promastigote binding to RPM. However, it did not prevent amastigote binding and the non-covalent binding of C3 onto the amastigote membrane could still be detected (1D). Non-specific binding of gold-labelled Protein A was not detected in the absence of specific antibody (plates not shown).

Parasite infectivity

As shown in Fig 1B, the in vivo infectivity of L. donovani LV9 promastigotes increased as the parasites passed from log to stationary phase of their growth cycle (Fig. 1A). Corresponding amastigote infectivity was 2.858 ± 0.193 LDU, approximately 4-fold higher than the most infective stationary phase promastigotes. An increase in infectivity was accompanied by a rise in the capacity of the stationary phase promastigotes to bind to RPM in vitro (Fig. 1C). This binding was largely CR3-mediated throughout, as demonstrated by the % inhibition observed upon M1/70 addition (Fig 1C, bar chart).

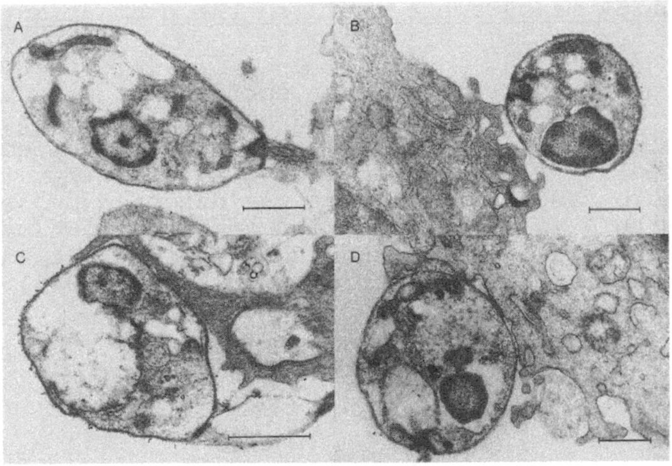

Plates 1A - D: Visualization of RPM-derived C3 on the surface of L. donovani promastigotes (1A) and amastigotes (1B) using specific anti-C3 antibody and gold-labelled protein A in the absence of an exogenous serum source. Deposition is also shown in the presence of NMS on the promastigote form (1C). SSH did not prevent limited C3 binding to amastigotes (1D). Bar=1 µm in A, C; 0.5 µm in B,D.

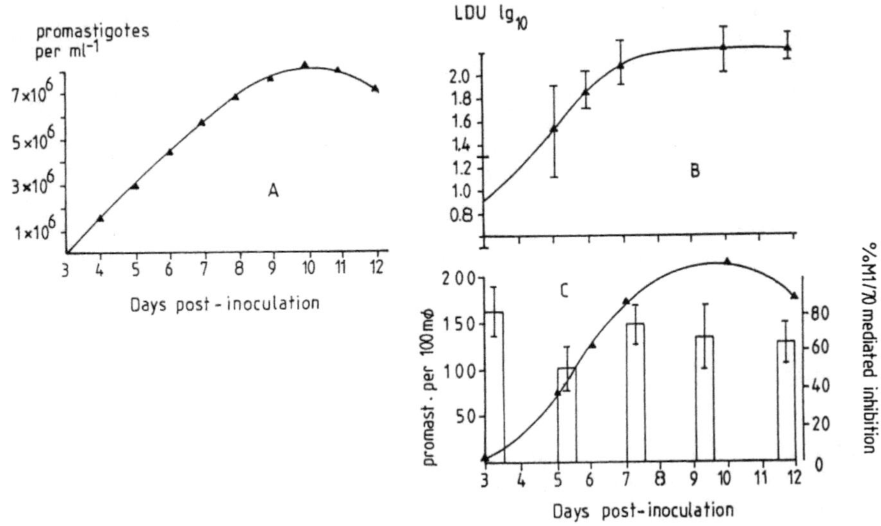

Figs. 1A - 1 C: Promastigote cultures passing from log to stationary phase of their growth cycle (1A) were assessed for in vivo infectivity (1B; mean values for 1 representative experiment, 3 - 5 mice per group). The in vitro binding capacity of these populations was also assessed (1C; mean values for same experiment, duplicate coverslips) and CR3 involvement was assessed by measuring M1/70-mediated inhibition.

ACP activation by L. donovani

Stationary phase promastigotes were found to deplete ACP components from a stock human serum source to a greater extent than equivalent numbers of log phase parasites, as measured by their capacity to inhibit the ACP-mediated haemolysis of rabbit erythrocytes (Fig. 2).

C3 deposition on promastigotes

The solid phase ELISA assay demonstrated that more C3 was deposited on the surface membrane of stationary phase promastigotes than on equivalent numbers of log stage parasites following incubation with human serum depleted of CP activity (Table 1). Parasite counts and viability assays confirmed that promastigotes were not lysed during the experiments: lower OD counts did not relate to a decrease in parasite numbers.

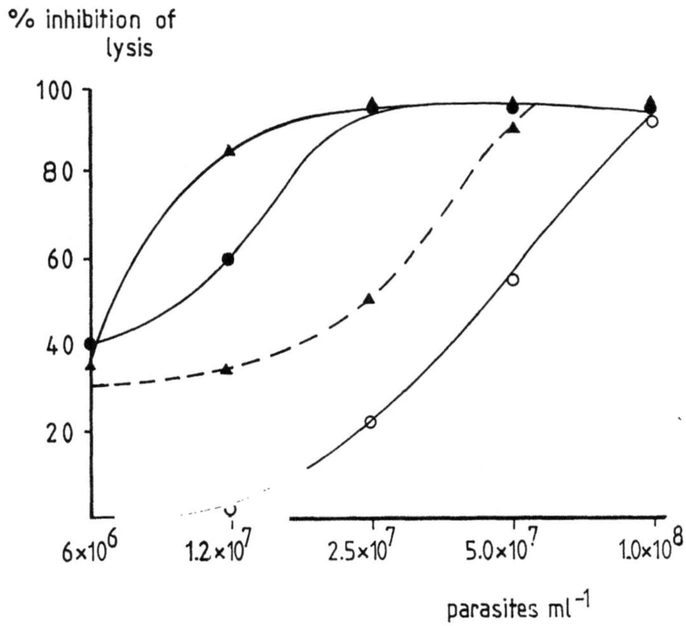

Figure 2: ACP fixation demonstrated by (i) ▲—▲ zymosan particles (positive control); (ii) ●—● stationary phase promastigotes; (iii) ▲---▲ log phase promastigotes and; (iv) o—o amastigotes.

No. of parasites per ml^{-1}	O.D. readings	
	log	stationary
$3.12x10^6$	0.085±.020	0.140±.038
$6.25x10^6$	0.184±.032	0.197±.021
$1.25x10^7$	0.295±.050	0.393±.037
$2.50x10^7$	0.538±.047	0.858±.045
$5.00x10^7$	1.245±.038	1.525±.047

Table 1: C3 deposition on surface of promastigote populations: OD values given are the means of triplicate samples. Zero value given by control samples prepared in absence of primary Ab layer.

DISCUSSION

In our earlier studies (Blackwell et al., 1985) we had demonstrated a role for C3-mediated binding to CR3 in both the serum-independent and serum-dependent entry of *L. donovani* promastigotes to host MPs *in vitro*. The extent to which this C3-dependent binding to MPs infuences infectivity of the parasite *in vivo* is complicated by the various observations (Pearson and Steigbigel, 1980; Mosser and Edelson, 1984) that *Leishmania* promastigotes may be lysed in nonimmune serum following ACP or CP activation. Differential survival of log and stationary phase promastigotes in normal human serum (Franke et al., 1985) suggested that the more infective stationary phase organisms are more resistant to complement-mediated lysis. No attempt was made, however, to quantify total ACP activity or, more specifically, the amount of C3 bound to the surface membrane of the different growth phase forms of the parasite. One interpretation of the earlier (Franke et al., 1985) findings might be that the membrane of stationary phase promastigotes simply acts as a less efficient C3 acceptor surface. This would explain their enhanced survival in serum but would be contrary to a role for C3 in mediating enhanced parasite entry into host MPs. An alternative hypothesis might predict equivalent C3 binding but a differential fate for bound C3 on the surfaces of log versus stationary phase organisms.

In the studies presented here we have attempted to clarify some of these issues relating to the different infectivity of log and stationary phase promastigotes in vivo by examining total ACP and C3-binding capacity in relation to CR3-mediated parasite uptake in vitro in a system not complicated by serum-dependent lysis of organisms. Our results show (Fig. 1) that the curves for increased infectivity in vivo as promastigotes in culture pass from log to stationary phases of growth are mimicked precisely by an increase in serum-independent binding of promastigotes to RPM in vitro. Moreover, the percentage of this increased binding inhibitable by the anti-CR3 antibody (M1/70) remained constant, suggesting that it might be matched by an equivalent increase in C3 bound to stationary phase organisms. This was confirmed by data showing both enhanced total ACP activity (Fig. 2) by stationary phase promastigotes in the Riches and Stanworth (1980) assay and a specific increase in bound C3 (Table 1) measured in the solid phase ELISA system. Hence, rather than being less efficient activators of the ACP, the surface membrane of stationary phase promastigotes appears to provide a more potent C3 binding surface which in turn promotes their CR3-mediated entry into host MPs. It is not yet known whether this difference in C3 binding capacity relates to differential expression of GP63 (Colomer-Gould et al., 1985), the potent C3 fixing antigen (Russell and Wilhelm, 1986) of Leishmania spp., on the different forms of the parasite or whether structurally and functionally different molecules are involved.

The actual efficiency of this CR3-mediated invasion depends on the levels of the specific ligand, iC3b, on the promastigote membrane. Deposition is initiated by the covalent binding of metastable C3b, which is converted to iC3b by fluid phase inactivators (Pangburn et al., 1983). Using an anti-C3 antibody and immunoelectron microscopy we have visualised both macrophage-derived and serum-enhanced deposition of C3 on the surface of stationary phase promastigotes (Wozencraft et al., 1986; Plates 1A and 1C). Inhibition of promastigote binding in the presence of the nucleophile SSH (Blackwell et al., 1985) confirms that the CR3-mediated binding is dependent upon covalently bound C3, but the anti-C3 antibody did not allow us to distinguish which form of C3 was present on the parasite surface. Using C3 iodination and gel electrophoretic techniques (Wozencraft and Blackwell 1986) we have been able to demonstrate C3b and iC3b on both log and stationary phase organisms. These have not been quantified precisely to determine whether the increased sensitivity of log phase parasites relates to relatively higher C3b deposition thus leading to formation of the terminal membrane attack complex (MAC). The data suggest, however, that the relative resistance of the stationary phase organisms might rely more on mechanisms which hinder MAC formation. The presence of polysaccaride side chains on bacterial membranes, for example, can prevent MAC formation by stearic hindrance (Joiner et al., 1986) while another protozoan parasite, Trypanosoma brucei, blocks MAC formation by an unknown mechanism (Devine et al., 1986). Antigen changes defined by lectin-binding studies do occur on the surface of L. major as their infectivity increases (Sacks and Perkins, 1984). We do not know, however, whether these are functionally relevant molecular changes mediating the enhanced infectivity of stationary phase promastigotes either by increased C3 binding or by blocking or preventing MAC formation.

In our studies presented here we have shown strong correlations between the increased infectivity of stationary phase promastigotes, their C3 binding capacity, and their CR3-mediated attachment to host MPs. These correlations do not, however, extend to the greater infectivity in vivo and in vitro of a numerically equivalent amastigote inoculum. While amastigotes show both RPM-derived and serum-dependent C3 deposition on their surface (Wozencraft et al., 1986; Plate 1B), this does not appear to mediate entry via CR3 (Blackwell et al., 1985). Our inability to inhibit C3 deposition with SSH (Plate ID) suggests that at least some C3 is bound noncovalently to the amastigote surface. The relatively lower performance of amastigotes in inhibiting ACP-mediated haemolysis of rabbit erythrocytes (Fig. 2) is also consistent with their inability to bind C3 covalently. Overall, these results indicate that the resistance of amastigotes to complement-mediated lysis (Mosser et al., 1985) and their mode of entry into host MPs involve quite different mechanisms and do not simply reflect an extension of the changes observed when promastigotes pass from the log to the stationary phase of the cell cycle.

The extent to which CR3-mediated promastigote invasion is important in initiating infection in vivo with other Leishmania spp has not yet been clearly defined. MP secretion of complement proteins and expression of CR3 may vary amongst the different MP sub-populations which different species of Leishmania invade. These functions are also affected by the state of MP activation (Wright et al., 1986) and hence by the immune status of the host. Nevertheless, our findings suggest that L. donovani, and possibly a range of other intracellular pathogens, exploit components of a major host defence mechanism to achieve successful invasion of target host cells.

ACKNOWLEDGEMENTS

This work was supported by project grants from the Medical Research Council, U.K. and the Wellcome Trust. Jenefer M. Blackwell is a Wellcome Trust Senior Lecturer.

REFERENCES

Blackwell JM (1985) Trans Roy Soc Trop Med Hyg 79: 606-612
Blackwell JM, Ezekowitz RAB, Roberts MB, Channon JY, Sim RB, Gordon S (1985) J Exp Med 162: 324-331
Bradley, DJ and J Kirkley (1977) Clin Exp Immunol 30: 119-129
Channon JY, Roberts MB, Blackwell JM (1984) Immunol 53: 345-355.
Colomer-Gould V, Quintao LG, Keithly J, Nogueira N (1985) J Exp Med 162: 902-916
Devine DV, Falk RJ, Balber AE (1986) Infect Immun 52: 223-292
Engels W, Endert J, Van Boven CPA (1985) J Immunol Methods 81: 43-53
Franke ED, McGreevy PB, Katz SP, Sacks DL (1985) J Immunol 134: 2713-2718
Joiner KA, Grossman N, Schmetz M, Leive L (1986) J Immunol 136: 710-715
Mosser DM, Edelson PJ (1984) J Immunol 132: 1501-1505
Mosser DM, Edelson PJ (1985) J Immunol 135: 2785-2789
Mosser DM, Wedgwood JF, Edelson PJ (1985) J Immunol 134: 4128-4131
Pangburn MK, Schreiber RD, Muller-Eberhard HJ (1983) J Immunol 131: 1930-1935
Pearson RD, Steigbigel RT (1980) J Immunol 125: 2195-2201
Riches DWH, Stanworth DR (1980) Immunology Lett 1: 363-366
Russell DG, Wilhelm H (1986) J Immunol 136: 2613-2620
Sacks DW, Perkins PV, (1984) Science 223: 1417-1419
Wozencraft AO, Blackwell JM (1986) Immunol: submitted
Wozencraft AO, Sayers G, Blackwell JM (1986) J Exp Med 164: 1322-1327
Wright SD, Detmers PA, Jong MTC, Meyer BC (1986) J Exp Med 163: 1245-1259

INTERACTION OF LEISHMANIA AND MEMBRANE-STABILIZING DRUGS WITH HUMAN PHAGOCYTIC CELLS

Arsalan Kharazmi

State Serum Institute, Department of Clinical Microbiology, Rigshospitalet, DK-2200, Copenhagen N, Denmark.

INTRODUCTION

Leishmaniasis inflicts more than one million people in the world and causes over 1000 deaths per year. Therapy of these patients poses a serious problem. Pentavalent antimonials at higher doses exhibit considerable toxicity, some patients with visceral leishmaniasis become unresponsive to conventional doses of antimonials, and the usual second line drugs such as pentamidine and amphotericin B are also toxic and difficult to administer. In order to devise more effective ways to control the disease, there is a great need for better understanding of the host-parasite interactions. Identification and characterization of the parasite products which are involved in attraction of the parasite by the host cells, and in binding to and ingestion of the parasite by these cells will be of particular interest in this context.

ROLE OF MEMBRANE-STABILIZING DRUGS

The initial step in the interaction of leishmania parasite with host target cell is binding to receptor(s) on the cell surface, followed by either actively penetrating into or being phagocytosed by the target cell. It is now more or less accepted that the actual ingestion of both promastigotes and amastigotes occurs via the "zipper" hypothesis (Silverstein 1977) involving the interaction of the molecules on the host cell surface with those on the parasite surface (Blackwell 1985; Chang & Chang 1986; Handman & Goding 1985). The actin and myosin microfilaments of the phagocyte provide the motor mechanism for parasite ingestion.

Apart from the precise molecular identity of the receptor-ligand interactions involving the known receptors on the macrophage, namely the FC (FCR), the complement (CR3), and the mannose-fucose (MFR) receptors on one hand, and the ligands on the parasite surface on the other hand, the cell membrane as a whole plays an important role in host-parasite interactions. Integrity and fluidity of cell membrane is essential in recycling and expression of receptors on the cell surface, and in interiorizing the parasite during the ingestion process. It is, therefore, tempting to postulate that pharmacological alterations of cell membrane by drugs which can change membrane fluidity may result in inhibition of either receptor-ligand binding or actual ingestion of the parasite by the phagocytic cell, or both.

NATO ASI Series, Vol. H11
Host-Parasite Cellular and Molecular Interactions
in Protozoal Infections. Edited by K.-P. Chang and D. Snary

It has been shown that lysosomotropic agents such as chloroquine, methylamine, and amantidine interfere with the flow of cell membrane occuring during endocytosis in cultured rat fibroblasts and that these drugs impair the transfer of extracellular material bound to the plasma membrane, from the phagosomal membrane to the lysosomes (Trouet, Tulkens & Schneider 1980). The binding of serotonin and opiates to their receptors on mouse brain cell membrane can be modulated by lipid fluidity (Heron et al. 1980). Chlorpromazine, a phenotiazine derivative, has been shown to alter erythrocyte membrane fluidity (Ogiso, Iwaki & Mori 1981; Suda et al. 1981). Another interesting aspect of some membrane stabilizing drugs such as chlorpromazine, trifluoperazine, and clopenthixol is that they are capable of modulating chemotaxis and oxidative burst of phagocytic cells (Rechnitzer, Kristiansen & Kharazmi 1985; Smith, Bowmann & Iden 1981).

Based on the involvement of cell membrane in leishmania-macrophage interactions and the role of membrane-stabilizing drugs on membrane-associated phenomena, the long range prospect for development of more effective and more useful drugs against leishmania would be to look for drugs which would interfere with parasite interactions at the host cell surface. Prevention of parasite enterance to the target cell will result in inhibition of parasite multiplication extracellularly and in more rapid elimination of the parasite by the immune system.

ROLE OF PARASITE PRODUCTS

The parasite-host relationship represented by leishmania-macrophage interactions have been studied in order to understand the mechanisms by which leishmania parasites invade, occupy and multiply in otherwise important cells of the host defence system. On entering the host, promastigotes are in danger of becoming agglutinated or lysed by factors proven to be in serum of many mammals. However, once inside the host macrophage, amastigotes are protected from antibodies and any other circulating substances that might be harmful to the parasite.

Internalization of Leishmania Parasites

Chemotaxis. In order to minimize the time between coming into the host skin and becoming inhabited in the macrophage, promastigotes must either actively search out for the macrophage, or the macrophage must move towards the promastigote, either movement involving chemotaxis.

It is known that parasite surfaces can activate complement to C5a which is chemotactic for macrophages (Bray 1983). The work carried out in our laboratory showed that the sonic extract prepared from L. mexicana amazonensis promastigotes exhibited chemotactic activity for human blood monocytes and neutrophils in vitro (Table 1 & Fig 1). The chemotactic activity of the sonicate was heat-labile and was lost by heating at 100°C. Serum was not required for the chemotactic activity, indicating a direct role for parasite products in attracting the host cells (Sørensen, Kharazmi & Nielsen 1986).

TABLE 1.
COMPARISON OF CHEMOTACTIC ACTIVITY OF L.M. AMAZONENSIS SONIC EXTRACT WITH CASEIN AND FMLP FOR HUMAN NEUTROPHILS AND MONOCYTES.
DATA PRESENTED ARE MEAN ± SEM OF 2 EXPERIMENTS, EACH PERFORMED IN TRIPLICATE FOR NEUTROPHILS AND DUPLICATE FOR MONOCYTES.

SONIC EXTRACT CONCENTRATION	PERCENTAGE OF CASEIN RESPONSE		PERCENTAGE OF FMLP RESPONSE	
(mg/ml)	NEUTROPHILS	MONOCYTES	NEUTROPHILS	MONOCYTES
0.15	9 ± 3	13 ± 1	23 ± 10	17 ± 0
0.3	18 ± 4	13 ± 1	50 ± 10	17 ± 0
0.6	38 ± 7	16 ± 1	106 ± 15	22 ± 2
1.2	47 ± 4	22 ± 1	136 ± 6	31 ± 1
2.4	80 ± 2	40 ± 2	231 ± 4	56 ± 1
4.8	86 ± 8	59 ± 7	244 ± 12	80 ± 4

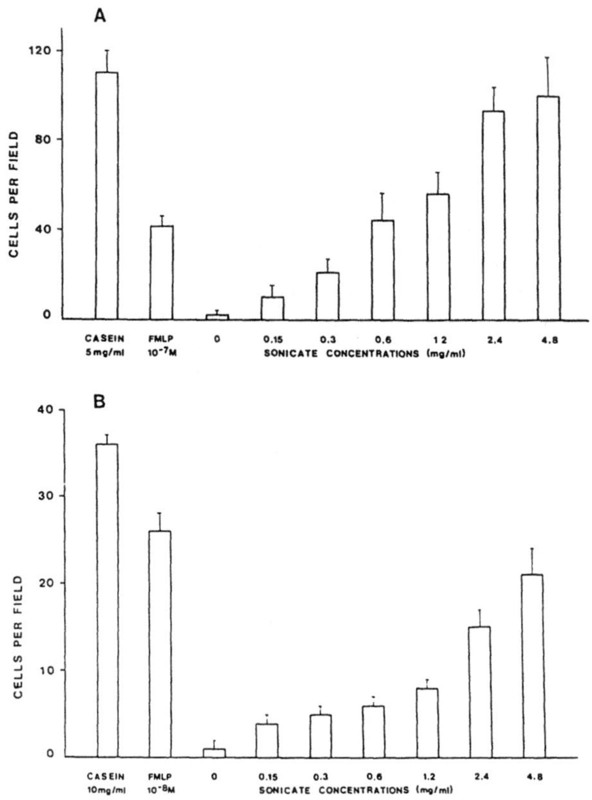

Fig 1.
Chemotactic activity of L.m. amazonensis sonic extract for human neutrophils (A) and monocytes (B) compared with casein and FMLP. The results are mean of 2 experiments each performed in triplicate. Bars represent SEM.

Binding. Once in the proximity of the macrophage, the parasite must bind in some way or other to the macrophage surface in order to become internalized. It is believed that the parasite becomes attached to the macrophage via some form of receptor-ligand binding. Herman (1980) argues that binding of cytophilic antibody via the FC receptor could establish an Fab site for the parasite attachment. Bray & Alexander (1983) consider activated complement on the surface of the parasite to mediate attachment. Leishmania can however be taken up by macrophages in the absence of serum (Blackwell et al. 1985), indicating that complement produced locally may play a role in the binding of leishmanias to macrophages.

Uptake. Apparently leishmania parasites are taken up by the macrophages by the sequential interactions between parasite ligands and host cell receptors, the "zipper" mechanism (Silverstein 1977). Pretreatment of macrophages with cytochalacin B, which is known to disrupt the regular arrangement of microfilaments associated with endocytosis, inhibits ingestion of various species of leishmania (Ardehali, Khoubyar & Rezai 1979). Pretreatment of the host cell with iodoacetamide, an irreversible inhibitor of glycolysis, prevents uptake of leishmania parasites (Zenian 1980). Both examples emphasize the role of phagocytic process in uptake.

CONCLUSIONS

Leishmaniases are a wide range of very variable diseases caused by different leishmania species and strains, and manifesting in different types of host reactions to the particular leishmania involved. Knowledge as to how the parasite in itself influences macrophage and neutrophil function, is necessary in order to understand the ways in which the different diseases manifest themselves. On the other hand membrane-stabilizing drugs play an important role in the membrane-associated phenomena such as binding and ingestion. It is, therefore, important to find out whether certain leishmanial factors are involved in the interaction of the parasite with the phagocytic defence system and what role membrane-stabilizing drugs play in this interaction. Such a knowledge will enable us to devise more effective ways to control the disease.

REFERENCES

Ardehali SM, Khoubyar K, Rezai HR (1979) Studies on the effect of the antiphagocytic agent cytochalasin B on the Leishmania-macrophage interaction. Acta Tropica 36: 15-21.

Blackwell JM (1985) Receptors and recognition mechanisms of Leishmania species. Trans. Roy. Soc. Trop. Med. Hyg. 79: 606-612.

Blackwell JM, Ezekowitz RAB, Roberts MB, Channon JY, Sim RB, Gordon S (1985) Macrophage complement and lectin-like receptors bind Leishmania in the absence of serum. J. Exp. Med. 162: 324-331.

Bray RS (1983) Leishmania: Chemotaxic responses of promastigotes and macrophages in vitro. J. of Protozool. 30(2): 322-329.

Chang CS, Chang KP (1986) Monoclonal affinity purification of a leishmania membrane glycoprotein and its inhibition of leishmania-macrophage binding. Proc. Nat. Acad. Sci. 83: 100-104.

Handman E, Goding JW (1985) The Leishmania receptor for macrophages is a lipid-containing glycoconjugate. EMBO J. 4: 329-336.

Herman R (1980) Cytophilic and opsonic antibodies in visceral leishmaniasis in mice. Infect. Immun. 28: 585-593.

Heron DS, Shinitzky M, Hershkowitz M, Samuel D (1980) Lipid fluidity markedly modulates the binding of serotonin to mouse brain membranes. Proc. Nat. Acad. Sci. 77: 7463-7467.

Ogiso T, Iwaki M, Mori K (1981) Fluidity of human erythrocyte membrane and effect of chlorpromazine on fluidity and phase separation of membrane. Biochem. Biophys. Acat. 649: 325-335.

Rechnitzer C, Kristiansen JE, Kharazmi A (1985) In vitro modulation of human neutrophil chemotaxis by cis(Z)- and trans(E)-clopenthixol and chlorpromazine. Acta Pathol. Microbiol. Immunol. Sec. C. 93: 199-203.

Silverstein SC (1977) Endocytic uptake of particles by mononuclear phagocytes and the penetration of obligate intracellular parasites. Am. J. Trop. Med. Hyg. 26(6) suppl.: 161-168.

Smith RJ, Bowmann BJ, Iden SI (1981) Effects of trifluoperazine on human neutrophil function. Immunol. 44: 677-684.

Suda T, Shimizu D, Maeda N, Shiga T (1981) Decreased viscosity of human erythrocyte suspension induced by chlorpromazine and isoxsusprine. Biochem. Pharmacol. 30: 2057-2064.

Sørensen AL, Kharazmi A, Nielsen H (1986) Leishmania promastigote sonicate exhibits chemotactic activity for human neutrophils and monocytes. Submitted.

Trouet A, Tulkens P, Schneider YJ (1980) Subcellular localization of infectious agents: Pharmacological and pharmacokinetic implications. In: The Host Invader Interplay. ed. Van den Bossche, North Holland H. p. 31-44.

Zenian A (1980) Leishmania tropica: Biochemical aspects of promastigote attachment to macrophages in vitro. Exp. Parasitol. 51: 175-187.

SUPPRESSION OF MACROPHAGE ACCESSORY FUNCTIONS BY LEISHMANIA DONOVANI

Neil E. Reiner[1,2], Winnie Ng[1] and W. Robert McMaster[3]

From the Departments of Medicine[1], Microbiology[2] and Medical Genetics[3], The University of British Columbia and Vancouver General Hospital, Vancouver, B. C., Canada

INTRODUCTION

Parasite-specific, T cell-dependent cell-mediated immune responses are important host-protective mechanisms which contribute to the resolution of infections due to protozoa of the genus *Leishmania*. For reasons that are not understood, visceralizing species of *Leishmania* such as *L. donovani* cause persistent disease and have findings consistent with parasite specific T-cell non-responsiveness (Ho, Koech, Iha, Bryceson, 1983). In order to dissect the basic mechanisms responsible for these defective host responses, we examined the effects of *L. donovani* infection upon macrophage accessory functions that are required for the induction of antigen-specific T-cell activation including: 1) the expression of major histocompatibility complex MHC class II immune response gene products and 2) the elaboration of interleukin 1 (IL-1) (Unanue, 1981).

MATERIALS AND METHODS

Amastigotes of *L. donovani* obtained from hamster spleens were used to infect monolayers of adherent resident or elicited peritoneal cells (>90% macrophages). In experiments concerned with modulation of MHC antigens, thioglycollate elicited macrophages were infected for 4 h and then either untreated or stimulated with recombinant murine interferon-gamma (IFN) (Genentech). H-2K, H-2D, and class II antigens were quantitated in an indirect ^{125}I-binding assay performed under saturating conditions using monoclonal anti-MHC antibodies in the first step and ^{125}I-conjugated F $(ab')_2$ fragments of goat anti-mouse-Ig in the second. In experiments concerned with modulation of IL-1 production, macrophage supernatants were assayed for monokine activity by co-stimulation of C3H/Hej thymocytes (Mizel, 1982).

RESULTS

Kinetics of induction of macrophage class II antigens. Stimulation of control cells with 12.5 U/ml of IFN for 96 h resulted in a progressive, time-dependent increase in I-E expression (Table 1). In contrast, leish-

NATO ASI Series, Vol. H11
Host-Parasite Cellular and Molecular Interactions
in Protozoal Infections. Edited by K.-P. Chang and D. Snary

mania-infected cells were maximally induced after 24 h of lymphokine treatment. Relative to control cells, induction of I-E on infected cells was suppressed by 67% and 56% respectively at 72 h and 96 h of stimulation.

Table 1

KINETICS OF INDUCTION OF MACROPHAGE MHC CLASS II (I-E) ANTIGENS BY IFN[a]

Cell Treatment		Binding ^{125}I-anti-immunoglobulin (CPM x 10^3) — Duration of Culture with IFN (hours)			
Leishmania	IFN (12.5 U/ml)	24	48	72	96
-	-	1.9	1.4	0.4	0.0
+	-	1.6	3.5	1.8	1.4
-	+	5.1	6.4	8.3	14.3
+	+	4.3	5.2	2.9	6.2

[a]Cells were >90% infected at all time points. The data shown are the mean ± S.D. In the IFN-stimulated state, the differences between infected and control cells at 72 h and 96 h were both significant at the 0.1% level.

Dose response relationship for induction of macrophage class II antigens by IFN. Leishmania-induced suppression of macrophage class II antigens was not related to an alteration of the IFN dose-response relationship. Thus, increasing the optimal IFN concentration (12.5 U/ml) for normal cells by as much as 8 fold failed to restore I-A expression by infected cells (IFN = 200 U/ml, I-A = 60% of control) to normal.

Modulation of macrophage class I and class II antigens during infection with L. donovani. The results shown in Table 2 indicate that parasite-induced suppression of MHC antigens was not selective for class II gene products. In this experiment, relative to normal cells, binding of anti-I-A and anti-I-E by IFN treated infected cells were suppressed by 70% ($p < 0.001$) and 47% ($p < 0.01$) respectively. Similarly, binding of anti-H-2K was suppressed by approximately 50% in both the basal ($p < 0.01$) and lymphokine treated ($p < 0.001$) states. In contrast, suppression of H-2D induction of IFN treated infected cells was less pronounced (15%).

Parasite dose-response relationship for suppression of macrophage MHC antigen expression. Examination of the relationship between the intensity of leishmania-infection and the degree of suppression of MHC antigens revealed that the macrophage parasite burden correlated with the level of suppression observed. This is shown for I-A in Figure 1. Similar results were obtained for class I antigens (data not shown).

Effects of inhibition of macrophage cyclooxygenase activity on suppression of MHC class II antigens by L. donovani. To examine the potential relationship between parasite-induced increases in macrophage prostaglandin

(PG)E$_2$ synthesis and suppression of class II antigens, cells were incubated with the cyclooxygenase inhibitor sodium meclofenamate beginning 1 h prior to infection. Meclofenamate inhibited PGE$_2$ synthesis by >90% (data not shown). As shown in Figure 2, this treatment significantly augmented I-A expression by IFN-stimulated infected cells ($p < 0.05$) compared to IFN stimulated, non-meclofenamate treated infected cells.

Table 2

MODULATION OF MACROPHAGE MHC CLASS I AND CLASS II ANTIGENS DURING INFECTION WITH *LEISHMANIA DONOVANI*[a]

Cell Treatment		Binding of ^{125}I-anti-immunoglobulin (CPM x 10^3)			
		Determinant			
Leishmania	IFN (12.5 U/ml)	I-A	I-E	H-2K	H-2D
-	-	0.2	0.9	1.1	9.3
+	-	0.1	0.9	0.6	8.3
-	+	13.7	5.6	4.2	17.0
+	+	4.4	2.4	2.9	14.5

[a]MHC antigens were assayed after 96 h of IFN stimulation. The data are the mean ± S.D. of triplicate determinations.

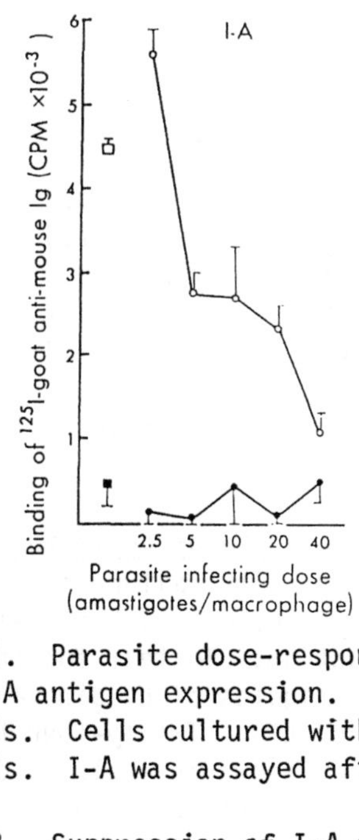

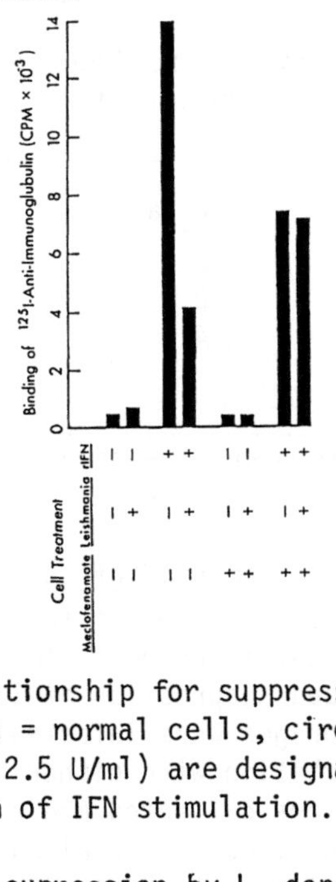

Figure 1. Parasite dose-response relationship for suppression of macrophage I-A antigen expression. Squares = normal cells, circles = infected cells. Cells cultured with IFN (12.5 U/ml) are designated by open symbols. I-A was assayed after 96 h of IFN stimulation.

Figure 2. Suppression of I-A antigen expression by *L. donovani* is prevented by inhibition of PGE_2 synthesis. Cells were cultured with meclofenamate for 1 h prior to addition of Leishmania. Four h later the cells were either untreated or stimulated with IFN (12.5 U/ml) for 96 h.

IL-1 activity in supernatants of resident peritoneal macrophages infected with L. donovani. Conditioned media from monolayers of resident macrophages infected with L. donovani at a parasite:cell ratio of 10:1 showed no increase in IL-1 content when compared to supernatants of control cells (Table 3). In contrast, IL-1 activity was increased in supernatants generated in the presence of Listeria ($p < 0.025$ vs. control) or lipopolysaccharide ($p < .005$ vs. control). Lower inocula of L. donovani (as low as 1 parasite/cell) also did not induce IL-1 production (data not shown). The failure of resident peritoneal macrophages to produce IL-1 in response to L. donovani was not related to the state of accessory cell activation since macrophages elicited with proteose-peptone also did not produce IL-1 in response to infection (data not shown). The absence of increased IL-1 activity in supernatants of leishmania infected cells was not related to inhibition of release of monokine. Thus, lysates of infected cells did not contain intracellular IL-1 in excess of that detected in lysates of control cells. Membrane IL-1 was also not increased on L. donovani infected macrophages (data not shown).

Table 3

INTERLEUKIN 1 ACTIVITY IN SUPERNATANTS OF RESIDENT PERITONEAL MACROPHAGES INFECTED WITH L. DONOVANI[a]

Macrophage Stimulus	Thymocyte Proliferation (CPM x 10^3)	Macrophage Stimulus	Thymocyte Proliferation (CPM x 10^3)
Nil	0.8[a]	Nil	0.5
Listeria	2.6	Listeria	1.5
L. donovani	0.6	L. donovani + Listeria	0.6
LPS	4.0	LPS	2.1
-	-	L. donovani + LPS	1.8

[a]The values shown are the mean of triplicate determinations. The S.D. were less than 10% of the mean.

Effects of leishmania-infection on the capacity of cells to respond to secondary stimulation for IL-1 production. The failure of macrophages to elaborate IL-1 in response to L. donovani could have been related to parasite-induced suppression of IL-1 production. To test for this possibility, macrophages were infected with amastigotes of L. donovani for 4 h prior to stimulation with either Listeria or LPS. Twenty-four h later supernatants were collected and assayed. As shown in Table 3, compared to Listeria stimulation of normal macrophages, the responses of leishmania -infected monolayers to Listeria were significantly suppressed ($p < 0.001$).

In contrast to the ability of L. donovani infection to suppress the IL-1 response to secondary stimulation with Listeria, suppression was not found when the second stimulus was LPS. Results identical to these were also obtained using peptone-elicited cells (data not shown).

DISCUSSION

The persistence of L. donovani (and other Leishmania spp.) within mammalian hosts suggests that these organisms have evolved means of subverting the immune response. A logical point at which to invoke such mechanisms would be at the early parasite-macrophage interface. The results presented above indicate that L. donovani is capable of both evading and suppressing the macrophage IL-1 response as well as suppressing macrophage MHC class II antigen expression. These two parasite-induced accessory cell abnormalities may have important consequences for defective induction of anti-leishmania T-cell immunity since both of these accessory cell functions play an obligatory role during antigen-specific T-cell activation (Unanue, 1981, Mizel, 1982). In this regard, the level of T-lymphocyte stimulation obtained during antigen presentation is proportional to the density of class II molecules (as well as IL-1) on antigen presenting cells (Kurt-Jones, Virgin, Unanue, 1985). Thus, T-cell activation is quantitatively dependent upon the amplication of class II molecules on accessory cells by IFN. Suppression by L. donovani of this macrophage response to IFN stimulation would be expected to limit the ability of parasitized accessory cells to activate leishmania-specific T-cells. The results of these studies suggest, therefore, that deficient parasite-specific T-lymphocyte responses in both experimental and clinical forms of leishmaniasis may be related to parasite-induced suppression of discrete macrophage accessory functions. Biologic immune response modifiers such as IL-1, colony-stimulating factors or IFN may be of value in reversing these abnormalities in-vivo.

References

Ho MD, Koech K, Iha DW, Bryceson AAM (1983). Immunosuppression in Kenyan visceral leishmaniasis. Clin exp Immunol. 51:207-214.

Unanue ER (1981) The regulatory role of the macrophage in antigenic stimulation. II Symbiotic relationship between lymphocytes and macrophages. Adv Immunol. 31:1-136.

Mizel SB (1982) Interleukin 1 and T cell activation. Immunol Rev. 63:52-72.

Kurt-Jones EA, Virgin HW, Unanue ER (1985) Relationship of macrophage Ia and membrane IL-1 expression to antigen presentation. J. Immunol. 135:3652-3654.

EXTRACELLULAR DEVELOPMENT OF ERYTHROCYTIC MALARIA PARASITES

William Trager

Department of Parasitology, The Rockefeller University, New York, NY 10021, USA

INTRODUCTION

The propagation of a parasite in vitro apart from its intact host has always led to extensive progress in understanding host-parasite relationships. This has been especially well illustrated with the explosive growth in work on the biochemistry, immunology and molecular biology of the human malaria parasite Plasmodium falciparum during the 10 years since its cultivation was first reported (Trager & Jensen 1976). In these cultures all the erythrocytic stages of the parasite, including the male and female gametocytes as well as the asexual schizogonic cycle, develop in human erythrocytes maintained in vitro. Just as with all other obligate intracellular parasitic protozoa that have been grown in vitro, the parasites develop in their living host cells. Can they be grown axenically in a non-living medium? How much of the intactness of the host cell do they require? Why should a eukaryotic organism having its own protein-synthesizing machinery need to depend on a living host cell for its development, and what is the nature of this dependence? Might it be that the host cell furnishes signals or triggers for differentiation of the parasite, as well as essential nutrients? The erythrocytic stages of malaria may provide particularly favorable material for attempts to answer these questions.

Much is now known about the complex process of entry of a malarial merozoite into its host erythrocyte (Aikawa & Miller 1983, Breuer 1985). For P. falciparum the initial attachment of a merozoite to an erythrocyte depends on interaction of a merozoite surface protein of 130 K dalton molecular weight with the erythrocyte glycophorins (Pasvol & Jungery 1983, Perkins 1984). The merozoite must then orient so that its apical end is against the membrane of the erythrocyte. A special junction is formed which then travels as a circle around the merozoite as it enters the forming parasitophorous vacuole of the red cell (Aikawa et al. 1978). For P. falciparum it has been shown that both ATP and spectrin of the host cell are essential to this process (Dluzewski et al. 1983, Olson & Kilejian 1982). At the same time the rhoptries or paired organelles of the merozoite produce a secretion of unknown nature that is presumed to be concerned in the endocytic invagination of the red cell membrane (Bannister 1977), and in its rapid transformation into a membrane of different character, the parasitophorous membrane (Langreth 1977).

It seemed reasonable to suppose that this complex process might serve as a trigger for the differentiation of the merozoite into a trophic parasite, which occurs soon after entry. That this is not the case was indicated by the following surprising observation. When schizonts of P. falciparum were incubated overnight at 37° in a medium of human erythrocyte extract supplemented with ATP and pyruvate, merozoites emerging from them developed into small extracellular rings (Trager & Lanners 1984). This did not occur in the absence of ATP and pyruvate. Electron micrographs of these early rings showed loss of characteristic merozoite organelles, as the inner membrane and the rhoptries, enlargement of the cell and assumption

NATO ASI Series, Vol. H11
Host-Parasite Cellular and Molecular Interactions
in Protozoal Infections. Edited by K.-P. Chang and D. Snary

of a ring or cup shape with elongation of the nucleus. A few of these forms showed tiny pigment grains, suggesting that feeding and digestion of hemoglobin had begun, even though the parasite was not surrounded by a parasitophorous membrane as it is when in its normal position in a red cell. Hence, it appears that differentiation of a merozoite into an early ring is triggered not by the complex process of entry into the erythrocyte but rather by the milieu in which the merozoite finds itself after entry.

In order to provide a cleaner system for quantitative work, it seemed desirable to attempt to do the experiments with an inoculum consisting only of recently emerged viable merozoites.

MATERIALS AND METHODS

Since malarial merozoites ordinarily have a short extracellular life (Bannister 1977), it is essential to have highly synchronized cultures that can be depended on to produce a harvest of segmenting schizonts at a predictable time. The methods for this continue to be varied slightly and still are subject to improvement. In recent work the following schedule has given fairly satisfactory results for both clones A-2 and D-4 of FCR-3/Gambia. Stock cultures are maintained in 2 flow vessels (Trager 1979). On a Wednesday morning harvested parasite suspension from such vessels is centrifuged, resuspended to a 25% hematocrit and 4 ml of the suspension is layered over 4 ml of 65% Percoll in the standard RPMI-1640-Hepes-10% serum culture medium (designated RPS) held in a 15 ml conical glass centrifuge tube. This is centrifuged 20 min at 2000 g. The brown band so obtained containing 50% or more schizonts is resuspended and washed by centrifugation for 10 min at 500 g in complete medium. The pellet of schizonts is then mixed with 5 ml of 8% erythrocyte suspension, placed in a petri dish and incubated for 3 hr in a candle jar at 37°C. This permits a 3 hr "window" for invasion of new cells. All the older parasites are then killed with sorbitol treatment (Lambros & Vanderberg 1979). The cells with 0 to 3 hr rings are returned to a flow vessel. Two days later they are subcultured and still maintained in flow vessels. On the 5th day (i.e. after 2½ cycles) when there is a high proportion of schizonts, harvests from the flow vessels are used to expand the cultures into 24 large petri dishes each with 10 ml of either 5% or 8% red cell suspension in RPS. These cultures are kept in candle jars and receive fresh medium manually at 9 AM and 5 PM of the following day.

One day still later, or a week from the day of synchronization, they are ready for use in preparation of a merozoite suspension. By 11 AM there are many segmenters and reinvasion is occurring. The contents of the plate are transferred to centrifuge tubes, centrifuged and made to 25% suspension in RPS. These are layered in 4 ml amounts over 4 ml of 65% or 60% Percoll in RPS in 15 ml centrifuge tubes and centrifuged 20 min at 2000 g. The brown bands obtained are resuspended in RP-15S (15% serum) at the rate of 2 bands per 5 ml. This is centrifuged 8 min at 500 g to pellet the schizonts. These are then resuspended in 7 to 10 x their volume of RP-15S and pooled in 1 to 1.5 ml amounts in small petri dishes. These are incubated in a candle jar at 37° for 1½ to 2 hr with swirling every 10 min (without removal from the candle jar). During this time the culture media to be tested are mixed and the culture vessels prepared.

Usually the contents of two petri dishes are pooled in one centrifuge tube and centrifuged 8' at 70 to 100 g. This suffices to sediment most of the schizonts leaving a cloudy grey-brown supernatant that contains merozoites and pigment clumps. This supernatant is removed and used to inoculate the experimental media, usually at the rate of 0.1 ml to 1 ml of medium.

For each experiment controls are prepared as a test of the invasive ability of the merozoites. These are best set up in wells of a 24-well Linbro plate. Each of 2 wells receives 5 µl of a 50% suspension of human red cells in RPS followed immediately by 0.05 ml of the merozoite suspension. The drops are well mixed with a Pasteur pipet and a small droplet of the mixture is used to prepare thin films stained with Giemsa stain. These serve for counts of the merozoites per 1000 red cells, from which the number of merozoites per ml can be estimated (see footnote (C) to Table 1). These slides are also used to determine the extent of contamination with schizonts or other stages. The plate is then incubated for ½ hr in a candle jar to allow for contact of merozoites and red cells. Then 0.5 ml of RP-15S is added to each well and the plate incubated overnight. The next day the contents of the wells are mixed and resuspended evenly and transferred to centrifuge tubes to centrifuge down the cells. These are used to prepare thin Giemsa-stained films from which the number of newly invaded rings per 10,000 cells can be counted. Division of this number by the number of merozoites per 10,000 cells on the 0-time slides gives the invasion rate.

The experimental tests are conveniently done either in the wells of a Linbro plate (0.5 ml) or in small petri dishes (1.5 or 2 ml), both held in a candle jar, or in 25 ml flasks (2 ml) that receive a current of a gas mix of 2% O_2, 5% CO_2, balance N_2. After incubation at 37° for 18-20 hr, or in some experiments for 6-7 hr, the tests are concluded by transferring this material to a centrifuge tube, centrifuging 15 min at 2000 g, removing most of the supernatant and resuspending the pellet evenly in a minimal volume. If relative counts are to be done, 5 µl of 50% red cell suspension can be added at this point (or alternatively it can be included in the original preparation). Thin films stained with Giemsa are then used to count the number of extracellular forms per 1000 red cells.

RESULTS AND DISCUSSION

Data on typical merozoite preparations obtained with 2 different clones (Trager et al. 1981) of the FCR-3/Gambia isolate are shown in Table 1. It is interesting to note that the 2 clones gave somewhat similar merozoite numbers, ranging from about 50 x 10^6 to 200 x 10^6 per ml. There was, however, a difference in the efficacy of sedimentation of the schizonts by the final low speed centrifugation. Whereas with clone A-2, 500 rpm usually reduced them below countable numbers, this was not the case with D-4 (see Exp. 84) and a speed of 700 rpm had to be used. This might be related to the knobby condition of the A-2 cells as compared with the knobless D-4 cell. Furthermore, it is worth noting that when schizonts were effectively removed, the invasive ability of the A-2 merozoites was 10 times higher than that of the D-4 merozoites, suggesting that the former may have a longer viability. When such merozoite suspensions were inoculated to the red cell extract medium, extracellular rings developed in appreciable numbers only if ATP and pyruvate were supplied (Table 2).

In the experiments shown in Table 1 the rings seen in the control slides used for the determination of invasion rate were all late rings and many showed fine pigment. This was different from most experiments with schizont inocula when the rings in control preparations tended to range from early to late. Such a difference would be expected since with the merozoite inocula invasion would have had to occur during the first half hour

TABLE 1
TYPICAL MEROZOITE SUSPENSIONS

Exp.No.	Parasite clone[a]	Schizonts separated by	Final centrifugation[b] (RPM)	MZ/ml[c] $\times 10^6$	Schizonts per 10,000 rbc[d]	Invasion rate[e]
E-48	A-2	Plasmagel	500	150	30	<1%
52	A-2	Plasmagel	500	24	0	4%
56	A-2	65% Percoll	500-1st ext.	80	0	3%
			500-2nd ext.	80	0	5%
57	A-2	65% Percoll	500	140	0	2%
59	A-2	65% Percoll	500	50	0	0
63	A-2	65% Percoll	500	60	8	5%
66	A-2	60% Percoll	500	76	0	2%
84	D-4	65% Percoll	500	200	100	8%
85	D-4	65% Percoll	1000	50	2	0.1%
86	D-4	65% Percoll	700	200	4	0.1%
87	D-4	65% Percoll	700	150	0	0.3%
89	D-4	60% Percoll	700	50	0	0.3%
90	D-4	60% Percoll	700	60	0	0.2%

a) Both are clones from the FCR-3/Gambia isolate: A-2 knobby, D-4 knobless

b) The schizont suspensions were incubated in a candle jar at 37° for 1½ hr and were then centrifuged 8 min at the indicated speed. The supernatant constituted the merozoite extract (MZ).

c) 0.05 ml MZ suspension was mixed with 5 µl 50% red cell suspension and stained slides prepared. Merozoites were counted per 1000 red cells and from this the no. per ml calculated. The accuracy of this method can be quite good as shown by the following result: In Exp. E-57 one well received 0.05 ml MZ suspension with 5 µl 50% red cells while another received 0.1 ml of the same suspension also with 5 µl 50% red cells. The counts turned out to be 270 and 560 merozoites respectively per 1000 red cells.

d) The count of contaminating schizonts seen in the slides used for the merozoite count.

e) The mixture of merozoites and red cells used for (c) was incubated ½ hr under culture conditions, then supplemented with 0.5 ml culture medium and incubated overnight. Slides were then prepared and the number of rings to 10,000 red cells counted. This number, divided by the number of merozoites per 10,000 red cells (from c) gave the invasion rate, the percent of merozoites that succeeded in invading a red cell and developing into a ring.

TABLE 2
EXTRACELLULAR DEVELOPMENT FROM MEROZOITE SUSPENSION (E-48)

Well No.	Medium[a]	No. of each stage[b] per 5000 red cells Merozoite	Extracell. ring[c]	Intracell. ring	Schizont[d]
A-5	RCM	40	12	2	2
A-6	RCM	37	14	0	4
B-1	RCM+ATP+Py	43	72	2	4
B-2	RCM+ATP+Py	22	78	2	6

a) RCM is the complete red cell extract medium. ATP as the dipotassium salt was supplied at 1.6 mM and sodium pyruvate (Py) at 3.6 mM.

b) Counts after 18 hr. Abnormally stained clearly degenerate organisms were not counted.

c) Not all of these were ring-shaped, but all were larger than merozoites with relatively more cytoplasm.

d) These were contaminating schizonts from the inoculum (see Table 1).

or so after mixing, resulting in rings about 18-20 hr old the next day. It accordingly seemed desirable to find out whether extracellular developing forms would be present earlier than on the following day. In several experiments sampling was done at 7 hr, as well as the next day. In E-57, for example, 2 wells with RCM + ATP + Py gave counts at 7 hrs of 104 and 170 extracellular rings/1,000 rbc as compared with counts in corresponding wells sampled at 18 hr of only 35 extracellular rings/1,000 rbc. These were derived from a merozoite inoculum of 800 per 1,000 rbc, so that at 7 hr 1/5 to 1/8 of the merozoites had begun to develop. Electron micrographs of parasites fixed at this time in another similar experiment (E-59) showed all gradations between merozoites and the early extracellular rings. The lower counts at 18 hr as compared to 7 hr show that a considerable proportion of the merozoites that initiate development are not able to maintain their viability.

Despite this result, it seemed preferable to continue to make most of the observations after about 20 hr, especially since one ultimate objective is more extensive extracellular development. Some experiments have shown a troubling variability in counts at this time, as illustrated in Table 3. The results in Dishes 4 and 6 and in Flask 9 suggest a favorable effect of CoA, but clearly more consistent results are needed.

Uptake of rhodamine 123 indicates the presence of a membrane potential (Tanabe 1983). Accordingly this can be used as an indicator of parasite viability. If the dye is used at the relatively high concentration of 10 µg/ml the whole parasites become brightly fluorescent when viewed by epifluorescence in a suitable microscope with the appropriate filters. At about 1/10 of this concentration the fluorescence tends to be localized in the parasites' mitochondria (Divo et al. 1985). In preliminary experiments rhodamine 123 was added at 10 µg/ml (as 10 µl of a solution with 1 mg/ml) after 18 hr incubation and the preparation returned to culture conditions for an additional 30 to 60 min. The material was then centrifuged and

TABLE 3

COUNTS OF THE NUMBERS OF DEVELOPING EXTRACELLULAR FORMS AFTER 20 HR AT 37° UNDER THE CONDITIONS AND IN THE MEDIA INDICATED (EXP. E-66)

Conditions	No.	Medium[a]	Extracellular forms/1000 rbc
Petri dishes in candle-jar	1	RCM+ATP+Py	22
1.5 ml medium	2	RCM+ATP+Py	28
0.15 ml MZ susp.	3	As above +CoA[b]	18
5 µl 50% rbc	4	As above +CoA	66
	5	Like 3,4 + DTT[c]	17
	6	Like 3,4 + DTT	72
Same as above but held	7	Same as 1,2	10
in flasks with a gas mix	8	Same as 1,2	38
(5% CO_2, 2% O_2) passing through	9	Same as 3,4	61 (independent repeat:78)
	10	Same as 3,4	27 (independent repeat:32)
	11	Same as 5,6	24
	12	Same as 5,6	24

a) See footnote (a) of Table 2.

b) Coenzyme A (CoA) added at 0.06 mM

c) Dithiothreitol added at 1 mM

washed once in a high-potassium buffer, and wet mounts were examined by epifluorescence. In experiment E-86 with a merozoite inoculum (see Table 1) rhodamine was added in this way to 3 culture dishes, one with only the RCM medium, one supplemented with ATP + Py and the third with ADP + Py. Only the latter two showed small round fluorescent bodies. Similar observations have been made in subsequent experiments with merozoite inocula. In the RCM medium either no fluorescent bodies were seen or else a few large schizonts, from contaminating material in the merozoite suspension. In the presence of the ATP supplement small round brightly fluorescent forms were seen. Similarly, in an experiment using a schizont inoculum, addition of rhodamine on the following day revealed in the RCM medium numerous bright intracellular schizonts and a few small fluorescent bodies, whereas in RCM + ATP + Py these small fluorescent bodies were numerous. Phase contrast examination and stained slides of corresponding material all support the inference that these small bodies that take up rhodamine correspond to the extracellular developing forms. We may also infer that these forms are still viable.

There are now several lines of evidence that partial extracellular development of erythrocytic stages of malarial parasites is possible. In work with the bird malaria P. lophurae, trophozoites of this species removed from their host erythrocytes with a hemolytic antiserum and complement, developed under appropriate conditions into schizonts in a 20 hr period of incubation at 40°C (Trager 1971, Trager & Jensen 1980). Essentia to this development was a duck erythrocyte extract preparation in a nutrient medium of high potassium content and supplemented with certain cofactors. Most striking was the effect of ATP with pyruvate (at 2 and 5 mM respectively) and CoA (at 0.05 mM). The requirement for CoA pointed to a major biosynthetic lesion (Trager & Brohn 1975) also indicated by the lack of the enzymes for biosynthesis of CoA in P. lophurae (Brohn & Trager 1975). Viable extracellular P. lophurae with normal ultrastructure were seen even after 3 days in vitro (Langreth & Trager 1973).

Most recently, erythrocytic P. knowlesi removed from their host monkey erythrocytes by nitrogen decompression have been found to develop from late trophozoites-early schizonts with 1 to 4 nuclei into schizonts with 8 nuclei after 4 hr incubation at 37°C in an "artificial intracellular medium" (Nillni et al. 1985). This medium contained Na^+ at 20 mM, K^+ at 120 mM, no Ca^{2+}, other ingredients as in Dulbecco's modified Eagle medium, with glucose at 10 mM and 10% heat inactivated dialyzed calf serum, and had a pH of 6.7. Both the relatively low pH and the high potassium were important to this 4 hr period of development. Even after 4 hr about 80% of the parasites carried on protein synthesis at rates comparable to those of infected erythrocytes. The interesting observation was made that certain proteins synthesized by the isolated parasites were secreted into the culture medium in keeping with other lines of evidence that some parasite proteins are released into the infected host cell (Aley et al. 1984, Kilejian 1979, Ginsberg et al. 1983).

It is important to note that in the work with both P. lophurae trophozoites and P. knowlesi schizonts the parasites removed from their host cells were surrounded by the parasitophorous membrane. This is closely adherent to the plasma membrane of the parasite and so far it has not been possible to obtain these stages in a viable condition but divested of the parasitophorous membrane. If this membrane plays an essential role, then further development of the isolated merozoites of P. falciparum cannot be expected unless such a membrane is provided. Of great interest would be comparative experiments done with very early rings that have been allowed to develop intracellularly for a few hours and have then been removed from their host cells with hemolytic antiserum or by nitrogen decompression.

Also of great interest would be studies on extracellular development of a species of Babesia since these organisms, though at first surrounded by an invaginated red cell membrane, rapidly lose it and then develop directly in the cytoplasm of the erythrocyte (Rudzinska et al. 1976). Babesia bovis would be especially suitable for such work since it grows readily in bovine red cells in culture and since its merozoites are capable of longer extracellular life than those of Plasmodium sp. (Levy et al. 1981).

The effect of ATP on the initial extracellular differentiation of P. falciparum merozoites suggests that "triggering" substances might be as important as nutrient requirements in promoting further development. Dluzewski et al. (1986) in experiments with resealed ghosts report that the protease inhibitor leupeptin not only inhibited invasion but also inhibited intracellular development of trophozoites of P. falciparum. They have also suggested (Dluzewski et al. 1983) that the need for ATP in invasion may be related to the activity of a protein kinase. A plasmodium protein kinase has been demonstrated in the rodent malaria parasites P. berghei and P. chabaudi (Wiser et al. 1983). Protein kinase activators such as the phorbol esters (May et al. 1986) would be worth investigating in relation to their possible effects on initial extracellular development.

Such studies might also have a bearing on what it is that makes a small percentage of erythrocytic merozoites develop into gametocytes rather than asexual forms. Since gametocytes of both sexes arise from cloned parasites in vitro as well as in vivo (Downs 1947, Bhasin & Trager 1984), merozoites that form gametocytes must have the same genetic endowment as those that form asexual schizonts. Some subtle environmental factor present in a small proportion of host erythrocytes could be responsible for directing development into the sexual forms. In line with this possibility is the recent finding of a stimulating effect of cortisone on gametocyte formation in vitro (Maswoswe et al. 1985).

REFERENCES

Aikawa M, Miller LH (1983) Structural alteration of the erythrocyte membrane during malarial parasite invasion and intraerythrocyte development. Ciba Found. Symp. 94:45-63

Aikawa M, Miller LH, Johnson J, Rabbege J (1978) Erythrocytic entry by malarial parasites: A moving junction between erythrocyte and parasite. J. Cell Biol. 77:72-82

Aley SB, Barnwell JW, Daniel W, Howard RJ (1984) Identification of parasite proteins in a membrane preparation enriched for the surface membrane of erythrocytes infected with Plasmodium knowlesi. Mol. Biochem. Parasitol. 12:69-84

Bannister LH (1977) Invasion of red cells by Plasmodium. In: Taylor AER, Mullers R (eds) Parasite invasion. Blackwell, Oxford, p 27-55

Bhasin VK, Trager W (1984) Gametocyte-forming and non-gametocyte-forming clones of Plasmodium falciparum. Am. J. Trop. Med. Hyg. 33:534-537

Breuer WV (1985) How the malarial parasite invades its host cell, the erythrocyte. Intl. Rev. Cytol. 96:191-238

Brohn FH, Trager W (1975) Coenzyme A requirement of malaria parasites: Enzymes of coenzyme A biosynthesis in normal duck erythrocytes and erythrocytes infected with Plasmodium lophurae. Proc. Natl. Acad. Sci. USA 72:2456-2458

Divo AA, Geary TG, Jensen JB, Ginsburg H (1985) The mitochondrion of Plasmodium falciparum visualized by rhodamine 123 fluorescence. J. Protozool. 32:442-446

Dluzewski AR, Rangachari K, Wilson RJM, Gratzer WB (1986) Protease inhibitors and their effects on invasion and development of Plasmodium falciparum. Abtract 27, Society of Protozooloy

Dluzewski AR, Rangachari K, Wilson RJM, Gratzer WB (1983) A cytoplasmic requirement of red cells for invasion by malarial parasites. Mol. Biochem. Parasitol. 9:145-160

Downs WG (1947) Infection of chicks with single parasites of Plasmodium gallinaceum Brumpt. Am. J. Trop. Med. 46:41-44

Ginsburg H, Krugliak M, Eidelman O, Cabantchik ZI (1983) New permeability pathways induced in membranes of Plasmodium falciparum infected erythrocytes. Mol. Biochem. Parasitol. 8:177-190

Kilejian A (1979) Characterization of a protein correlated with the production of knob-like protrusions on membranes of erythrocytes infected with Plasmodium falciparum. Proc. Natl. Acad. Sci. USA 76:4650-4653

Lambros C, Vanderberg JP (1979) Synchronization of Plasmodium falciparum erythrocytic stages in culture. J. Parasitol. 65:418-420

Langreth SG (1977) Electron microscope cytochemistry of host-parasite membrane interactions in malaria. Bull. WHO 55:171-178

Langreth SG, Trager W (1973) Fine structure of the malaria parasite Plasmodium lophurae developing extracellularly in vitro. J. Protozool. 20:606-613

Levy MG, Erp E, Ristic M (1981) Cultivation of Babesia. In: Ristic M, Kreier JP (eds) Babesiosis, Academic Press, NY, p 207-223

Maswoswe SM, Peters W, Warhurst DC (1985) Corticosteroid stimulation of the growth of Plasmodium falciparum gametocytes in vitro. Ann. Trop. Med. Parasitol. 79:607-616

May WS, Lapetina EG, Cuatrecases P (1986) Intracellular activation of protein kinase C and regulation of the surface transferrin receptor by diacylglycerol is a spontaneously reversible process that is associated with rapid formation of phosphatidic acid. Proc. Natl. Acad. Sci. USA 83:1281-1284

Nillni EA, Schmidt-Ullrich R, Mikkelsen RB, Wallach DFH (1985) Extracellular development of Plasmodium knowlesi erythrocytic stages in an artificial intracellular medium. Mol. Biochem. Parasitol. 17:219-237

Olson JA, Kilejian A (1982) Involvement of spectrin and ATP in infection of resealed erythrocyte ghosts by the human malarial parasite, Plasmodium falciparum. J. Cell Biol. 95:757-762

Pasvol G, Jungery M (1983) Glycophorins and red cell invasion by Plasmodium falciparum. Ciba Found. Symp. 94:174-195

Perkins ME (1984) Surface proteins of Plasmodium falciparum merozoites binding to the erythrocyte receptor, glycophorin. J. Exp. Med. 160: 788-798

Tanabe K (1983) Staining of Plasmodium yoelii-infected mouse erythrocytes with the fluorescent dye rhodamine 123[1]. J. Protozool. 30:707-710

Trager W (1971) Malaria parasites (Plasmodium lophurae) developing extracellularly in vitro: Incorporation of labeled precursors. J. Protozool. 18:392-399

Trager W (1979) Plasmodium falciparum in culture: Improved continuous flow method. J. Protozool. 26:125-129

Trager W, Brohn FH (1975) Coenzyme A requirement of malaria parasites: Effects of coenzyme A precursors on extracellular development *in vitro* of *Plasmodium lophurae*. Proc. Natl. Acad. Sci. USA 72:1834-1837

Trager W, Jensen JB (1976) Human malaria parasites in continuous culture. Science 193:673-675

Trager W, Jensen JB (1980) Cultivation of erythrocytic and exoerythrocytic stages of plasmodia. In: Kreier JP (ed) Malaria, Academic Press, NY, Vol. 2, p 271-313

Trager W, Lanners HN (1984) Initial extracellular development *in vitro* of merozoites of *Plasmodium falciparum*. J. Protozool. 31:562-567

Trager W, Tershakovec M, Lyandvert L, Stanley H, Lanners N, Gubert E (1981) Clones of the malaria parasite *Plasmodium falciparum* obtained by microscopic selection: Their characterization with regard to knobs, chloroquine sensitivity, and formation of gametocytes. Proc. Natl. Acad. Sci. USA 78:6527-6530

Wiser MF, Eaton JW, Sheppard JR (1983) A *Plasmodium* protein kinase that is developmentally regulated, stimulated by spermine, and inhibited by quercetin. J. Cell. Biochem. 21:305-314

ACKNOWLEDGEMENTS

This work was supported by a grant from the National Institute of Allergy and Infectious Diseases and also received assistance through a Biomedical Research Support Grant from the National Institute of Health, U.S. Public Health Service.

THE RELATIVE ROLES OF N- AND O-LINKED CARBOHYDRATE IN THE INVASION OF HUMAN RED CELLS BY MEROZOITES OF PLASMODIUM FALCIPARUM

G. Pasvol,[1] C. Hodson,[2] M. J. A. Tanner,[3] and C. I. Newbold[1]

[1]Tropical Medicine Unit, Nuffield Department of Clinical Medicine, John Radcliffe Hospital, Oxford, OX3 9DU, UK. [2]National Blood Transfusion Service, Lancaster Centre Quernmore Rd., Lancaster, LA1 3JP. [3]Department of Biochemistry, University of Bristol, Medical School, University Walk Bristol, BS8 1TD.

INTRODUCTION

Identification of the molecular determinants involved when merozoites of Plasmodium falciparum invade human red cells remains extremely controversial. Thus the relative roles of the major red cell sialoglycoproteins (the glycophorins) and the major anion transport protein (band 3), if they do play a role in invasion, remain undecided. Equally, the identity of parasite molecules to which putative red cell receptors bind is unclear. The bulk of evidence indicates that red cell sialoglycoproteins (α, β, γ and δ - see Anstee, 1981) might all have a role to play in merozoite invasion (for a review see Pasvol, 1984). However the nature of the specific components involved in these molecules is unclear. Thus, for example, whilst the outer sugar rich T1 peptide of α has been found to inhibit invasion in one laboratory (Perkins, 1984), this has not been confirmed by another (Breuer et al, 1983). Moreover, even where there is experimental agreement, interpretations differ. For example, although it is agreed that the T6 peptide of α (which includes the transmembrane and an outer portion of the molecule close to the membrane) inhibits invasion, it has been argued that the underlying mechanism may be via a non-specific hydrophobic, rather than a specific, interaction. It is quite conceivable that even trypsin treatment, which cleaves α, β and γ, but not δ (and which in all hands, though to varying degrees, reduces invasion) may be acting on another as yet unidentified surface molecule. Thus at present, no portion of the protein backbone of the glycophorins has been identified as being a specific red cell 'receptor' for merozoites.

NATO ASI Series, Vol. H11
Host-Parasite Cellular and Molecular Interactions
in Protozoal Infections. Edited by K.-P. Chang and D. Snary

The Possible Role of Sugars

Attention has therefore been diverted to the possible specific role that sugars on both the sialoglycoproteins and band 3 might play in invasion. A major pointer to their importance, and one on which all are agreed, is that neuraminidase treatment of human red cells reduces invasion by P. falciparum (Miller et al, 1977). However, once again, the mechanism by which removal of sialic acid acts has not been resolved. Whilst some believe that this is purely a charge effect (Friedman et al, 1984), others implicate a specific role for sialic acid either on its own or in conjunction with other sugars (see Pasvol, 1984). Very recently the role of sialic acid has been taken a step further by data which suggest that parasite strains may differ in their requirement of red cell surface sialic acid for successful invasion (Mitchell et al, 1986). The extreme resistance to invasion of a particular type of red cell (Tn) which retains 50% of normal sialic acid content points to a specific, rather than a charge effect being involved (Pasvol et al, 1982). These Tn cells, because of an acquired defect in galactosyl transferase, have O-linked sugars which lack the outer galactose and sialic acid residues. Moreover Cad cells in which there is an extra N-acetyl galactose (GalNAc) residue on each O-linked sugar have also been found to resist invasion despite a full complement of sialic acid (Cartron et al, 1983).

Finally, the most controversial aspect in terms of the role of sugars in invasion has been the effect of N-acetyl glucosamine (GlcNAc), because of the possibility of toxicity of this sugar on parasite maturation. GlcNAc was first said to inhibit invasion without toxicity (Weiss et al, 1981). This sugar subsequently appeared to be toxic but only at concentrations higher than that at which it inhibited invasion (Jungery et al, 1983). When it was coupled to bovine serum albumin to avoid the problem of toxicity and was still inhibitory at much lower concentrations, it was argued that this might be due to the hydrophobic nature of the linker molecule (Hermentin et al, 1984). A further paper from another group supported a specific role for this sugar (Gupta et al, 1985) whilst another refuted it (Howard et al, 1982). It would thus appear that the role of carbohydrate in the invasion of red cells by P. falciparum remains unresolved.

METHODS AND RESULTS

The Effect of Endoglycosidases

Because GlcNAc is located primarily on the large complex N-linked sugars of the glycophorins and band 3, we have turned our attention to these large and prominent sugars. Enzymes which specifically cleave these sugars from red cell glycopeptides have recently become available. The endoglycosidases D and H used in previous experiments by Howard et al (1982) to investigate their effect on invasion leave these complex N-linked sugars on red cells intact. We have used three endoglycosidases, namely endoglycosidase F from Flavobacterium meningosepticum (endo F) which cleaves both high mannose and complex glycoproteins (Elder and Alexander, 1982) and two forms of endo-β-galactosidase one from Flavobacterium keratolyticus (endo-β1) (Kitamikado et al, 1981) and another from Bacteroides fragilis (endo-β2) (Scudder et al, 1983). Endo-β-galactosidase cleaves sugar chains with repeating units of N-acetyllactosamine such as those comprising the major portion of the band 3 oligosaccharide.

Since extreme variability has been found in the ability of endoglycosidases to cleave different complex sugars, it was first necessary to establish both whether these enzymes would act on glycophorin and band 3 sugars, and to devise suitable conditions for treatment of red cells which would leave them intact for invasion experiments. Both 1% mercaptoethanol and non-ionic detergent which facilitate cleavage from other glycoproteins were for obvious reasons unsuitable for red cells. Moreover, prolonged incubation in 0.1 M phosphate buffer at pH 6.0 was found to reduce invasion considerably in sham-treated controls. As a compromise, the treatment buffer was supplemented with glucose (4g/l), the concentration of enzyme increased appropriately and incubation carried out for 4 hours at 37°C in the presence of 50 mM EDTA and 1 mM PMSF. Between 5 and 10 μl endo F was used to treat 500 μl of packed red cells (Tanner et al, 1986). Similar conditions were used for treatment with endo-β1 using 250 μg enzyme per 100 μl packed red cells. Treatment of red cells with endo-β2 was carried out with crude bacterial culture supernatant initially in the absence of EDTA or PMSF. The effects of these enzymes on red cell membrane proteins were monitored by surface labelling of proteins with ^{125}I by the iodogen method (Markwell and Fox, 1978). It can be seen that both the dimeric and monomeric forms of α were reduced in molecular weight by enzyme treatment (Fig 1). Prolonged exposure of the film failed to demonstrate any detectable parent molecule. Treatment with endo-β1 and β2 showed condensation of band 3 into a tighter band and minimal advance of the front of the band to a lower molecular weight. Treatment with endo-β2 in addition showed disappearance of the a dimer and evidence of possible endo F activity as well indicating protease contamination.

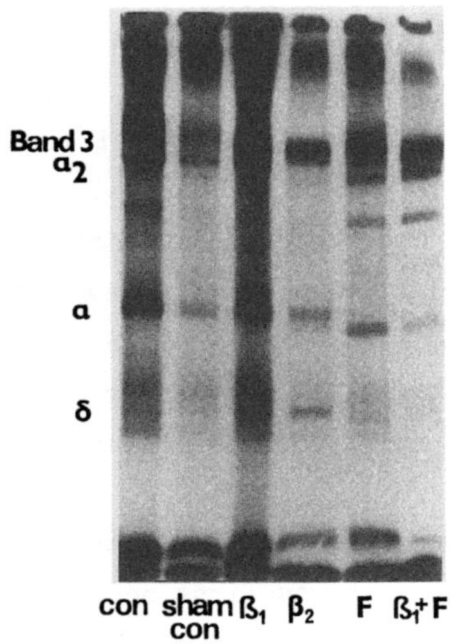

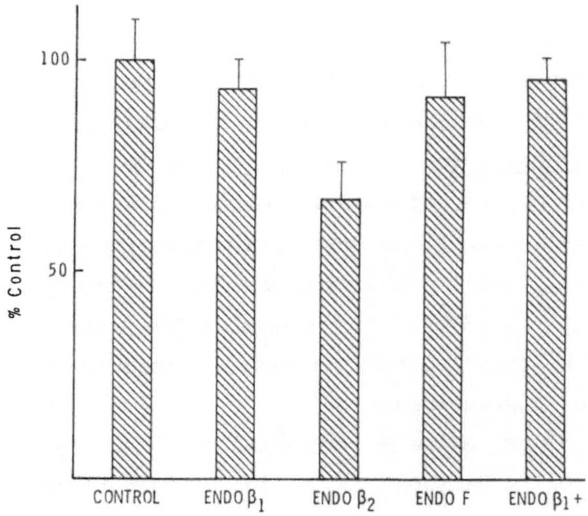

FIGURE 1. SDS PAGE of endoglycosidase treated red cells surface labelled with ^{125}I. Con=control; sham con=sham control, β_1 and β_2=endo β galactosidases 1 and 2 (see text); F=endoglycosidase F; B_1+F=combination of endo-β-galactosidase 1 and endoglycosidase F.

FIGURE 2. Invasion by P. falciparum of red cells treated with endoglycosidases as in figure 1. Results represent the mean of results of 3 experiments ± standard deviation (invasion of controls in the 3 experiments were 10.4, 10.8 and 14.9 ring-form parasites per 100 red cells).

When invasion by merozoites of P. falciparum (Strain T996) was assessed by standard methods it was found that there was no significant reduction in invasion of the treated cells when compared to the sham-treated controls. (Fig. 2). The exception to this was cells treated with endo-β_2 where a significant reduction (33%) in invasion was detected. However in this case ^{125}I surface labelling indicated the possible presence of contaminating protease. Subsequently when the same enzyme was partially purified and the digestion carried out in the presence of EDTA and PMSF no reduction in invasion was seen (G. Pasvol unpublished). Furthermore no invasion inhibition was detected with the combination of endo-β_1 and endo F where cleavage of the N-acetyllactosamine residues might enhance subsequent cleavage by endo F of the large band 3 oligosaccharide. It thus appeared unlikely that these complex sugars on the glycophorins containing GlcNAc glucosamine might be a major target for invading merozoites.

Tn red cells. These cells enabled us to examine the role of the O-linked sugars in merozoite invasion. We obtained Tn red cells from three donors (Bain, Dal. and Sch.). (Results on invasion of cells from Bain. have previously been reported, Pasvol et al, 1982). On this occasion cells expressing the Tn determinant (specified by GalNAc) were separated from non-expressing cells from the same donor using agglutination by the GalNAc-specific lectin from Helix pomatia (Hp). Thus in addition to the total cell population, Hp +ve and Hp -ve cells from the same donor could be tested. Cells from a blood O individual and a blood group A individual (Find.) (expressing GalNAc) served as controls. It can be seen in the three Tn donors that was considerable heterogeneity in expressing the Tn antigen, sialic acid content, agglutination with a polyclonal Tn antibody and a monoclonal antibody R1.3 (Table 1).

TABLE 1

SOME CHARACTERISTICS OF Tn CELLS

	Control (gp. O)	Control Find. (gp. A)	Bain.	Dal.	Sch.
% Hp +ve Cells	0	~100	95	35	92
Sialic Acid (% Normal)	100	100	40	28	74
Hp +ve Cells					
Anti Tn[a]	-ve	-ve	++++	++++	+
R1.3[a,b]	C	C	-ve	-ve	C
Intact tetra-saccharides relative to control (%)[c]	100	100	15	0	60

[a] Agglutination results with a rabbit anti-Tn antibody. ++++=Strong +=weak C=complete.

[b] R1.3 is a monoclonal antibody which recognises the outer portion of α and δ and is sialic acid dependant.

[c] These figures are approximate and speculative, assuming that 70% of total red cell sialic acid is located on α and δ.

Surface labelling of these cells with ^{125}I showed a marked difference in the molecular mass of $\alpha 2$ and δ of Bain., some blurring of $\alpha 2$ and δ of Dal. whereas only a minimal shift in the molecular weight of α of Sch. was detected (Fig. 3).

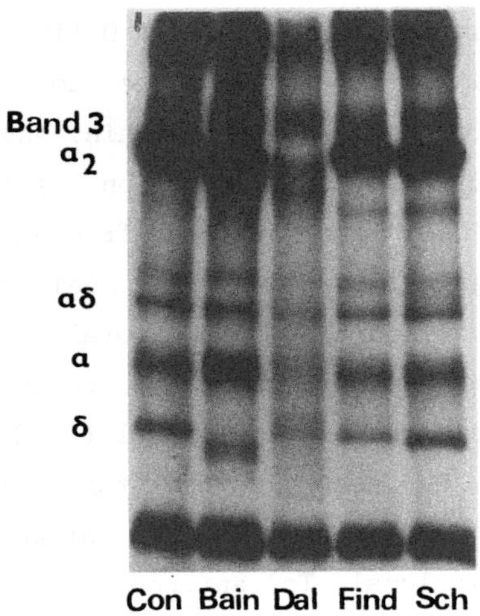

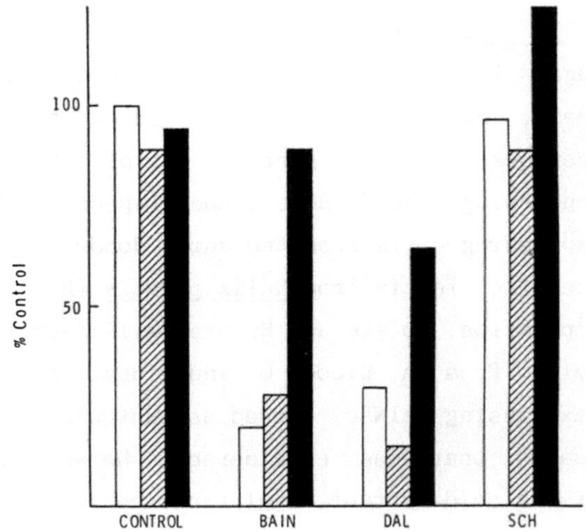

FIGURE 3. SDS PAGE of various Tn and control cells surface labelled with ^{125}I. Con=control; Bain., Dal. and Sch. are red cells from three different Tn donors; Find. is a control ABO Blood group A donor.

FIGURE 4. Invasion by P. falciparum of red cells from various Tn donors. Open bars represent the total red cell population, hatched bars the Helix pomatia (Hp) +ve population (i.e. Tn cells) and the solid bars represent Hp -ve population (see text). Hp -ve in control (Find.) represents a very small number of unagglutinated cells. The results are the mean of two counts expressed as a percentage of control (in this experiment 11.9 ring-form parasites per 100 red cells).

However, the most dramatic results were seen with the relative rates of invasion of these cells (Fig. 4). Parasites readily invaded total and Hp +ve control blood group A cells (Find.), whereas invasion of Bain. and Dal. cells (total or Hp +ve) were dramatically reduced. Cells not agglutinated by Hp from Bain. and Dal. (i.e. Tn -ve) showed a return towards control rates of invasion. However most striking were the cells from Sch., 92% of which are Tn +ve, and yet when Tn positive cells were lectin purified, their invasion rates were normal.

It would thus appear that Tn positivity in itself is not sufficient to render cells resistant to invasion. When used in invasion experiments both the source and further characteristics of Tn cells need to be determined before cells used in different laboratories can be compared. This view is strengthened by recent results from our laboratory where cells which have clearly been serologically typed as Cad positive were found to be fully susceptible to invasion (G. Pasvol unpublished) in contrast to the data presented by Cartron et al (1983) who found such Cad cells from another donor to be resistant. The difference in the two results may be because of a difference in the location of the Cad determinant in the two cell types. For example, it may be present on the glycophorins in the cells described by Cartron and on another as yet undefined molecule (e.g. glycolipid) in our study. It is thus evident that red cells assigned to a particular serotype might represent heterogenous populations. Detailed biochemical studies are therefore required before a precise molecular defect can be incriminated in resistance to invasion in these cells.

DISCUSSION

Our data have shown that removal of the N-linked sugars on the glycophorins by endoglycosidase F leaves red cells fully susceptible to invasion by P. falciparum. Successful removal of these sugars was monitored by ^{125}I surface labelling and together with the invasion data would suggest that these sugars on glycophorin play no specific role in invasion. Treatment of red cells with endo-β-galactosidase considerably 'tightens' the diffuse band of band 3 on PAGE and was also found to have no effect on invasion. Since no distinctive reduction of band 3 to a lower molecular weight was detected with endo-β it was not possible to monitor successful removal of the sugar as easily as with endo F. Nevertheless extensive treatment with the enzyme was found to have no effect. Recent data using the erythroglycan isolated from band 3 has suggested that this sugar might act as one of the 'receptors' for P. falciparum during invasion (Friedman et al, 1985). However, previous studies where reagents were purified and used in solution in different laboratories, have often revealed extremely variable results. This may be because of contamination (e.g. with proteases), non-specific interactions or for other unexplained reasons. Since endo-β removes only the outer portion of the single large N-linked sugar on band 3, the inner core portion containing most of the GlcNAc remains intact and thus a role for this portion of the molecule cannot be excluded.

Overall the data presented here do not support a major role for the N-linked sugars in the invasion of red cells by P. falciparum.

In contrast, modification of the O-linked sugars on glycophorin produced a dramatic effect on invasion. Current data suggest that there are 15 O-linked tetrasaccharides on α, 11 on δ and a variable number on β and γ. Tn red cells occur in an acquired condition where, because of a relative deficiency of galactosyl transferase, galactose is not added to the N-acetyl galactosamine residue. N-linked sugars and glycolipids remain unaffected. However the Tn phenotype appears to be an extremely heterogeneous condition as the percentage cells expressing Tn, the sialic acid content of these cells compared to normal, and a number of other parameters vary from one Tn individual to the next. None of these characteristics (particularly the percentage cells expressing Tn) appears to be correlated with invasion. The role of the O-linked saccharides is also supported by very recent data indicating that the O-linked tetrasaccharide isolated from α at a concentration as low as 1 mM was the most potent carbohydrate inhibitor of invasion (Hermentin et al, 1986).

How can all these data be reconciled? This must remain purely speculative, particularly at the present time when the role played by almost every determinant which has been proposed as being a specific receptor for P. falciparum on red cells has been questioned. Moreover the role of carbohydrate in itself need not be specific. Decreased invasion of red cells with naturally-occurring or artificially-induced alterations in carbohydrate might reflect an underlying change in protein tertiary structure. However if a specific interaction is implicated, it is possible that a lectin-like molecule on the merozoite surface recognises a group of sugars on the red cell in a manner analagous to the recognition of a cluster of sugars by wheat germ agglutinin (Bhavanandan and Katlic, 1979) or phytohaemagglutin E (Yamashita et al, 1983). In these models the lectin recognises a cluster of particular sugar residues, but binding can be inhibited by a number of related sugar structures including monosaccharides. One might speculate that the cluster of tetrasaccharides on amino acid positions 10 to 15 on α and δ would be strong candidates. Such speculations await more detailed analyses of the number and positions of intact or defective tetrasaccharides present on the glycophorins of the various Tn phenotypes. These might answer the critical questions as to the nature of the specificity of merozoite-red cell interaction.

ACKNOWLEDGEMENTS

We would like to thank Mr L McShine at the Stichting Rode Kruis Bloedbank, Groningen-Drenth, Netherlands, for the Sch. cells, and Mr G Ingles at the Glasgow and West of Scotland Transfusion Centre for the Dal cells. We would also like to thank Miss H Goodburn for her technical, and Miss P Wright for her secretarial assistance. GP is a Wellcome Senior Lecturer and he, CIN and MJAJ are supported by the Wellcome Trust and GP and CIN by the Rockefeller Foundation. We would also like to thank Drs G Luzzi, C F Gilks, K Marsh and D Lee for their critical comments.

REFERENCES

Anstee DJ (1981) The blood group MNSs-active sialoglycoproteins. Semin Hematol 18:13-31.

Bhavanandan VP, Katlic AW (1979) The interaction of wheat germ agglutinin and sialoglycoproteins.The role of sialic acid. J Biol Chem 254:4000-4008.

Breuer WV, Kahane I, Baruch D, Ginsburg H, Cabantchik ZI (1983) Role of internal domains of glycophorin in Plasmodium falciparum invasion of human erythrocytes. Infect Immun 42:133-140.

Cartron JP, Luilier OPM, Soulier JP (1983) Susceptibility to invasion by Plasmodium falciparum of some human erythrocytes carrying rare blood group antigens. Br J Haematol 55:639-647.

Elder JH, Alexander S (1982) Endo-β-N-acetylglucosaminidase F: Endoglycosidase from Flavobacterium meningosepticum that cleaves both high-mannose and complex glycoproteins. Proc Natl Acad Sci USA 79:4540-4544.

Friedman MJ, Blankenberg T, Sensabaugh G, Tenforde TS (1984) Recognition and invasion of human erythrocytes by malarial parasites: Contribution of sialoglycoproteins to attachment and host specificity. J Cell Biol 98:1672-1677.

Friedman MJ, Fukuda M, Laine RA (1985) Evidence for a malarial parasite interaction site on the major transmembrane protein of the human erythrocyte. Science 228:75-77.

Gupta SK, Schulman S, Vanderberg JP (1985) Stage-dependent toxicity of N-acetyl-glucosamine to Plasmodium falciparum. J Protozool 32:91-95.

Hermentin P, Paulsen H, Kolar C, Enders B (1984) Plasmodium falciparum: Carbohydrates as receptor sites of invasion. Exp Parasitol 58:290-306.

Hermentin P, Neunziger G, Paulsen H, Kolar C, Seiler FR, Dahr W, Enders B (1986) Toxicity of "hydrophobic groupings" and the role of carbohydrates in Plasmodium falciparum infection. Behring Inst Mitt 80:64-73.

Howard RJ, Haynes JD, McGinniss MH, Miller LH (1982) Studies on the role of red blood cell glycoproteins as receptors for invasion by Plasmodium falciparum merozoites. Mol Biochem Parasitol 6:303-315.

Jungery M, Pasvol G, Newbold CI, Weatherall DJ (1983) A lectin-like receptor is involved in invasion of erythrocytes by Plasmodium falciparum. Proc Natl Acad Sci USA 80:1018-1022.

Kitamikado M, Ito M, Li Y-T (1981) Isolation and characterization of a keratan sulfate-degrading endo-β-galactosidase from Flavobacterium keratolyticus. J Biol Chem 256:3906-3909.

Markwell MAK, Fox CF (1978) Surface-specific iodination of membrane proteins of viruses and eucaryotic cells using 1,3,4,6,-tetrachloro-3α, 6α-diphenylglycoluril. Biochemistry 17:4807-4817.

Miller LH, Haynes JD, McAuliffe FM, Shiroishi T, Durocher JR, McGinniss MH (1977) Evidence for differences in erythrocyte surface receptors for the malarial parasites Plasmodium falciparum and Plasmodium knowlesi. J Exp Med 146:277-281.

Mitchell GH, Hadley TJ, McGinniss MH, Klotz FW, Miller LH (1986) Invasion of erythrocytes by Plasmodium falciparum malaria parasites: evidence for receptor heterogeneity and two receptors. Blood 67:1519-1521.

Pasvol G (1984) Receptors on red cells for Plasmodium falciparum and their interaction with merozoites. Philos Trans R Soc Lond [Biol] 307:189-200.

Pasvol G, Jungery M, Weatherall DJ, Parsons SF, Anstee DJ, Tanner MJA (1982) Glycophorin as a possible receptor for Plasmodium falciparum. Lancet 2:947-950.

Perkins M (1984) Binding of glycophorins to Plasmodium falciparum merozoites. Mol Biochem Parasitol 10:67-78.

Scudder P, Uemura K, Dolby J, Fukuda MN, Feizi T (1983) Isolation and characterization of an endo-β-galactosidase from Bacteroides fragilis. Biochem J 213:485-494.

Tanner, JA, Anstee, DJ, Mallinson, G, Ridgwell, K, Martin, PG, Avent, ND, Parsons, SF. "Effect of endoglycosidase F preparations on the surface components of the human erythrocyte". Carbohydrate Research, 1986. (In Press).

Weiss MM, Oppenheim JD, Vanderberg JP (1981) Plasmodium falciparum: Assay in vitro for inhibitors of merozoite penetration of erythrocytes. Exp Parasitol 51:400-407.

Yamashita K, Hitoi A, Kobata A (1983) Structural determinants of Phaseolus vulgaris erythroagglutinating lectin for oligosaccharides. J Biol Chem 258:14753-14755.

ANTIGENS OF PLASMODIUM FALCIPARUM MEROZOITES: IDENTIFICATION, ISOLATION AND FUNCTIONAL CHARACTERIZATION

Hans-G. Heidrich, Werner Strych, Friedrich Lottspeich and Arja Miettinen-Baumann

Max-Planck-Institut für Biochemie, D-8033 Martinsried, Federal Republic of Germany

INTRODUCTION

Special problems are encountered in the attempt to develop a vaccine against malaria. The parasite stages which offer a suitable target are restricted to these which are not protected by a host cell, i.e., are free in the human blood circulation for at least a short period of time, and thus open to immunological attack by antibodies. The isolation and production of antigens from these stages is therefore a major task in the development of a vaccine for active immunization against the malaria parasite. It has already been shown that such antigens can produce partial protection in animals (Perrin et al. 1985). Merozoites are the asexual blood stage of the parasite which are released from the infected host cell at the end of the developmental cycle. They immediately recognize uninfected erythrocytes, make contact with them, orient themselves on the erythrocyte membrane and finally invade the host cell, all within about 30 sec. About 18-20 merozoites then develop in the erythrocyte. Specific triggers for invasion and development, most of them probably surface antigens, must be present in the merozoite.

It has been shown that some merozoite surface antigens stem

NATO ASI Series, Vol. H11
Host-Parasite Cellular and Molecular Interactions
in Protozoal Infections. Edited by K.-P. Chang and D. Snary

from a high molecular weight glycoprotein precursor of 185 000-195 000 Da present in schizonts (Perrin et al. 1980; Hall et al. 1983; Holder and Freeman 1984). During maturation of schizonts to merozoites this precursor is processed to various smaller proteins which are then found on the merozoites surface (Heidrich et al. 1983; Freeman and Holder 1983).

Neutralization of surface antigens which act as functional triggers, by antibodies would prevent invasion and development thus blocking the asexual erythrocytic cycle of the parasite. The time for immunological attack is short so that it is probably necessary to have high titers of antibody in the human blood. However, since the cycle is repeated periodically, it should be possible to eliminate all parasites over a period of time.

Here we describe and discuss the isolation of P. falciparum merozoites from cell culture and the purification and characterization of some of their surface antigens. The characterization of an 80 000 Da surface antigen is uescribed in detail as an example for other antigens which we are investigating, in order to demonstrate our working strategy.

RESULTS AND DISCUSSION

Starting material for all the experiments described were naturally released merozoites from Plasmodium falciparum (FCB1 strain) which were isolated from cultures using a sieving procedure with nylon filters (Heidrich et al. 1983). The yield of merozoites was raised to almost 10^{11} per experiment by using highly synchronous cultures. Merozoites were routinely examined in the electron microscope for their morphological integrity and were found to be unaltered and to possess a thick surface coat.

Detergent extracts from isolated merozoites, from isolated ^{35}S-methionin metabolically labelled merozoites or from merozoites, surface iodinated with $Na^{125}I$, were made with

Triton X-100, Zwittergent or other detergents. A cocktail of aprotinin, phenylmethane sulfonyl fluoride (PMSF), antipain, bestatin and pepstatin was routinely used as a protease blocker in the isolation medium (Heidrich et al. 1986).

Merozoite antigens were separated chromatographically. The first chromatography was carried out on Sephacryl S300, which groups antigens according to their molecular masses, using freshly extracted, unfrozen material after a centrifugation at

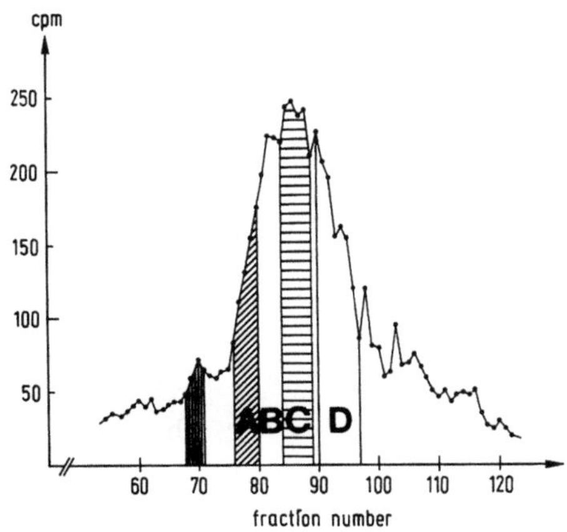

FIGURE 1. Chromatography of a Triton X-100 extract from freshly isolated merozoites on Sephacryl S300: separation of antigens according to size. The various pools were processed for ion-exchange chromatography shown in Figure 2.

100 000 g for 60 min containing about 20 mg unlabelled merozoite total protein, to which about 10^6 cpm of a similar Triton extract from ^{35}S-methionin labelled merozoites was added as a tracer. The separation profile is shown in Figure 1. The high molecular mass proteins $>$130 000 Da were found in fractions 58-66, in particular the 185 000-190 000 Da (from schizont extracts) and the 145 000-165 000 Da triplet proteins. Subsequent fractions contained proteins with decreasing molecular masses. Aliquots of fractions were precipitated with trichloroacetic acid, identified in gels and suitable pools of the column fractions processed for ion-exchange chromatography.

Ion-exchange chromatography was performed with a Pharmacia FPLC-System which gives highly reproducible gradients and results, and allows formation of both linear and step gradients. In this report only a few linear gradient separations are shown as examples (Figure 2). The homogeneity of some purified antigens is demonstrated in Figure 3. The 80 000 Da merozoite antigen, derived from a high molecular mass

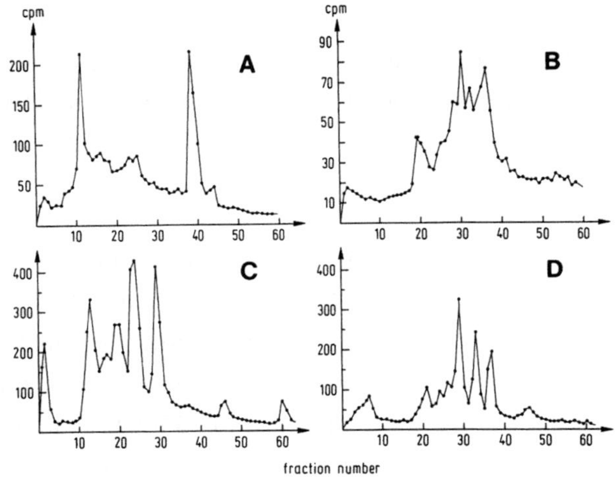

FIGURE 2. Ion-exchange chromatography of the various pools from Figure 1 in a Pharmacia FPLC-System with the resin MonoQ: separation of antigens according to charge. The various peaks were identified on gels.

precursor present in the schizont stage of the parasite, does not bind to the anion-exchange resin MonoQ under the conditions used and is found in the effluent of the column. Some antigens are still slightly contaminated; however, such antigens can be

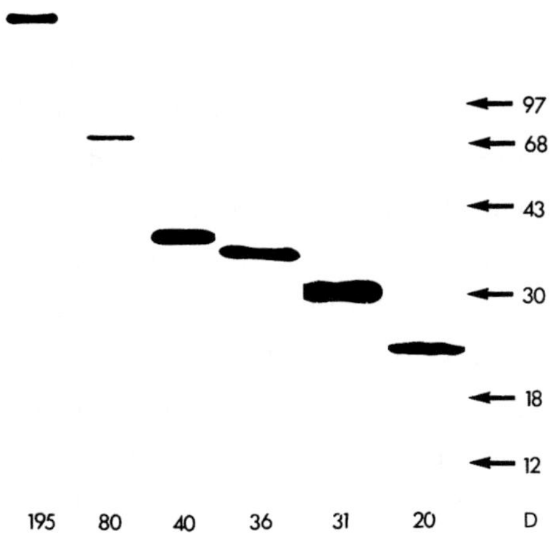

FIGURE 3. PAGE of various isolated merozoite antigens. Apparent molecular masses of antigens are shown at the bottom of the figure, that of molecular markers at the side (⟵).

purified using step gradients. The yield of the various purified antigens varied between 10-20 µg per separation. Most of the purified antigens precipitated with human immune serum showing that their epitopes were intact. Monoclonal antibodies directed against the 185 000 Da PSA precursor molecule (McBride et al. 1985) recognized the 80 000 Da antigen showing that it is very probably a processing product of the precursor.

For the production of antisera against various purified antigens, rabbits were injected subcutaneously at three-week intervals with 10-20 μg of the isolated antigen emulsified with Freund's complete adjuvant, followed by boosters with 10-20 μg of the antigen emulsified in Freund's incomplete adjuvant. Blood was drawn by heart punctation one week after the last injection and serum prepared. Pre-immune serum was prepared from blood of the same animal. In immunoprecipitation experiments sera recognized the corresponding antigens from unfractionated Triton X-100 or Zwittergent merozoite extracts providing additional confirmation that the epitopes of the isolated antigens were intact. The serum directed against the 80 000 Da antigen reacted exclusively with the 185 000 Da precursor in ^{35}S-methionine labelled extracts from schizonts. In extracts from iodinated merozoites it recognized only the 80 000 Da processing fragment (Figure 4). In other words, the

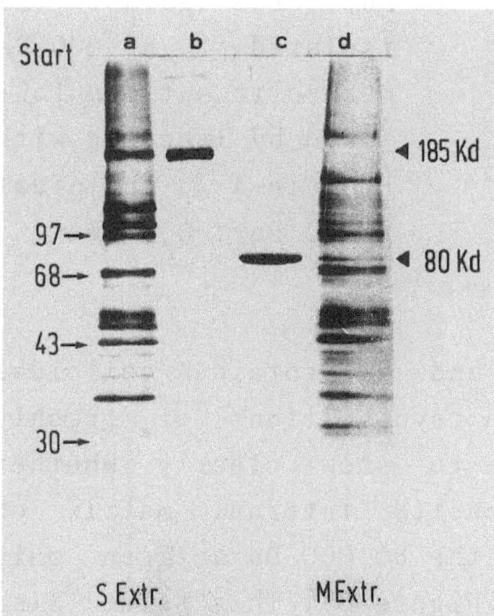

FIGURE 4. Immunoprecipitation of ^{35}S-methionin labelled schizont extract (lanes a and b) and merozoite extract (lanes c and d) with the antiserum against the isolated 80 000 Da processing fragment.

antiserum recognizes both the 185 000 Da precursor, present in schizonts, but not in merozoites, and the 80 000 Da processing product which is present in the merozoites, but not in schizonts. That none of the other merozoite surface antigens is recognized by this antiserum is evidence that none of these antigens is derived from the 80 000 Da fragment but are rather direct processing products from the 185 000 Da precursor.

Sera were tested in a modified in vitro inhibition test (Dejardins et al. 1979) using ^{3}H-hypoxanthine as radiotracer. The Protein A-Sepharose purified IgG-fraction of the serum against the 80 000 Da antigen did neither show invasion nor growth inhibition of parasite cultures when tested in several experiments and compared to control serum. The concentrations of IgG in these experiments were between 15 µg and 250 µg per ml of culture.

Sera were used to identify the position of the various antigens in the merozoite internal matrix or on the merozoite

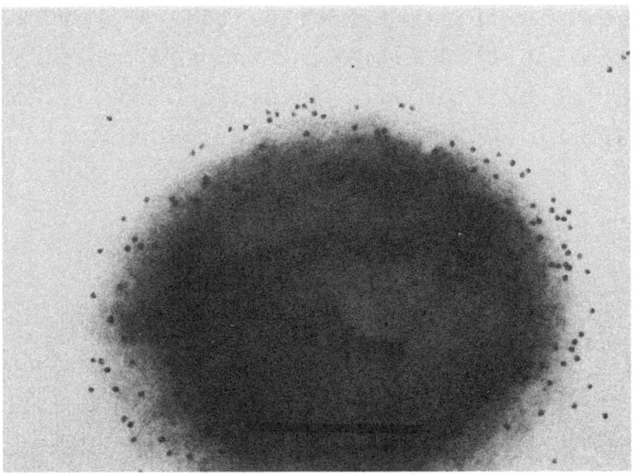

FIGURE 5. Electron micrograph of an ultracryo-section of freshly isolated merozoites reacted with antiserum against the isolated 80 000 Da merozoite antigen followed by labeling with Protein-A / colloidal gold. Bar: 0,5 µm.

surface. Using immunofluorescence and a Protein-A/colloidal gold labeling technique on ultrathin cryo-sections of freshly isolated merozoites, it was possible to show clearly whether antigens were derived from the merozoite internal matrix or from its surface. The serum against the 80 000 Da antigen only reacted with the merozoite surface (Figure 5). This serum also reacted strongly with the 80 000 Da radiolabelled antigen extracted from freshly isolated surface-iodinated merozoites in immunoprecipitation experiments (see above and Figure 4). This proves that the 80 000 Da merozoite antigen is a surface antigen (Heidrich et al. 1986).

The nucleotide sequence of the 185 000 Da glycoprotein precursor from a few strains of P. falciparum was analyzed (Mackay et al. 1985; Holder et al. 1985) so that the amino acid sequence of this high molecular mass precursor could be deduced. In order to fit our isolated processing fragments into the known sequence of the precursor we sequenced some of the merozoites surface antigens shown to stem from the 185 000 Da glycoprotein. Such analyses provide important information about, for example, the processing mechanism during schizogony, the putative functional properties of the various fragments and the use of the deduced glycosylation sites. N-terminal amino acid sequence analyses were performed on a gas phase sequencer. The phenylthiohydantoin amino acid derivatives were analyzed by a HPLC system which separates all component isocratically (Lottspeich 1985). In this way the position of the 80 000 Da isolated antigen within the precursor molecule sequence was defined unequivocally. Its N-terminal end lies at the position in the precursor molecule at which a signal peptidase starts the 185 000 Da protein. The N-terminus has the sequence VTHESYQELV. THE 80 000 Da antigen thus comprises the N-terminal end of the large precursor molecule. It contains the repetitive sequences of the precursor and should, according to several potential glycosylation sites found in the deduced amino acid sequence, be heavily glycosylated. However, the 80 000 Da fragment has never been found to be glycosylated in isolated merozoites from culture (Heidrich et al. 1985) showing that the potential glycosylation sites in this region of the precursor are not used. Other isolated fragments are glycosylated and these results (not shown here) fit well with the deduced amino acid sequence. We are convinced that we have also found the precusor's putative membrane anchor sequence which is thought to be taken into the erythrocyte during invasion (McBride and Heidrich, submitted).

ACKNOWLEDGEMENTS

This work received financial support from UNDP/World Bank/ WHO Special Programme for Research and Training in Tropical Diseases and from the Deutschen Bundesministerium für Forschung und Technologie.

REFERENCES

Perrin LH, Merkli B, Gabra MS, Stocker JW, Chizzolini C, Richle R (1985) Immunization with a Plasmodium merozoite antigen induces partial immunity in monkeys. J Clin Invest 75: 1718-1721

Perrin LH, Ramirez L, Er-Hsiang L, Lambert PH (1980) Plasmodium falciparum: characterization of defined antigens by monoclonal antibodies. Clin Exp Immunol 41: 91-96

Hall R, McBride J, Morgan G, Tait A, Zolg JW, Walliker D, Scaife J (1983) Antigens of the erythrocytic stages of the human malaria parasite Plasmodium falciparum detected by monoclonal antibodies. Mol Biochem Parasitol 7: 247-265

Holder A, Freeman RR (1984) The three major antigens on the surface of Plasmodium falciparum merozoites are derived from a single high molecular weight precursor. J Exp Med 160: 624-629

Heidrich HG, Strych W, Mrema JEK (1983) Identification of surface and internal antigens from spontaneously released Plasmodium falciparum merozoites. Z Parasitenkd 69: 715-725

Freeman RR, Holder AA (1983) Surface antigens of malaria merozoites: a high molecular weight precursor is processed to an 83 000 m.w. form expressed on the surface of Plasmodium falciparum merozoites. J Exp Med 185: 1647-1653

Heidrich HG, Matzner M, Miettinen-Baumann A, Strych W (1986) Immunoelectron microscopy shows that the 80 000 Da antigen of Plasmodium falciparum merozoites is localized in the surface coat. Z Parasitenkd 72: 681-683

McBride JS, Newbold CJ, Anand R (1985) Polymorphism of a high molecular weight schizont antigen of the human malaria parasite Plasmodium falciparum. J Exp Med 161: 160-180

Dejardins RD, Canfield CJ, Haynes JD, Chulay JD (1979) Quantitative assessment of antimalarial activity in vitro by a semiautomated microdilution technique. Antimicrob Ag Chemotherapy 1 : 710-718

Mackay M, Goman M, Bone N, Hyde JE, Scaife J, Certa U, Stunnenberg H, Bujard J (1985) Polymorphism of the precursor for the major surface antigens of Plasmodium falciparum merozoites: studies at the genetic level. EMBO J 4: 3823-3829

Holder AA, Lockyer MJ, Odink KG, Sandhu JS, Riveros-Moreno V, Nicholls SC, Hillmann Y, Davey LS, Tizard MLV, Schwarz RT, Freeman RR (1985) Primary structure of the precursor to the three major surface antigens of Plasmodium falciparum merozoites. Nature 317: 270-773

Lottspeich F (1985) Microscale isocratic separation of phenylthiohydantoin amino acid derivatives. J Chromatogr 32 : 321-327

Heidrich HG, Strych W, Prehm P (1984) Spontaneously released Plasmodium falciparum merozoites from culture possess glycoproteins. Z Parasitenkd 70: 747-751

STUDIES ON THE PLASMODIUM FALCIPARUM 195,000 MOLECULAR MASS MEROZOITE SURFACE ANTIGEN PRECURSOR

A.A. Holder, Y. Hillman, S.C. Nicholls and L.S. Davey

Department of Molecular Biology, Wellcome Biotech, Beckenham, Kent BR3 3BS, UK

INTRODUCTION

One of the proteins synthesized by the asexual blood stage of malaria which has been studied in some detail is the precursor to the major merozoite surface antigens (PMMSA), a glycoprotein that is synthesized throughout schizogony and present on the merozoite surface in processed form (Holder & Freeman, 1982, 1984; Howard et al., 1985). Complete or partial nucleotide sequences of the gene for this protein from a number of Plasmodium falciparum strains have been reported (Holder, et al., 1985; Mackay et al., 1985; Cheung et al., 1985; Weber et al,1986). Some features of the protein and its gene will be discussed here, in particular studies we have performed with a West African (Wellcome) strain of the parasite.

MATERIALS AND METHODS

The methods of antigen analysis, the preparation of specific antibodies, the determination of the gene structure and expression of parts of the gene have been described previously (Holder & Freeman, 1982, 1984; Freeman & Holder, 1983; Holder et al., 1985). Other specific DNA fragments were expressed in the open reading frame vector, pXY460 (a derivative of pXY410, Winther et al., 1985) that produces a hybrid β-galactosidase protein. The fusion proteins, containing the sequence of interest fused to the N-terminus of galactosidase, were prepared by affinity chromatography (Steers, Cuatrecasas & Pollard, 1971) and inoculated into animals to raise specific antisera.

NATO ASI Series, Vol. H11
Host-Parasite Cellular and Molecular Interactions
in Protozoal Infections. Edited by K.-P. Chang and D. Snary

RESULTS AND DISCUSSION

Analysis of the gene, composition and structure of the protein

The coding region of the PMMSA gene has an average A-T content of 76% (Holder et al., 1985), the two regions that stand out from this pattern are the short repetitive sequence and the sequence for a proline rich region (nucleotides 631 to 730 and 2641 to 2700) where the AT content drops to 46%. The high AT content is reflected in the codon usage bias for each amino acid. In the third position when there is an even choice A or T is present in 86.4% of the codons used. This pressure is maintained where alternative first base codons can be used, for example, for arginine AGA is used for 71% of the codons while CGN is used for the remaining 29% and for leucine UUR is used for 76.8% of the codons, while CUN is used for the remaining 23.2%. Some amino acids occur at low frequency, for example, there is only one trytophan (UGG) and the frequencies of arginine (CGN, 1.5%), alanine (GCN, 3.1%) and glycine (GGN, 3.5%) are low compared with composite data bases (Doolittle, 1981), while other amino acids such as lysine (AAR, 13%) and asparagine (AAY, 9.3%) are more abundant. This apparent bias against amino acids with GC rich codons must operate within the functional and conformational restraints operating on the protein structure.

An analysis of the protein for internal homologies using the Diagon method of Staden (1982) is shown in Fig. 1. Although there are no extensive internal duplications, there are numerous tetrapeptide, 9 pentapeptide and 1 hexapeptide sequence occuring more than once in the sequence outside of the repeat block. Although this repetition may be fortuitous it is possible that short functional sequence repeats may be conserved, for example as a recognition sequence for a processing protease.

Antibodies raised by immunization with one part of the molecule may recognise the related structures elsewhere in the molecule. Animals immunised with fusion proteins produced by expression of parts of the gene in a heterologous system such as *Escherichia coli* can produce cross reactive antibodies. The reaction of antibodies on Western blots with parasite-derived mono-

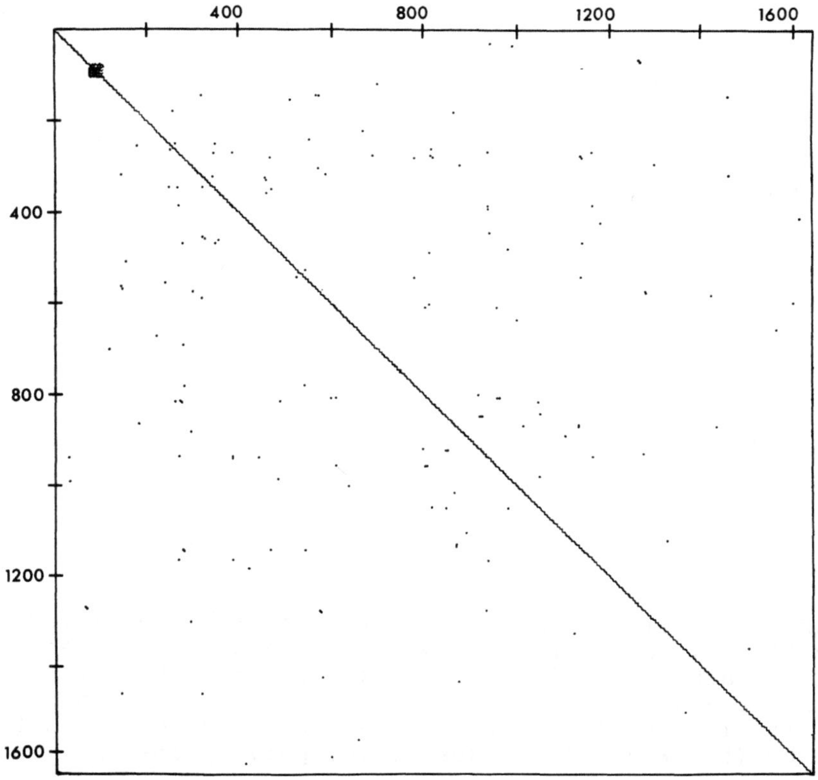

FIGURE 1. Dot matrix comparison of the Wellcome PMMSA protein sequence with itself to locate duplicated tetrapeptide sequences.

clonal antibody 89.1-affinity purified protein is shown in Fig.2. In addition to the 195K precursor and the 83K terminal product derived from the N-terminus of the protein, this preparation also contains 110K and 150K species representing possible processing intermediates extending from the N-terminus (Holder & Freeman, 1984). Rabbits immunised with protein expressed by pME1, an E. coli product containing amino acids 154 to 1048 (Holder et al., 1985), produced antibodies reacting with all species, and pME2 and pME3 products (containing amino acids 1047 to 1640 and 1047 to 1447, respectively) induced antibodies against the two larger species. However, the product of pME13, a construct coding for approximately the last 100 amino acids of PMMSA before the hydrophobic tail, expressed as a β-galactosidase fusion protein, induced antibodies recognising the 150K and 83K species as

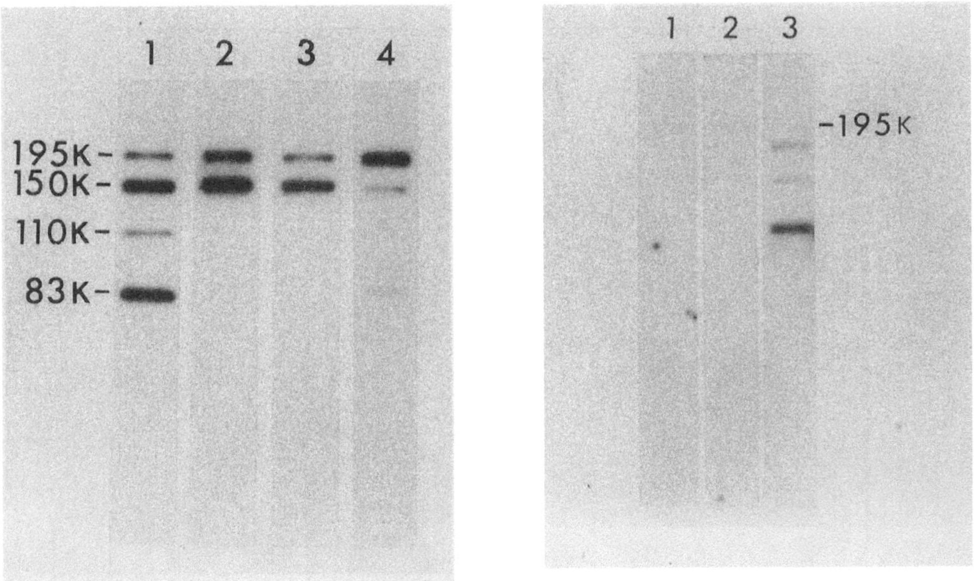

FIGURE 2. Reaction of antibodies with affinity purified PMMSA and specific fragments of it on Western blots. (A) probed with anti-pME1 (track 1), anti-pME2 (track 2), anti-pME3 (track 3) and anti-pME13 (track 4). (B) probed with the monoclonal antibodies 111.2 (track 1) and 111.4 (track 2) and with polyclonal anti-PMMSA serum (track 3). The sizes of the polypeptides are shown.

well as the 195K polypeptide (Fig. 2A, track 4).In contrast, monoclonal antibodies raised against the native protein and specific for this region, appear to recognise only the non-reduced intact 195K PMMSA (Fig.2B, tracks 1 and 2). It is possible that determinants not normally exposed to the immune system can be immunogenic when presented in a different way (for example as an E. coli product) and that these determinants may be shared by different regions of the molecule.

Western blotting was used to analyse the polypeptides in extracts of schizonts or merozoites which reacted with antibodies raised against the purified PMMSA or the pME2 product (Fig. 3). In schizont extracts that had not been reduced with dithiothreitol prior to electrophoresis, the rabbit polyclonal anti-PMMSA detected the 195K, 150K, 83K and two species of 38K and 36K. In the merozoite extract there was some residual 195K together with the 83K and 36K species and material at the dye front (<30K). In reduced extracts the polypeptides of 38K and 36K and the material

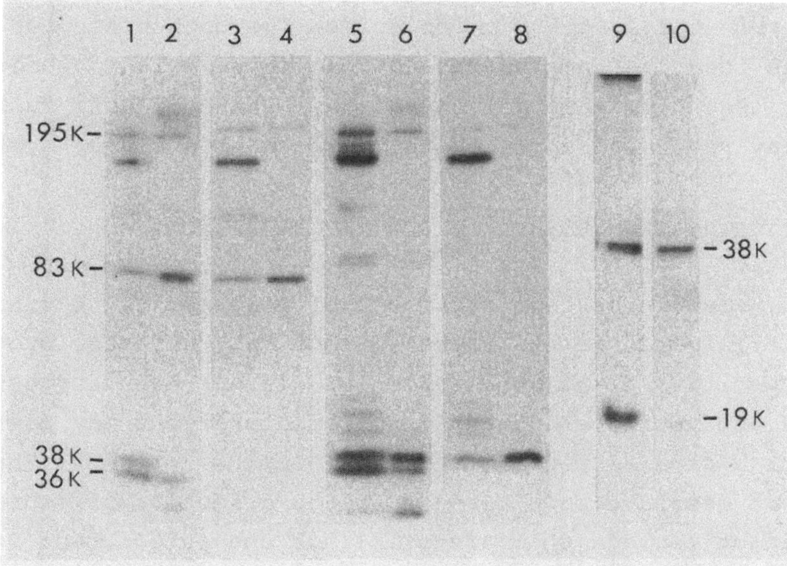

FIGURE 3. Reaction of antibodies with polypeptides in extracts of schizonts (tracks 1, 3, 5 and 7) or merozoites (tracks 2, 4, 6 and 8) electrophoresed through a 7.5% polyacrylamide gel. The samples in tracks 3, 4, 7 and 8 were reduced prior to electrophoresis, and those in tracks 1, 2, 5 and 6 were not. The polypeptides were transferred to nitrocellulose and then probed with anti-PMMSA (tracks 1 to 4) or anti-pME2 (tracks 5 to 8). Tracks 9 and 10 are a repeat of tracks 6 and 8 probed with anti-pME2, after electrophoresis of the extract on a 12.5% polyacrylamide gel. The molecular weights of specific bands are indicated.

at the front were no longer reactive with the antibodies. In identical extracts anti-pME2 antibodies detected the 195K and 150K species but not the 110K or 83K species. Molecules of 38K, 36K and <30K were also detected in the non-reduced samples and the reaction with the reduced polypeptides was reduced or abolished, except for a 38K species present in both schizont and merozoite extracts and not recognised by the anti-PMMSA serum. This reduction insensitive species corresponds to a fragment derived from the middle of the PMMSA polypeptide (Holder et al., 1987) and probably equivalent to the 45K species reported by Lyon et al (1986) in the CAMP strain. When the merozoite extract was fractionated on a 12.5% polyacrylamide gel and probed with anti-pME2 the low molecular weight, reduction -sensitive antigen was estimated at 19K. The distribution of cysteines in the PMMSA sequence is assymetric with over half located within the last 100 amino acids. The data presented here support the idea that antibodies against this region, either monoclonal (see Fig. 2B), polyclonal anti-PMMSA or anti-fusion protein, may recognise epitopes that are constrained by cystines.

Repetitive sequences: structure and expression.

The analysis of the gene structure of polypeptides synthesized by the malaria parasite has revealed that many of these proteins contain repetitive amino acid sequences. It has been suggested that these repetitive sequences in malaria antigens may be responsible for a strain specificity of the immune response. The mechanisms that generate such repeats or the selective pressures that retain or replace them are still not understood, and it is not clear whether this selection operates at the level of the

A. Amino acid sequence N-terminal to the repetitive region

Group 1: FQKEKMVLNEGTSGTAVTTSTPG
Group 2:..FHKEKMILNEEEITTKGA

B. Amino acid sequence of the repetitive region

Group 1:

WELL	SKG SVA SGG SGG SVA SGG SVA SGG SVA SGG SVA SGG
MAD20	SSG SVT SGG SVA SVA --- SVA SGG --- SGG SVA SGG
SGE2	SGG SVT SGG SGG SVA --- SVA SGG --- SGG SVA SGG

Group 2:

CAMP	SAQ SGT SGT SGT SGT SGT SGT SGT SAQ SGT SGT SAQ SGT SGT SAQ SGT SGT SGT SGT
K1	SAQ SGT SGT SGT SGP SGP SGT --- --- --- --- --- --- --- --- --- --- --- ---

C. Amino acid sequence C-terminal to the repetitive region

Group 1: SGNSRRTNPSDNSSDSD
Group 2: SPSSRSNTLPRSNTSSGASPPADASDSD....

FIGURE 4. A comparison of the repetitive amino acid region and the flanking sequences for the PMMSA from different isolates.

gene or the protein, possibly an adaptation to parasitism and interaction with the host immune system.

In the P. falciparum strains investigated so far the repetitive sequences and the conserved sequences around them close to the start of the PMMSA coding sequence, fall into two groups, as shown in Fig.4. In Group 1 the repetitive nucleotide structure is based on the two elements, TCA GTT GCT and TCA GGT GGC(T) whereas in Group 2 the structural elements are AGT GCT CAA and AGT GGT A(C)CA. One part of the Group 2 amino acid repeat (Gly Thr Ser Gly Thr) also occurs in the Group 1 sequences just before the repetitive sequence.

We expressed the repeat sequence from the Wellcome strain as a fusion protein with β-galactosidase to determine whether this structure is immunogenic and accessible to antibody on the native protein. DNA fragments from a Rsa 1 digest of pPfG1 were treated briefly with Bal 31 exonuclease, repaired with DNA polymerase Klenow fragment and then cloned into the Sma 1 site of the open reading frame vector, pXY460. Transformants were screened for functional β-galactosidase expression by growth on agar containing X-gal. Blue colonies were picked and lysates of these cells were screened by Western blotting. One of the transformants produced a hybrid β-galactosidase that reacted on a Western blot with a rabbit antiserum raised against PMMSA. This E. coli clone (pME 14) contained a plasmid with nucleotides 623 to 747 of the PMMSA gene, covering the repetitive sequence, fused to the β-galactosidase gene. Antiserum raised to this hybrid protein was assayed by immunoprecipitation of ^{35}S-methionine labelled polypeptides from extracts of blood stage parasites or of ^{125}I surface labelled polypeptides from merozoite extracts, solid phase radioimmunoassay against purified PMMSA and Western blotting against purified PMMSA or merozoite extracts. In each case the antibodies reacted with the PMMSA and, in the merozoite extracts, with the 83K species that is the terminal processing fragment from this region of the molecule (Fig. 5). These results indicate that the repetitive sequence is immunogenic and that this sequence is accessible to antibody in parasite extracts.

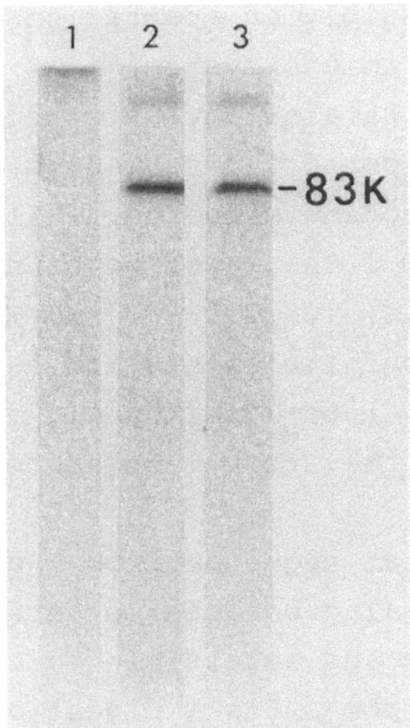

FIGURE 5. Radioiodinated merozoite surface protein immunoprecipitated by normal serum (track 1), anti-pME 14 (track 2) and anti-pME 1 (track 3); electrophoresed on a 12.5% polyacrylamide gel. The molecular weight of the specific band is indicated.

The genetic basis of antigenic diversity.

It has been proposed that the PMMSA is responsible for much of the antigenic diversity in *P. falciparum* and it has been shown that monoclonal antibodies against this protein can differentiate between different strains, defining 7 serotypes (McBride, Newbold and Anand, 1985). The repetitive sequence is one variable immunogenic region. The group 1 repeat sequences have been found in both African (Wellcome, SGE 2) and Asian (MAD 20, T9/94) isolates. Outside of the repeat region the K1 (Asian) strain gene differs from the Wellcome (African) strain by only 1.2% (19 out of 1523 amino acids, Mackay et al., 1986). A greater degree of diversity was detected in Southern blotting experiments using DNA from 4 cloned lines of a single Thai isolate, (Schwarz *et al.* ,1986). Variable and constant regions were defined and the alleles divided into two groups. More definitive information was obtained by comparing the partial sequence of the gene from the CAMP strain with other known sequences (Weber *et al*, 1986), defining alternating blocks of conserved and variable amino acid sequence throughout the molecule.

Can these distinct structural elements recombine to generate antigenic diversity? In the haploid blood stage there is no evidence of more than one allele in the genomic DNA, excluding a multigene family like that of the *Trypanosoma brucei* variant surface glycoprotein (Borst & Cross, 1982). Genomic rearrangements bringing together alternative blocks of sequence to assemble the complete gene (as, for example, in immunoglobulin gene expression) are unlikely, based on the analysis of genomic DNA sequences and digests. At present there is evidence for only two major groups of PMMSA alleles in the population and it is possible that at meiosis homologous recombination within the conserved regions, or gene conversion events, may lead to a reassortment of constant and variable regions within each allele. These processes, together with genetic drift with mutations in codons for amino acids that are not necessary for the structural or functional integrity of the protein, may account for the observed diversity in this gene and its product.

ACKNOWLEDGEMENTS

We thank Dr P.Talbot for providing the plasmid pXY 460.

REFERENCES

Borst P, Cross GAM (1982) Molecular basis for trypanosome antigenic variation. Cell 29:291-303

Cheung A, Shaw AR, Leban J, Perrin LH (1985) Cloning and expression in *Escherichia coli* of a surface antigen of *Plasmodium falciparum* merozoites. EMBO J 4:1007-1012

Doolittle RF (1981) Similar amino acid sequences: chance or common ancestry? Science 214:149-159

Freeman RR, Holder AA (1983) Surface antigens of malaria merozoites. A high molecular weight precursor is processed to an 83,000-mol wt form expressed on the surface of *Plasmodium falciparum* merozoites. J exp Med 158:1647-1653

Holder AA, Freeman RR(1982) Biosynthesis and processing of a *Plasmodium falciparum* schizont antigen recognized by immune serum and a monoclonal antibody. J exp Med 156:1528-1538

Holder AA, Freeman RR (1984) The three major antigens on the surface of Plasmodium falciparum merozoites are derived from a single high molecular weight precursor. J exp Med 160:624-629

Holder AA, Lockyer MJ, Odink KG, Sandhu JS, Riveros-Moreno V, Nicholls SC, Hillman Y, Davey LS, Tizard MLV, Schwarz RT, Freeman RR (1985) Primary structure of the precursor to the three major surface antigens of Plasmodium falciparum merozoites. Nature 317:270-273

Holder AA, Sandhu JS, Hillman Y, Davey LS, Nicholls SC, Lockyer MJ (1987) Processing of the precursor to the major merozoite surface antigens of Plasmodium falciparum. (submitted)

Howard RF, Stanley HA, Campbell GH, Langreth SG, Reese RT (1985) Two Plasmodium falciparum merozoite surface polypeptides share epitopes with a single M_r 185,000 parasite glycoprotein. Mol Biochem Parasitol 17:61-77

Lyon JA, Geller RH, Haynes JD, Chulay JD, Weber JL(1986) Epitope map and processing scheme for the 195,000-dalton surface glycoprotein of Plasmodium falciparum merozoites deduced from cloned overlapping segments of the gene. Proc Natl Acad Sci USA 83:2989-2993

Mackay M, Goman M, Bone N, Hyde JE, Scaife J, Certa U, Stunnenberg H, Bujard H (1985) Polymorphism of the precursor for the major surface antigens of Plasmodium falciparum merozoites: studies at the genetic level. EMBO J 4:3823-3829

McBride JS, Newbold CI, Anand R (1985) Polymorphism of a high molecular weight schizont antigen of the human malaria parasite Plasmodium falciparum. J Exp Med 161:160-180

Schwarz RT, Riveros-Moreno V, Lockyer MJ, Nicholls SC, Davey LS, Hillman Y, Sandhu JS, Freeman RR, Holder AA(1986) Structural diversity of the major antigen of Plsmodium falciparum merozoites. Mol Cell Biol 6:964-968

Staden R (1982) An interactive graphics program for comparing and aligning nucleic acid and amino acid sequences. Nucl Acids Res 10:2951-2961

Steers E, Cuatrecasas P, Pollard HB (1971) The purification of β-galactosidase from Escherichia coli by affinity chromatography. J Biol Chem 246:196-200

Weber JL, Leininger WM, Lyon JA (1986) Variation in the gene encoding a major merozoite surface antigen of the human malaria parasite Plasmodium falciparum. Nucl Acids Res 14:3311-3323

Winther MD, Allen G, Bomford RH, Brown F (1986) Bacterially expressed antigenic peptide from foot-and-mouth disease virus capsid elicits variable immunologic responses in animals. J Immunol 136:1835-1840

ASPECTS OF THE POSTTRANSLATIONAL MODIFICATION OF A MAJOR PLASMODIUM FALCIPARUM MEROZOITE SURFACE ANTIGEN

R.T. Schwarz, M.J. Lockyer* and A.A. Holder*

Laboratoire de Chimie Biologique, Universite des Sciences et Techniques de Lille, 59655 Villeneuve D'Ascq Cedex, France.
*Department of Molecular Biology, Wellcome Biotech, Beckenham, Kent, BR3 3BS, UK.

INTRODUCTION

The precursor to the major merozoite surface antigens (PMMSA) of *Plasmodium falciparum* is a candidate for the development of an antimalarial vaccine (Holder *et al*., 1985). A detailed knowledge of the protein and gene structure, especially the extent of sequence variation between different isolates, is therefore very important. The biosynthesis and posttranslational modification of this molecule are also of great interest. The protein is cleaved into a number of discrete fragments on the merozoite surface (Holder & Freeman, 1984) and there is evidence that the PMMSA protein is acylated. A glycolipid anchor may be the mode of attachment of this molecule to the cell surface (Haldar *et al*., 1985; Schwarz *et al*., 1986), by analogy with a number of protozoal (Ferguson *et al*., 1985, Etges *et al*., 1986) and other cell surface molecules (reviewed by Low *et al*., 1985). We have investigated some aspects of the posttranslational modification of the PMMSA protein by incorporating radiolabelled precursors.

MATERIALS AND METHODS

Cultures of *P. falciparum* in RPMI 1640 medium were labelled with either ^{35}S methioninine, 9,10(n)-^{3}H myristic acid or 1-^{3}H ethan-1-ol-2-amine hydrochloride for 3 hrs. Labelling with D-[2-^{3}H]mannose and D-[6-^{3}H]glucosamine hydrochloride was done for the same period of time in RPMI medium in which D-glucose was re-

NATO ASI Series, Vol. H11
Host-Parasite Cellular and Molecular Interactions
in Protozoal Infections. Edited by K.-P. Chang and D. Snary

placed by 20 mM fructose. Lysates and immunoprecipitates were prepared and analyzed by sodium dodecyl sulfate polyacrylamide gel electrophoresis (SDS-PAGE) as described previously (Schwarz *et al.*, 1986).

Larger amounts of tritium labelled PMMSA were prepared by immunoaffinity chromatography (Holder & Freeman, 1984). The eluate containing the ethanolamine-labelled PMMSA was dialyzed against water, lyophilized, taken up in 60 µl water and precipitated with 5 volumes of acetone using 50 µg bovine serum albumin as carrier. The pellet was delipidated by successive extractions with chloroform/methanol/water (1:1:0.3; v/v). After extraction of the pellet with a mixture of 200 µl acetone and 50 µl 0.1 M HCl, it was hydrolyzed in 6M HCl at 110 ^{0}C *in vacuo* for 20 hrs. The sample was dried over NaOH *in vacuo*, taken up in 20 µl water, spotted on cellulose thin layer plates (MN Cel 300) and chromatographed using 2-propanol/2-butanone/1 M HCl (60:15:25; v/v/v) as solvent.

TABLE 1. Synthesis of ^{35}S methionine labelled trichloroacetic acid precipitable material in the presence of either D-glucose or D-fructose. Aliquots of 25 µl of labelled lysates were spotted on Whatman GFC-filters and baked at 100 ^{o}C for 10 min. The filters were washed successively with ice-cold trichloroacetic acid (once with 10%, twice with 6%) and then with ethanol/ether, (1:1 v/v). Radioactivity was determined using a liquid scintillation counter.

Concentration (mM)	Maximal incorporation (%)	
	Glucose	Fructose
0.0	2.1	2.1
0.5	16.1	2.3
1.0	32.4	2.5
2.0	64.8	3.5
5.0	86.9	21.9
10.0	100.0	41.4
20.0	77.3	74.5
40.0	67.2	94.5

RESULTS AND DISCUSSION

To prevent interference of glucose with the uptake and activation of radioactive carbohydrates it was replaced by fructose and the capacity of this sugar to support the formation of ^{35}S labelled proteins was studied. The data in Table 1 demonstrate that label was incorporated in comparable amounts in the modified medium and this medium (containing 20mM fructose) was used for the incorporation of labelled sugars. Figure 1 shows that the PMMSA molecule could be labelled with ^{3}H mannose, ^{3}H glucosamine, ^{3}H myristic acid and ^{3}H ethanolamine, suggesting that the polypeptide is biosynthetically modified, with sugar and acyl moieties, consistent with glycolipid addition. A feature of this type of

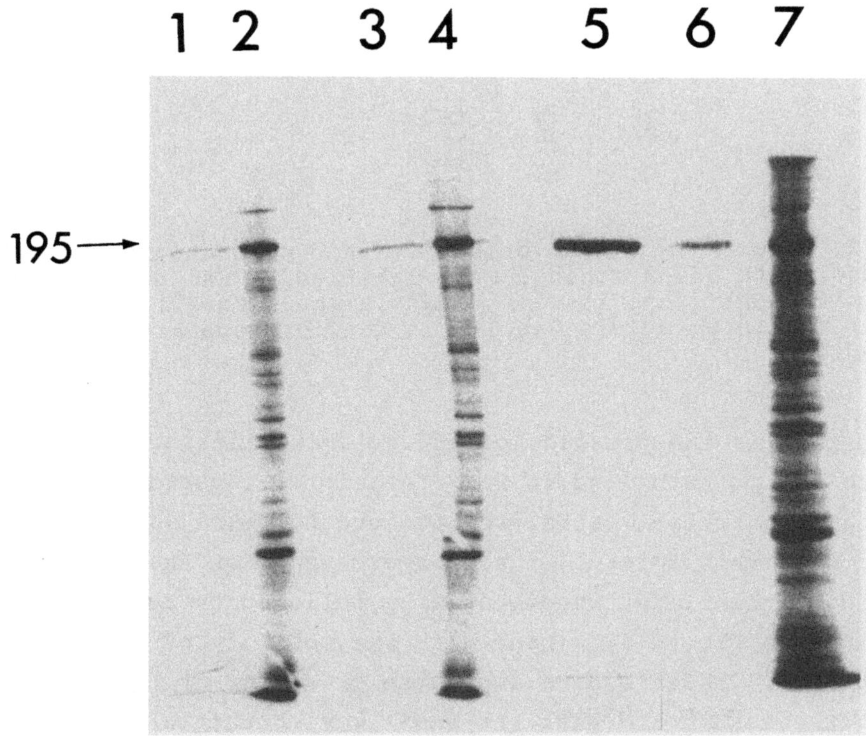

FIGURE 1. Sodium dodecyl sulfate-polyacrylamide gel electrophoresis of ^{3}H mannose- (lane 1), ^{3}H glucosamine- (lane 3), ^{3}H myristic acid- (lane 5), and ^{3}H ethanolamine- labelled PMMSA (lane 6) immunoprecipitated from extracts of P. falciparum (Wellcome strain). ^{35}S methionine-labelled proteins were run in parallel (lanes 2,4,7).

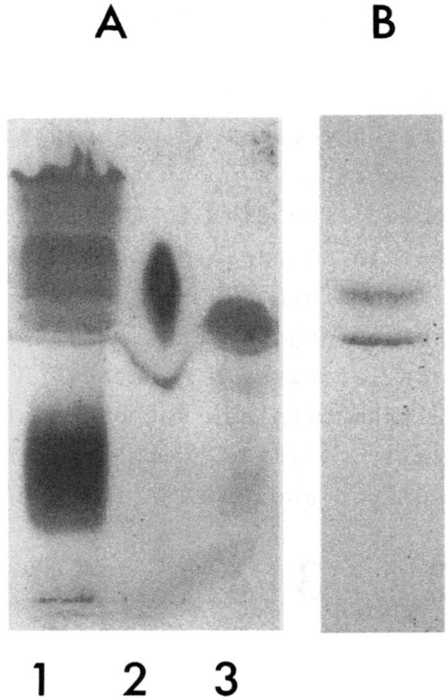

FIGURE 2. Panel A is a ninhydrin - stained chromatogram of a 6 M HCl hydrolyzate of ^{3}H ethanolamine-labelled PMMSA (co-precipitated with BSA) (lane 1) with ethanolamine (lane 2) and serine (lane 3) run in parallel. Panel B is the corresponding autoradiograph of lane 1.

modification is the presence of ethanolamine linking the glycolipid to the protein, and it was, therefore, important to establish that the tritium label from the ethanolamine was still associated with this moiety. An acid hydrolyzate of the PMMSA was analyzed by thin layer chromatography followed by autoradiography . As shown in Figure 3, a band with the mobility of ethanolamine was detected. These results and those of Haldar et al. (1985) support the assumption that the PMMSA may contain a lipid-anchor. The nucleotide sequence of the PMMSA gene (Holder et al., 1985) predicts a C-terminal hydrophobic amino acid sequence which may be replaced by a glycolipid linked through ethanolamine to the new terminal amino acid. Further work is needed to fully characterize this post translational modification.

ACKNOWLEDGEMENT.

R.T.S. thanks the Deutsche Forschungsgemeinschaft for a Heisenberg Fellowship during part of this work.

REFERENCES

Etges R, Bouvier J, Bordier C (1986) The major surface protein of Leishmania promastigotes is anchored in the membrane by a myristic acid-labelled phospholipid. EMBO J 5:597-601

Ferguson MAJ, Low MG, Cross GAM (1985) Glycosyl-sn-1,2-dimyristyl phospatidylinositol is covalently linked to Trypanosoma brucei variant surface glycoprotein. J Biol Chem 260: 14547-14555

Haldar K, Ferguson MAJ, Cross GAM (1985) Acylation of a Plasmodium falciparum merozoite surface antigen via sn-1,2 diacyl glycerol. J Biol Chem 260:4969-4974

Holder AA, Freeman RR (1984) The three major antigens on the surface of Plasmodium falciparum merozoites are derived from a single high molecular weight precursor. J exp Med 160:624-629

Holder AA, Lockyer MJ, Odink KG, Sandhu JS, Riveros-Moreno V, Nicholls SC, Hillman Y, Davey LS, Tizard MLV, Schwarz RT, Freeman RR (1985) Primary structure of the precursor to the three major surface antigens of Plasmodium falciparum merozoites. Nature 317:270-273

Low MG, Ferguson MAJ, Futerman AH, Silman I (1986) Covalently attached phosphatidylinositol as a hydrophobic anchor for membrane proteins. Trends Biochem Sci 11:212-215

Schwarz RT, Riveros-Moreno V, Lockyer MJ, Nicholls SC, Davey LS, Hillman Y, Sandhu JS, Freeman RR, Holder AA (1986) Structural diversity of the major antigen of Plasmodium falciparum merozoites. Mol Cell Biol 6:964-968

EXPORT OF PLASMODIUM FALCIPARUM PROTEINS TO THE HOST ERYTHROCYTE MEMBRANE: SPECIAL PROBLEMS OF PROTEIN TRAFFICKING AND TOPOGENESIS

Russell J. Howard[1], Shigehiko Uni[2], Jeffrey A. Lyon[3], Diane W. Taylor[4], Wendell Daniel[1] and Masamichi Aikawa[2]

[1]Malaria Section, Laboratory of Parasitic Diseases, NIAID, NIH, Bethesda, MD 20892, USA; [2]Institute of Pathology, Case Western Reserve University, Cleveland, OH 44106, USA; [3]Department of Immunology, Walter Reed Army Institute of Research, Washington DC 20012, USA; and [4]Department of Biology, Georgetown University, Georgetown, DC 20057, USA

INTRODUCTION

Asexual blood-stage malaria parasites induce several morphological, antigenic and functional changes of the infected erythrocytes membrane. In the human malaria Plasmodium falciparum these changes include the following: 1. expression of knob-like protrusions of the host cell membrane and underlying electron dense material (EDM), together called a 'knob'. 2. expression of the capacity of infected erythrocytes to cytoadhere specifically to capillary endothelial cells. Knobs bear the ligand(s) responsible for cytoadherence. 3. expression on the cell surface of a very large ($M_r \sim 300,000$) malarial protein called the cytoadherence protein, or, PfEMP1 (P. falciparum erythrocyte membrane protein 1). A body of indirect evidence, summarized below, suggests that PfEMP1 either bears the ligand(s) responsible for recognition of endothelial cells, or, is located in close physical proximity to the cytoadherence ligand(s). 4. insertion of another very large malarial protein ($M_r \sim 300,000$) under the surface membrane at the EDM of knobs. This protein can be distinguished from PfEMP1 not only by its submembrane location, but by specific reaction with monoclonal antibodies. We designate this molecule PfEMP2. 5. insertion of a histidine-rich protein, called the knob-protein or PfHRP1 (P. falciparum histidine-rich-protein 1), into the EDM at knobs. Like PfEMP2, PfHRP1 appears to be entirely submembranous at knobs. P. falciparum-infected erythrocytes also secrete malarial proteins into the extracellular medium (culture fluid or plasma). Another histidine-rich-protein distinct from

NATO ASI Series, Vol. H11
Host-Parasite Cellular and Molecular Interactions
in Protozoal Infections. Edited by K.-P. Chang and D. Snary

PfHRP1, PfHRP2, is released from intact infected cells into the extracellular medium.

All of these changes reflect the capacity of the intraerythrocytic malaria parasite to modify the properties of the host erythrocyte membrane at a distance. The host membrane is not only physically distant from the site of malarial protein synthesis (the parasite cytoplasm), but separated topologically by an additional membrane called the parasitophorous vacuole membrane (PVM). Originally derived from an invagination of erythrocyte membrane and pinching off around the invading merozoite, the PVM ensheaths the intracellular parasite and increases in size as the parasite grows. The barriers that one must consider when asking how malarial proteins reach the host erythrocyte membrane are the following: 1. the parasite plasmamembrane (PPM), 2. the parasitophorous vacuole (the space between the PPM and PVM), 3. the PVM, 4. the host erythrocyte cytoplasm and 5. the erythrocyte surface membrane (RBCM). Recently we described results of cryo-thin-section immunoelectron-microscopy which unambiguously established the fact that malarial proteins are inserted into and through the host erythrocyte membrane. Our results also established the presence of diverse 'organelles' in the host erythrocyte cytoplasm that appear to be involved in the trafficking of these proteins between the PVM and RBCM. Numerous questions emerge from these data, two of which are dealt with at length in this report.

What are the pathways whereby membrane-bound malarial proteins are exported from the parasite to the host erythrocyte membrane?

How do these pathways predict the topogenesis of malarial membrane proteins expressed on the cytoplasmic face of the host erythrocyte membrane or alternatively, on the external face of this membrane? We also consider the pathways whereby malarial proteins can be secreted as water soluble molecules from intact infected cells.

RESULTS

Parasite-derived Components Exported to the RBCM

Below we briefly review the nomenclature and properties of each parasite component that has been shown to be exported to the RBCM.

Electron Dense-material (EDM)

Movement of EDM from the asexual P. falciparum parasite to the host cell membrane has been suggested from transmission electron microscopy (Aikawa, Uni, Andrutis, and Howard, 1986). EDM was seen both under the PPM and in the parasitophorous vacuole. Unit membrane vesicles budding of the PVM were also seen with EDM on the cytoplasmic side of these membranes. The EDM associated with these vesicles was identical in size, shape and electron density to that located under knobs. Vesicles bearing EDM were also seen fusing with the RBCM, thereby completing a pathway of movement from PVM to RBCM (Aikawa et al., 1986).

PfHRP1

P. falciparum histidine-rich-protein 1, PfHRP1, or the knob-protein/knob-associated HRP (Kilejian, 1979), is a protein of unusually high histidine content that is only expressed by P. falciparum parasites which bear knobs (K^+ phenotype) (Kilejian, 1979; Hadley, Leech, Green, Daniel, Wahlgren, Miller and Howard, 1983; Vernot-Hernandez and Heidrich, 1984; Leech, Barnwell, Aikawa, Miller and Howard, 1984). PfHRP1 reacts specifically with a mouse IgG monoclonal antibody (McAb 89) that was used to define the subcellular location of this protein by cryo-thin-section immunoelectronmicroscopy (Taylor, Parra, Chapman, Stearns, Rener, Aikawa, Uni, Panton, Aley and Howard, 1986). This HRP is located under knobs at the EDM and in association with spherical organelles in the host erythrocyte cytoplasm (Taylor et al., 1986). In different K^+ parasites PfHRP1 ranges in apparent M_r on SDS-PAGE between M_r 80,000 and 120,000. Using McAb89 we have shown that PfHRP1 is not expressed at the external surface of the host cell membrane. Furthermore, K^+ parasites which fail to cytoadhere (ie K^+B^-; knob positive, binding negative) express PfHRP1 at knobs like the K^+B^+ parasites. This HRP presumably has a structural and/or functional role at knobs other than a direct role in cytoadherence.

PfHRP2

P. falciparum histidine-rich-protein 2, PfHRP2, has been sequenced (as deduced from cloned genomic DNA) and shown to contain 34%, histidine (H), 37% alanine (A) and 10% aspartic acid (D) with multiple tandem repeats AHH or AHHAAD (Wellems & Howard, 1986). PfHRP2 is secreted from intact infected cells, of K^+ and K^- phenotypes, as a water soluble protein (Howard, Uni, Aikawa, Aley, Leech, Lew, Wellems, Rener and Taylor, 1986). It is defined immunochemically by specific reaction with rabbit antiserum to the peptide AHH(AHHAAD)$_2$ (Rock, Marsh, Maloy, Taylor, Wellems, Saul and Howard, 1986) and reaction with a panel of mouse monoclonal antibodies (McAb87, Howard et al, 1986 and McAbs 2G12 & 1D6, Rock et al, 1986). By immunoelectronmicroscopy with these antibodies we localized PfHRP2 to the cytoplasm of the malaria parasite, to electron dense spherical accretions in the host erythrocyte cytoplasm and under/at the erythrocyte membrane (Howard et al., 1986 and Panton et al., unpublished data). The apparent M_r of PfHRP2 varies in different parasites from M_r60,00-85,000, although the M_r deduced from sequencing is significantly less (M_r 30,000-45,000).

PfEMP1

P. falciparum erythroycte-membrane-protein 1, PfEMP1, or the cytoadherent protein, was identified as a M_r~300,000 malarial surface protein that either mediated cytoadherence or is physically closely-associated with the cytoadherence ligand(s) on infected cells.

1. PfEMP1 is labeled by ^{125}I-lactoperoxidase labeling under conditions wherein hemoglobin and spectrin are not labeled
2. PfEMP1 is expressed on the surface of K^+B^+ infected cells but is not expressed on the surface of K^- cells.
3. Aotus monkey antisera which specifically block or reverse cytoadherence of particular isolates of P. falciparum, specifically react on immunoprecipitation of detergent-solubilized material with PfEMP1 of the same isolates. Antisera which do not affect cytoadherence of the particular isolate do not immunoprecipitate PfEMP1 from that isolate.

4. The sera described in 3 show the same specificity for reaction with PfEMP1 when added to intact infected cells and solubilized immune-complexes are purified and analyzed.
5. Cytoadherence is exquisitely sensitive to trypsin treatment of infected cells (0.1-1 μg/ml of trypsin, 5 min, 23°C, ablates cytoadherence). The same conditions of trypsin treatment quantitatively cleave the ^{125}I-surface labeled PfEMP1.

These results are from Leech, Barnwell, Miller and Howard 1984; Aley, Sherwood & Howard, 1984 and R. J. Howard, unpublished data.

Monospecific antibody reagents for PfEMP1 are unfortunately not yet available. PfEMP1 does not react with McAbs 8B7.4 and 4H9.1 which react with PfEMP2 (described next). Thus, although PfEMP1 and PfEMP2 share the properties of size range (M_r~300,000), antigenic diversity, size heterogeneity and localization at knobs, only PfEMP1 is expressed at the external face of the host erythrocyte membrane.

PfEMP2

P. falciparum-erythrocyte-membrane-protein 2, PfEMP2, is defined by its reaction with a pair of McAbs (4H9.1, an IgM and 8B7.4, an IgG) which react specifically with different epitopes on this antigen (Howard, Lyon, Uni, Saul, Aley, Klotz, Panton, Sherwood, Marsh, Aikawa and Rock, 1987). PfEMP2 has been identified as a M_r~300,000 malarial protein by immunoprecipitation with these McAbs. The apparent M_r of PfEMP2 varies with different *P. falciparum* isolates, and, the molecule is antigenically diverse in that while the PfEMP2 of all isolates tested reacts with McAb8B7.4, not all isolates react with McAb 4H9.1. PfEMP2 is not extracted from infected cells by Triton X-100 and other detergents or peturbants which fail to disrupt the erythrocyte (and parasite?) cytoskeleton. PfEMP2 was localized by cryo-thin-section immunoelectronmicroscopy (Howard et al., 1987). These results, together with indirect immunofluorescence studies and immunoelectronmicroscopy of saponin-treated membranes from infected cells, show that PfEMP2 is located within the parasite, within the host erythrocyte cytoplasm and as a submembrane constituent of the EDM at knobs.

Thus, several properties including association with the erythrocyte surface membrane are shared by PfEMP2 and the cell surface PfEMP1 molecule that appears to be involved in cytoadherence. However, PfEMP2 is distinguished from PfEMP1 by the following criteria: (1). PfEMP1, as identified by cell surface ^{125}I-labeling, does not react with McAbs 8B7.4 or 4H9.1. (2). In contrast to PfEMP1, PfEMP2 is not ^{125}I-radiolabeled under conditions of specific cell surface labeling (ie when spectrin and hemoglobin are not labeled). (3) In diverse parasites the apparent M_r of PfEMP1 and PfEMP2 are different, although both molecules migrate in the same range (M_r220,000 to >M_r300,000). (4) Whereas PfEMP1 can be shown to react with antibodies added to intact, non-fixed erythrocytes (indicating it to be exposed on the cell surface), PfEMP2 is not accessible to antibodies added to intact cells. The two McAbs do not react with intact infected cells, nor does a polyclonal rabbit antiseurm raised by immunization with affinity-purified PfEMP2 (see Howard et al, 1987).

The function of PfEMP2 is unknown. Since it is located at the EDM under knobs we presume it has some structural and/or functional role in knobs.

DISCUSSION

The cytoplasm of the normal human erythrocyte does not contain the machinery for protein synthesis, membrane synthesis, nor the machinery for subcellular protein trafficking. The demonstration that erythrocytes infected with asexual malaria parasites express malarial protein antigens not only on the erythrocyte surface (Howard, Barnwell & Kao, 1983; Leech, Barnwell, Miller, and Howard, 1984), but on the cytoplasmic side of the cell surface membrane (Taylor et al., 1986; Howard et al., 1987), and that malarial proteins are secreted 'through' this membrane into the extracellular medium (Howard et al., 1986) implies that several special problems of protein trafficking and topogenesis had been solved by the malaria parasite. The parasite plasamembrane (PPM), parasitophorous vacuole membrane (PVM) and erythrocyte surface membrane (RBCM) are clearly not barriers to the movement of malarial proteins. Nor is the considerable distance between the parasite and RBCM a barrier preventing transit of malarial proteins to the cell surface.

The problem of distance between the parasite and RBCM has been solved in a variety of ways. For the secreted malarial protein PfHRP 2, 'packets' of the protein, presumably representing aggregates of many molecules, are widely scattered in the host erythrocyte cytoplasm. These 'packets' bear no obvious relationship to any membranes. Transport of these 'packets' may be directed to the RBCM through an organelle system not yet evident (for example, microfilaments). Alternatively, PfHRP2 may simply diffuse through the erythrocyte cytoplasm and bind to the RBCM prior to its secretion. With the examples of PfHRP1 and PfEMP2, both of which appear to remain tightly membrane bound throughout their subcellular movements, cryo-thin-section immunoelectron microscopy has revealed a hitherto unknown system of organelles involved in this transit process. Maurers clefts, long known from transmission electron microscopy, do not appear to play a major role in export of either PfHRP1 or PfEMP2. Instead, circular electron dense organelles seen in cross-section (presumably spheres), appear to bear these proteins from the PVM to the RBCM. Whorls of membrane, often concentric and linked to one another by an electron dense area, also appear to be important in the transit of these molecules across the erythrocyte cytoplasm.

The identification of malarial organelles in the host erythrocyte cytoplasm and evidence that at minimum these organelles perform functions of protein trafficking has unusual biological implications. The asexual malarial parasite is generally considered to be bounded by its PPM. Accordingly, we must reach the extraordinary conclusion that this intracellular organism has either directed the de novo synthesis of organelles beyond its PPM, or, has exported organelles into its external environment (ie the erythrocyte cytoplasm). Alternatively, the extent to which the malaria parasite has created new structures beyond what is considered to be its limiting membrane, might suggest that the complex of malaria parasite and host erythrocyte should not be thought of as two separate entities. It is clear already that the malaria parasite has sufficiently altered the properties of the host erythrocyte membrane to create a membrane with entirely new properties, unlike those of normal erythrocytes or the parasite PPM.

We are beginning to learn the different ways in which malarial proteins are asymmetrically integrated into the RBCM. PfEMP1 and the variant antigen of P. knowlesi (Howard et al., 1983) for example, expose antigenic determinants on the external face of the RBCM. In contrast, PfEMP2 and PfHRP1 may expose little or none of their polypeptide sequence on the external face, but are exposed on the internal face of this membrane. The term 'protein topogenesis' was coined to describe this process of asymmetric integration of proteins into membranes (Blobel, 1980). It is reasonable to assume that the mechanisms for protein topogenesis within the malaria parasite, for both membrane and secreted proteins, are analogous to those much-studied in other organisms. Consequently, we will not discuss the multiple mechanisms of spontaneous or co-translocational protein insertion into membrane, (see for example Blobel, 1980; Engelman & Steitz, 1981; Wickner & Lodish, 1985), but will focus on how a particular orientation of a malarial protein into membrane shortly after, or during its synthesis within the parasite, could produce an asymmetrically integrated malarial protein at the special membranes of interest to us: PPM, PVM and RBCM.

Trafficking of EDM from Parasite to Knobs. The schematic diagram in Figure 1 shows how EDM synthesized within the malaria parasite can pass 'through' the topological barriers of PPM and PVM to form electron-dense accretions on the cytoplasmic face of Maurers clefts, thence, form EDM under knobs. This directional flow of material was suggested by electron microscopy and by studies on the effects of 2-F-L-histidine on *P. falciparum*. The latter studies, discussed in detail elsewhere in this volume (Aikawa, Uni, Andrutis, Kirk, Cohen and Howard, 1987) showed that 2-F-L-histidine induced a blockade of this process of EDM transport and confirmed indirectly the link between EDM under the parasite membrane and EDM under knobs.

Trafficking of the P. falciparum Glycoprotein and S-Antigen. The major *P. falciparum* glycoprotein (M_r~200,000) is expressed on the PPM of mature trophozoite-and schizont-infected erythrocytes. It is not expressed on the RBCM (Howard, Lyon, Diggs, Haynes, Leech, Barnwell, Aley, Aikawa and Miller, 1984). Since other malarial proteins are exported beyond the PPM (PfEMP1, PfEMP2, PfHRP1, PfHRP2), there must be special 'signal sequences' which direct a malarial protein to the

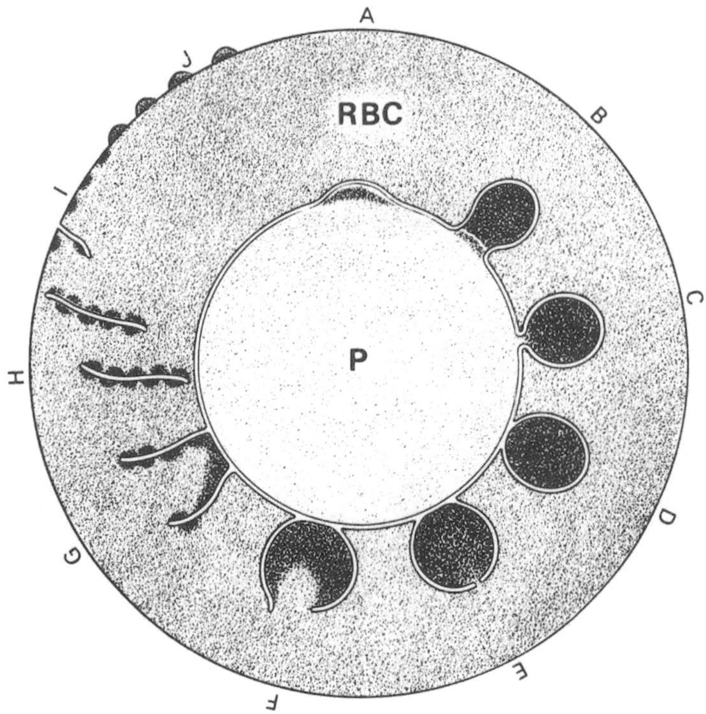

FIGURE 1. Schematic diagram showing a hypothetical route of transport of EDM from within the parasite cytoplasm to the cytoplasmic face of the RBCM under knobs. Transmission electron micrographs showing several of the stages drawn here have been published (Aikawa et al., 1986). The sequential steps are indicated clockwise A-J. A, EDM synthesized within the parasite cytoplasm (P) accumulates under the parasite plasmamembrane; B, the parasite plasmamembrane and parasitophorous vacuole membrane are pushed out into the erythrocyte cytoplasm; C, pinching-off and fusion of the parasite plasmamembrane; D, the EDM is now sealed within a membrane and lies in the lumen of the parasitophorous vacuole; E, fusion of the parasitophorous vacuole membrane and the membrane which surrounds the EDM; F, the EDM is now in the erythrocyte cytoplasm and surrounds long unit membrane clefts (Maurer's clefts) that are contiguous with the parasitophorous vacuole membrane; G, the EDM condenses to form smaller accretions the same size, electron density and shape as those seen under knobs; H, the Maurer's cleft bearing EDM moves through the erythrocyte cytoplasm to the erythrocyte surface membrane. The EDM does not appear to produce a knob protrusion on these unit membrane vesicles (see Aikawa et al., 1986); I, the cleft fuses with the surface membrane, placing the EDM on the cytoplasmic face of this membrane; J, the step of interaction of EDM with the surface membrane and cytoskeleton to produce a knob protrusion is drawn schematically as a final separate step. This step emphasizes the fact that knob production appears to be a unique property of interaction of EDM with the surface membrane.

appropriate trafficking system. The scheme shown in Figure 2 depicts the route of trafficking of this glycoprotein from presumed site of synthesis (rough endoplasmic reticulum, RER) to the PPM. This pathway is typical for a membrane protein destined to the plasmamembrane of eukaryotic cells (see Blobel, 1980; Engelman & Steitz, 1981; Wickner & Lodish, 1985).

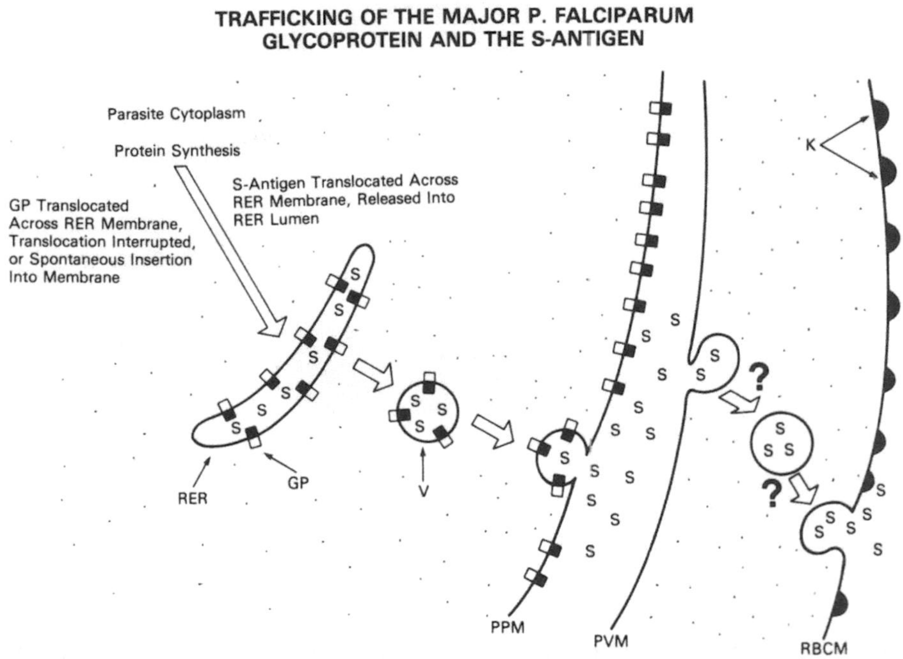

FIGURE 2. Schematic diagram showing trafficking of the malarial glycoprotein to the PPM and the S-antigen to the parasitophorous vacuole. : M_r ~200,000 P. falciparum glycoprotein, designated to indicate an asymmetric disposition in membranes; S: the P. falciparum S-antigen; RER: rough endoplasmic reticulum; V: unit-membrane bounded secretory vesicle; PPM: parasite plasmamembrane; PVM: parasitophorous vacuole membrane; RBCM: erythrocyte surface membrane; K: knobs.

The pathway of export of the other malarial protein depicted in Figure 10, the S-antigen, is also typical for a secreted water-soluble protein. The S-antigen is known to accumulate in the parasitophorous vacuole prior to rupture of the infected cell and merozoite reinvasion (Coppel, Crewman, Lingelbach, Brown, Saint, Kemp and Anders, 1983). At this time it is released into the extracellular fluid (Winchell, Ling and Wilson, 1984). Translocation (either spontaneous or co-translational) of the S-antigen into the lumen of the RER would then lead directly to its release, via secretory vesicles, into the parasitophorous vacuole. If

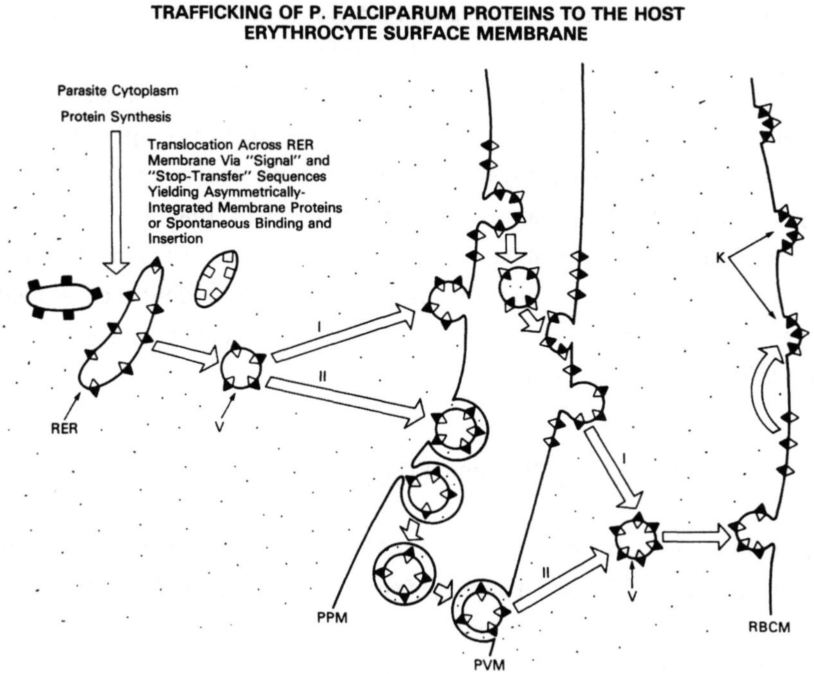

FIGURE 3. Schematic diagram showing trafficking of membrane proteins from the intracellular P. falciparum parasite to the RBCM. Three simple asymmetric dispositions of newly-synthesized malarial proteins are indicated at the RER: exclusively on the cytoplasmic membrane face; trans-membrane; exclusively on the membrane face within the lumen. Only the subsequent trafficking of the transmembrane protein is indicated. Two different schemes for crossing the topological barriers of PPM and PVM are presented (indicated I & II). Scheme I involves fusion of the vesicle V with the PPM, vesicle blebbing from the PPM, fusion with the PVM and subsequent blebbing from the PVM to produce a vesicle within the erythrocyte cytoplasm. Scheme II involves blebbing of the PPM around the original vesicle without fusion, transport of this bilamellar complex to the PVM where fusion of the outer membrane would release the vesicle into the erythrocyte cytoplasm. Thereafter the two schemes are indistinguishable. The alternative predictions from these schemes which would allow them to be distinguished experimentally are described in the text. The pathways shown would account for the trafficking of PfEMP1 to the external face of the RBCM and trafficking of PfEMP2 and PfHRP1 to the EDM under knobs. It should be noted that the protein asymmetry in membrane created within the parasite is identical to that expressed at the RBCM.

vesicles which cross the parasitophorous vacuole would express a different set of protein domains on their external surface.

Resolution of these questions will require detailed ultrastructural analysis of not only the presence or absence of malarial proteins on

blebbing of the PVM occurs as drawn in this figure, the S-antigen could conceivably be located as well within vesicles in the erythrocyte cytoplasm. Fusion of such vesicles with the RBCM would also deliver the S-antigen to the extracellular medium. The last two steps are conjectural for this particular antigen, but do demonstrate how a typical pathway of secretion from a eukaryotic cell could lead to secretion from a malaria-infected erythrocyte. The only malarial protein that we have studied that is secreted from intact cells, PfHRP2, does not appear to traverse the erythrocyte cytoplasm within a membrane-bound vesicle so cannot utilize this particular route.

Trafficking of PfEMP1, PfEMP2 and PfHRP1

Figure 3 shows two schemes for trafficking of membrane-bound proteins, such as PfEMP1, PfEMP2 and PfHRP1, from parasite cytoplasm to the RBCM. In both schemes, the protein asymmetry in membrane that was generated within the parasite is identical to that at the RBCM. One scheme (I) involves multiple vesicle fusion and blebbing events in order to cross the topological barriers of PPM and PVM. The other (scheme II) accomplishes this movement by maintaining the original vesicle and blebbing of the PPM around the vesicle, followed by transfer of this bi-membrane complex to the PVM where fusion of the outer membrane releases the vesicle.

Experiments can be designed to distinguish these alternative schemes. Scheme I predicts that proteins destined for the RBCM are also expressed on the PPM and that the proteins will have the same asymmetry in both membranes. Scheme II predicts that such proteins should not be identified as surface membrane components on the PPM. Scheme I also predicts that this class of malarial proteins are (transitory?) constituents of the PVM and that their asymmetry is reversed in this membrane compared with the RBCM: protein domains facing 'out' on the PVM vace 'inwards' when on the RBCM. These proteins do not become constituents of the PVM if Scheme II is applicable. Finally, in scheme II all membrane vesicles bear these proteins in the same asymmetrical orientation. Purification of vesicles should yield vesicles with only one set of protein domains expressed on the outside. In Scheme I, the

different membranes, but also the asymmetrical disposition of these proteins on different membranes. The development of reliable methods for purification of the PPM, PVM, RBCM and vesicle membranes from malaria-infected cells would rapidly advance this field.

Trafficking of PfHRP2

The pathway of movement of PfHRP2 cannot as yet be accounted for by extrapolation of the typical modes of protein trafficking seen in diverse eukaroytic cells to the malaria-infected erythrocyte. PfHRP2 has been identified in packets within the erythrocyte cytoplasm that do not appear to be defined by an outer membrane. Similar 'packets' are also seen within the parasite cytoplasm. Furthermore, PfHRP2 is secreted as a water-soluble protein from intact infected cells into the extracellular fluid (Howard et al., 1986). On the basis of these observations alone there is no simple explanation for the obvious capacity of PfHRP2 to cross three topological membrane barriers, the PPM, PVM and RBCM. If PfHRP2 was surrounded by a membrane while in transit from parasite cytoplasm to the RBCM, the scheme II shown in Figure 3 could transport it as a water soluble protein within the lumen of a vesicle. This does not appear to be the case. We are currently investigating whether PfHRP2 is able to translocate spontaneously across membranes.

Additional Questions. Since studies on the subcellular trafficking of malarial proteins and their topogenesis have just begun, it is appropriate to conclude by listing questions that can now be attacked with the reagents and methodologies currently available.

1. How does the malaria parasite direct the biogenesis of organelles involved in protein trafficking within the erythrocyte cytoplasm? Are these organelles controlled by the malaria parasite? Do these organelles also function in post-translational protein modification and other enzymatic functions?

2. What are the signals on different malarial proteins which direct them to different subcellular sites? Do different malarial proteins directed to the same site (eg PfEMP2 and PfHRP1) move in association or separately?

3. Does the movement of membranes and proteins from the parasite and PVM to the RBCM imply a reverse pathway of inward membrane movement? Or, are membranes shed from the RBCM to maintain the surface area of this

membrane? We note that in diverse eukaryotic cells in which vesicles bearing membrane proteins and/or secreted proteins move to the cell surface, an analogous system of vesicle movement in the opposite direction carries membrane and often receptor-ligand complexes back into the cell. The schemes shown in Figures 2 and 3 could operate also in reverse, thereby delivering components which bound to the RBCM to the parasitophorous vacuole (as soluble molecules) or the PPM (for membrane-bound molecules). The recent discovery of a transferrin receptor on P. falciparum-infected erythrocytes and movement of iron-loaded transferrin-receptor complex in to the parasite (Jungery, 1986) could be accounted for by these pathways of membrane movement.

4. Are there receptor-ligand interactions between different organelles involved in trafficking across the erythrocyte cytoplasm and the RBCM? What directs insertion of the delivered malarial proteins into the appropriate site at the RBCM (knobs or intervening membrane)? What is the mechanical basis for organelle movement (are microfilaments involved) or is it simply diffusion?

ACKNOWLEDGEMENTS

The authors are grateful to Mrs. Wilma Davis for typing of the manuscript and editorial assistance.

REFERENCES

Aikawa M (1966) The fine structure of the erythrocytic stages of three avian malarial parasites, Plasmodium fallax, P. lophurae and P. cathemerium. Am J Trop Med Hyg 15:449-71

Aikawa M (1971) Plasmodium: The fine structure of malarial parasites. Exp. Parasitol. 30:284-320

Aikawa M, Uni S, Andrutis AT, Howard RJ (1986) Membrane-associated electron-dense material of the asexual stages of Plasmodium falciparum: Evidence for movement from the intracellular parasite to the erythrocyte membrane. Am J Trop Med Hyg 35:30-36

Aikawa M, Uni S, Andrutis AT, Kirk KL, Cohen LA, Howard RJ (1986) Inhibition of knob formation on Plasmodium falciparum-infected erythrocytes by 2-fluoro-L-histidine (NATO ASI series: current volume)

Aley SB, Sherwood JA, Howard RJ (1984) Knob-positive and knob-negative Plasmodium falciparum differ in expression of a strain-specific malarial antigen on the surface of infected erythrocytes. J. Exp. Med. 160:1585-1590

Blobel G (1980) Intracellular protein topogenesis. Proc Natl Acad Sci USA 77:1496-1500

Coppel RL, Crewman AF, Lingelbach KR, Brown GV, Saint RB, Kemp DJ, Anders RF (1983) Isolate-specific S-antigen of Plasmodium falciparum contains a repeated sequence of eleven amino acids. Nature 306:751-756

Engelman DM, Steitz TA (1981) The spontaneous insertion of protein into and across membranes: The helical haripin hypothesis. Cell 23:411-422

Gumbiner B, Kelly RB (1982) Two distinct intracellular pathways transport secretory and membrane glycoproteins to the surface of pituitary tumor cells. Cell 28:51-59

Hadley TJ, Leech JH, Green TJ, Daniel WA, Wahlgren M, Miller LH, Howard RJ (1983) Plasmodium falciparum: a comparison of knobby (K+) and knobless (K-) parasites of two strains. Mol Biochem Parasitol 9:271-278

Howard RJ, Barnwell JW, Kao V (1983) Antigenic variation in Plasmodium knowlesi malaria: Identification of the variant antigen on infected erythrocytes. Proc Natl Acad Sci USA 80:4129-4133

Howard RJ, Lyon JA, Diggs CL, Haynes JD, Leech JH, Barnwell JW, Aley SB, Aikawa M, Miller LH (1984) Localization of the major Plasmdoium falciparum glycoprotein on the surface of mature intraerythrocytic trophozoites and schizonts. Mol Biochem Parasitol 11:349-362

Howard RJ, Lyon J, Uni S, Saul AJ, Aley SB, Klotz F, Panton LJ, Sherwood JA, Marsh K, Aikawa M, Rock EP (1987) Trafficking of a $M_r \sim 300,000$ Plasmodium falciparum protein (PfEMP2) from the intraerythrocytic parasite to the cytoplasmic face of the host cell membrane (submitted)

Howard RJ, Uni S, Aikawa M, Aley SB, Leech JH, Lew AM, Wellems TE, Rener J, Taylor DW (1986) Secretion of a malarial histidine-rich protein (PfHRP2) from Plasmodium falciparum-infected erythrocytes. J Cell Biol 103:1269-1277

Jungery M, Rodriguez-Lopez MH (1986) Plasmodium falciparum-derived transferrin receptors on the surface of infected erythrocytes. J Cell Biochem (Suppl) 10A:163

Kilejian A (1979) Characterizatin of a protein correlated with the production of knob-like protrusions on membranes of erythrocytes infected with Plasmodium falciparum. Proc Natl Acad Sci USA 76:4650-4653

Klotz FW, Cooper MD (1986) A feline thymocyte antigen defined by a monoclonal antibody (FT_2) identifies a subpopulation of non-helper cells capable of specific cytotoxicity. J Immunol 136:2510-2514

Leech JH, Barnwell JW, Aikawa M, Miller LH, Howard RJ (1984) Plasmodium falciparum malaria: association of knobs on the surface of infected erythrocytes with a histidine-rich protein and the erythrocyte skeleton. J Cell Biol 98:1256-1264

Leech JH, Barnwell JW, Miller LH, Howard RJ (1984) Identification of a strain-specific malarial antigen expressed on the surface of Plasmodium falciparum-infected erythrocytes. J Exp Med 159:1567-1575

Lodish HF, Braell WA, Schwartz AL, Strous GJM, Zilberstein A (1981) Synthesis and assembly of membrane and organelle proteins. In, Membrane Research, Classic Origins and Current Concepts. Int Rev Cytol Suppl. 12. AL Muggleton-Harris Ed (New York, Academic Press) pp 247-309

Palade G (1975) Intracellular aspects of the process of protein synthesis. Science 189:347-358

Rock EP, Marsh K, Maloy WL, Taylor DW, Wellems TE, Saul AJ, Howard RJ (1986) Comparative analysis of the Plasmodium falciparum histidine-rich proteins PfHRP1, PfHRP2 and PfHRP3 in malaria parasites of diverse origin. (submitted)

Rothman JE (1986) The compartmental organization of the Golgi apparatus. Sci American 253:74-89

Rudzinska MA, Trager W (1968) The fine structure of trophozoites and gametocytes in Plasmodium coatneyi. J Protozool 15:73-88

Taylor D, Parra M, Chapman GB, Stearns ME, Rener J, Aikawa M, Uni S, Panton LJ, Aley SB, Howard RJ (1986) Plasmodium falciparum histidine-rich protein 1 (PfHRP1): Its export from the parasite and insertion into the host erythrocyte skeleton under knobs (submitted)

Vernot-Hernandez JP, Heidrich HG (1984) Time-course of synthesis transport and incorporation of a protein identified in purified membranes of host erythrocytes infected with a knob-forming strain of Plasmodium falciparum. Mol Biochem Parasitol 12:337-350

Wellems TE, Howard RJ (1986) Homologous genes encode two distinct histidine-rich proteins in a cloned isolate of Plasmodium falciparum. Proc Natl Acad Sci USA (in press)

Wickner WT, Lodish HF (1985) Multiple mechanisms of protein insertion into and across membranes. Science 230:400-407

Winchell EJ, Ling IT, Wilson RJM (1984) Metabolic labeling and characterizatin of S-antigen, the heat-stabile, strain-specific antigens of Plasmodium falciparum. Mol Biochem Parasitol 10:287-296

INHIBITION OF KNOB FORMATION ON *PLASMODIUM FALCIPARUM*-INFECTED ERYTHROCYTES BY 2-FLUORO-L-HISTIDINE

[1]Masamichi Aikawa, [1]Shigehiko Uni, [2]Annette T. Andrutis, [3]Kenneth L. Kirk, [3]Louis A. Cohen and [2]Russell J. Howard

[1]Institute of Pathology, Case Western Reserve University, Cleveland, Ohio 44106, [2]Laboratory of Parasitic Diseases, National Institute of Allergy & Infectious Diseases, and [3]Laboratory of Chemistry, National Institute of Arthritis, Diabetes and Digestive and Kidney Diseases, National Institutes of Health, Bethesda Maryland 20892

INTRODUCTION

Knob protrusions and associated submembrane electron dense material (EDM) appear on the membrane of erythrocytes infected with *Plasmodium falciparum* (Aikawa et al. 1972, Luse and Miller 1971). These knobs form focal junctions with the endothelial cells of capillaries, resulting in the sequestration of infected erythrocytes along the vascular endothelium (Luse and Miller 1971, Udeinya et al. 1981, Oo et al. 1986). Parasites that do not produce knobs (K- phenotype) neither attach to endothelial cells *in vitro*, nor do they sequester *in vivo* (Barnwell et al. 1983). These findings indicate the presence of cytoadherence properties at knobs and the importance of knobs for parasite survival *in vivo*.

Knob bearing (K+) parasites synthesize a protein of unusually high histidine content (Pf HRP 1) that is not expressed by knob negative (K-) variants (Kilejian 1979). Pf HRP 1 has Mr 80,000-110,000 in different parasites (Hadley et al. 1983, Leech et al. 1984, Vernot-Hernandez and Heidrich 1984) and has been localized to purified membranes (Vernot-Hernandez and Heidrich 1984) and cytoskeletal material (Leech et al. 1984) from K+ infected erythrocytes. Recently, using a specific monoclonal antibody and immuno-electron microscopy we demonstrated export of Pf HRP 1 from the intraerythrocytic parasite to the surface membrane and incorporation of this protein into the EDM at knobs (unpublished data). EDM, identical in appearance to that under knobs, was observed by electron microscopy at the parasite plasma membrane as well as at unit membrane-bounded Maurer's clefts (Aikawa et al., 1986). These clefts presumably transport the EDM and associated Pf HRP 1 to the surface membrane knobs. Asexual *P. falciparum* parasites also express at least two other HRP. Pf HRP 2 (apparent Mr 60,000-80,000 on SDS-PAGE) is secreted as a water-soluble protein from intact infected cells of both K+ and K- parasites (Howard et al. 1986B). Pf HRP 3 (Wellems and Howard, 1986), of Mr 30,000-45,000 is not yet well characterized with respect to subcellular location. Both Pf HRP 2 (Wellems and Howard, 1986), and Pf HRP 3 (Stahl et al. 1985, Wellems and Howard 1986) have been sequenced and shown to contain 34% and 28% histidine respectively.

In view of the unusual histidine content of Pf HRP 1 and its location at knobs, together with the existence of two other HRP in asexual

NATO ASI Series, Vol. H11
Host-Parasite Cellular and Molecular Interactions in Protozoal Infections. Edited by K.-P. Chang and D. Snary

parasites, we tested whether histidine analogues might affect growth of *P. falciparum in vitro* (Howard et al. 1986A). 2-Fluoro-L-histidine (2-F-L-histidine) was particularly effective, compared with 24 other analogues of L-histidine, in inhibition of *in vivo* growth and inhibition of incorporation of ^{3}H- amino acids into malarial proteins. 2-F-L-histidine has been shown to inhibit protein synthesis by bacteria, viruses and other eukaryotic cells (De Clercq et al. 1978).

The current study examines the effects of 2-F-L-histidine on the appearance and subcellular distribution of EDM in K+ *P. falciparum* and the relationship between concentration of 2-F-L-histidine and the number of knobs expressed at the cell surface.

MATERIALS AND METHODS

An isolate of knob-positive *P. falciparum*, Malayan Camp-strain was obtained from non-splenectomized *Aotus triviratus* monkeys (Leech et al. 1984). Infected blood (20-30% parasitemia) was cryopreserved as ring-stages and stored over liquid nitrogen (Leech et al. 1984). The cryopreserved parasites were thawed and cultured at 37°C in RPMI 1640 medium containing 0.025 mM histidine (25% of the normal concentration) and supplemented with 30 mM HEPES, pH 7.2, 10 mg/ml hypoxanthine, 2 mg/ml D(-) glucose, 50 ug/ml gentamicin, 15% (v/v) horse serum, 3 ml/100 ml of 7.5% (w/v) $NaHCO_3$ and various concentrations of 2-F-L-histidine or 5-Fluoro-L-histidine (5-F-L-histidine). Culture was performed for 18-24 hrs, when the parasite in control cultures lacking histidine analogues had matured to the trophozoite or early schizont stage. Flat-bottomed 96-well culture trays (Costar, Cambridge, MA) were used for culture, each well containing 100 ul of cell suspension plus 200 ul of medium containing histidine analogue to give final concentrations of 0, 0.025, 0.05, 0.1, 0.125, 0.5 or 1.0 mM (Howard et al. 1986A).

For electron microscopy the cells from four wells treated with the same analogue concentration were pooled. The samples were fixed in 2% glutaraldehyde, 50 mM sodium phosphate, pH 7.4 plus 4% w/v sucrose (1 hr at 4°C), pelleted by centrifugation and resuspended in 20 mM sodium phosphate pH 7.4, 15 mM NaCl. The cells were post fixed in 1% OsO_4, dehydrated in a graded series of ethyl alcohol solutions, embedded in Epon 812 and sectioned with a diamond knife on a Porter-Blum Mt-2 ultramicrotome. After staining with 1% uranyl acetate and lead citrate the sections were examined with a JEOL 100CX electron microscope.

RESULTS

Electron Dense Material (EDM) in Untreated Infected Erythrocytes

Since the presence of EDM in *P. falciparum*-infected erythrocytes has been described previously (Aikawa et al. 1986), only a brief description is provided here for comparison with the results using histidine analogues. EDM measuring about 380 nm in height and about 190 nm in width was seen between the parasite plasma membrane and the parasitophorous vacuole membrane with trophozoite stages (Fig. 1, inset). The EDM generally gave a uniformly electron dense appearance, less electron-dense than the cup-shaped material under knobs. The area of EDM and frequency of these bodies decreased as parasites matured into

TABLE 1
EFFECT OF 2-F-L-HISTIDINE ON EDM
ASSOCIATED WITH THE PARASITOPHOROUS VACUOLE

Concentration of 2-F-L-histidine (mM)	Infected Erythrocytes with EDM (%)	Area of EDM per Infected Erythrocyte (um^2)*
0	67	0.021 ± 0.003
0.025	74	0.029 ± 0.004
0.05	80	0.032 ± 0.004
0.1	87	0.038 ± 0.004**
0.25	94	0.053 ± 0.007**

TABLE 2
EFFECT OF 2-F-L-HISTIDINE ON KNOBS

Concentration of 2-F-L-histidine (mM)	Infected Erythrocytes with Knobs (%)	Number of Knobs per Infected Erythrocyte*
0	62	9.6 ± 1.4
0.025	56	3.8 ± 0.8**
0.05	36	2.4 ± 0.3**
0.1	24	1.8 ± 0.3**
0.25	17	1.7 ± 0.2**

* Mean values ± standard error calculated from electron photomicrographs (see Materials and Methods). Values refer to erythrocyte sections.

** Values significantly different ($p<0.01$, t-test) compared with 0 mM 2-F-L-histidine.

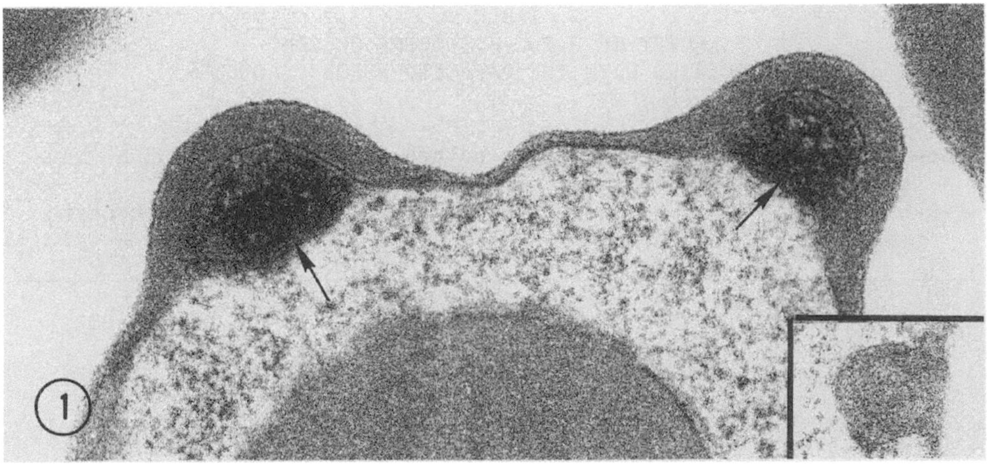

Figure 1. Electron micrograph of EDM (arrow) associated with the parasitophorous vacuole in a P. falciparum infected cell treated with 0.25 mM 2-F-L-histidine. The density of EDM increased after 2-F-L-histidine treatment. X50,000. Inset: EDM in an untreated infected erythrocyte. X50,000.

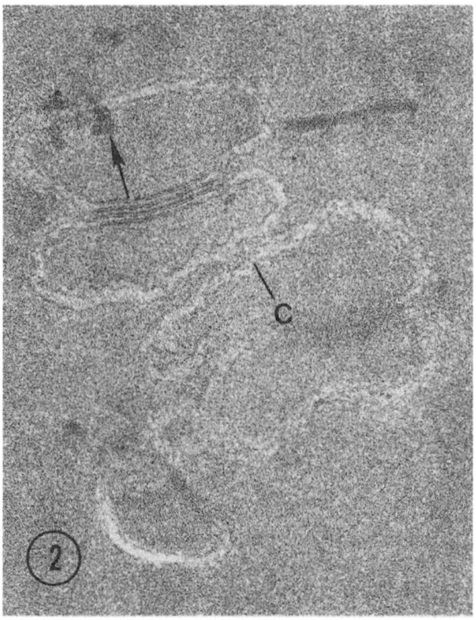

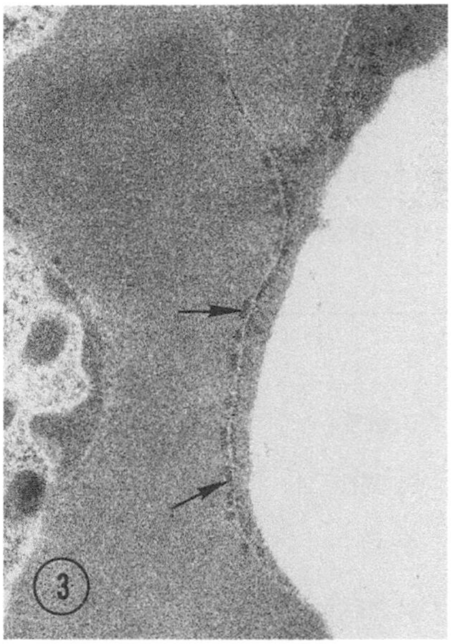

Figure 2. Maurer's clefts (C) in the erythrocyte cytoplasm of a trophozoite-infected cell without treatment. EDM (arrow) associated with these membranes is similar in size & density to that seen under knobs at the cell surface. X55,000.

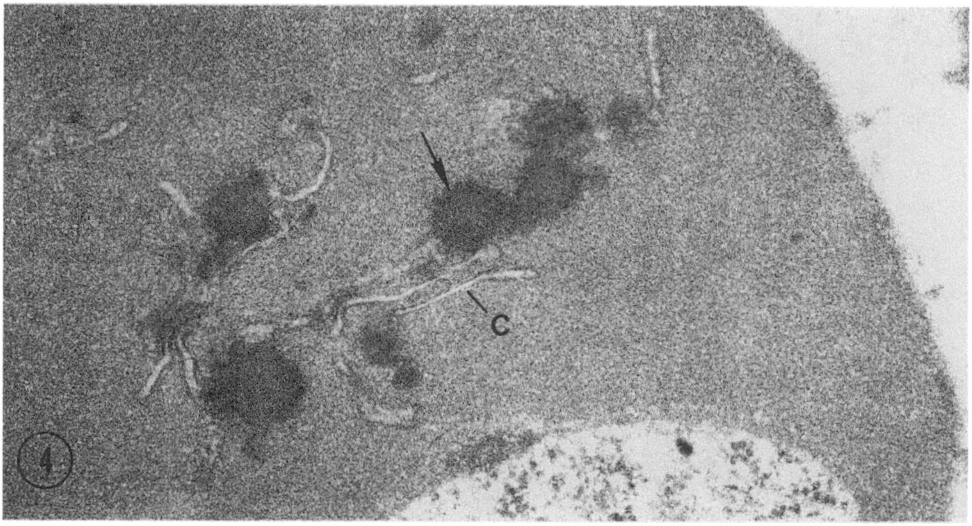

Figure 4. Electron micrograph of EDM (arrow) associated with Maurer's clefts (C) after treatment with 0.25 mM 2-F-L-histidine. The EDM is denser & larger than in control. X40,000.

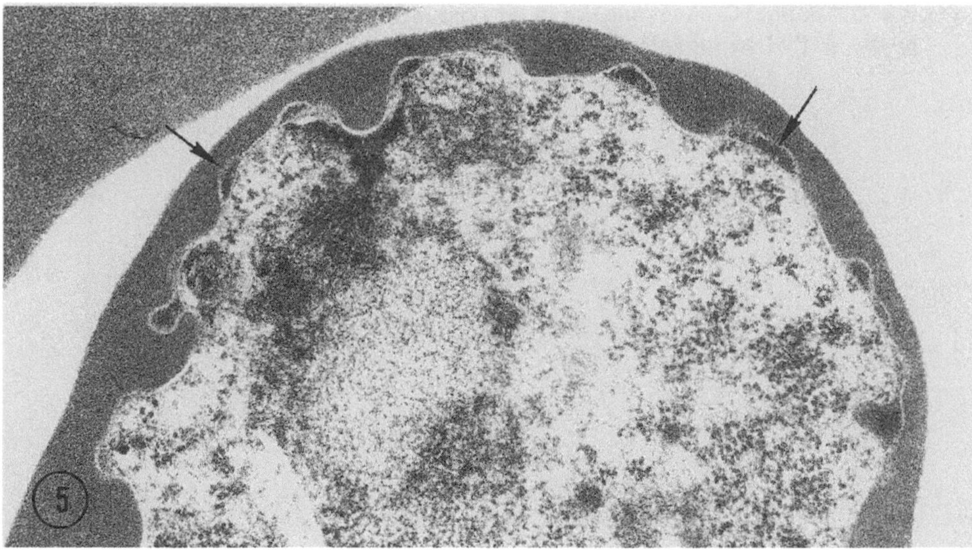

Figure 5. An infected erythrocyte after treatment with 0.25 mM 2-F-L-histidine showing a cell surface membrane devoid of knobs & extensive EDM (arrow) in the parasitophorous vacuole. X29,000.

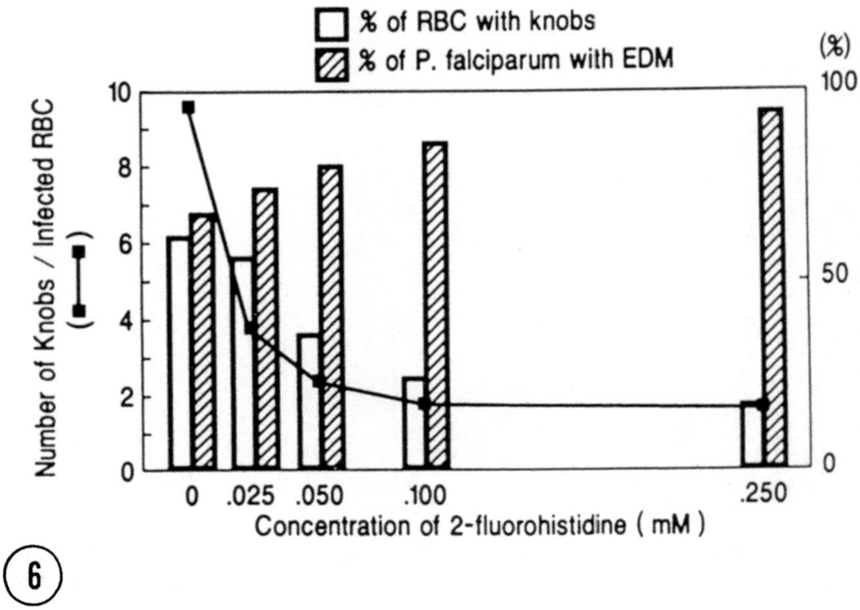

Figure 6. Quantitative analysis of the effects of 2-F-L-histidine on knobs & EDM associated with the parasitophorous vacuole.

schizonts.

EDM was also associated with unit membrane clefts (Maurer's clefts) in the host erythrocyte cytoplasm of K+ parasites. This EDM was of the same density and morphology as that seen under knobs at the cell surface (Fig. 2). K- P. falciparum parasites which lack EDM at the infected cell surface, also lacked this EDM associated with Maurer's clefts. Ring and early trophozoite stages of K- parasites also lacked this EDM. From observation of continuity between clefts bearing EDM and the parasitophorous vacuole membrane and continuity between these clefts and the cell surface membrane, we proposed that the EDM is transported from the parasite plasma membrane via Maurer's clefts to the inner face of knobs present at the erythrocyte membrane.

Effects of 2-F-L-histidine on EDM

2-F-L-histidine increased the amount of EDM in the parasitophorous vacuole. The surface area of EDM per cell and frequency of EDM with different concentrations of 2-F-L-histidine are shown in Table 1. Both of these values increased compared to control cells with the lowest concentration tested (0.025 mM) and showed additional increments with higher levels of this analogue. Figure 1 shows the appearance of this EDM with 0.25 mM 2-F-L-histidine. Compared with control cells (Fig. 1, inset), we also noted that the density of this EDM increased with increasing levels of 2-F-L-histidine. The structurally related histidine analogue, 5-F-L-histidine, was shown to have no effects of P. falciparum growth or protein synthesis at levels up to 1.0 mM (Howard et al. 1986A). 5-F-L-histidine also had no effect on the appearance of EDM at the same concentration.

2-F-L-histidine also increased the amount and density of EDM associated with Maurer's clefts in the host erythrocyte cytoplasm (Figs. 3, 4). With 0.05 mM 2-F-L-histidine, there was an increase in this EDM such that the entire length of a cleft in this section was covered with this material. The EDM was on the cytoplasmic face of this membrane and of the same size and density as the submembrane EDM at knobs. At high concentrations of 2-F-L-histidine the size of EDM associated with the cytoplasmic clefts increased to include deposits as large as 500 nm diameter.

Effects of 2-F-L-histidine on Knobs

In the culture lacking histidine analogue, 62% of infected erythrocytes seen in thin section bore knobs at the cell surface. The knob-positive erythrocyte had a mean value of 9.6 knobs per cell in these sections (Table 2).

Addition of 0.025 mM 2-F-L-histidine decreased the percentage of knob-positive infected erythrocytes to 56% and decreased the number of knobs per cell to 3.8 (mean value) (Table 2). As the concentration of 2-F-L-histidine increased the number of infected cells bearing knobs and number of knobs both decreased. Figure 5 shows the absence of surface membrane knobs and submembrane knob-associated EDM, with a trophozoite-infected cell treated with 0.25 mM 2-F-L-histidine. There were extensive deposits of EDM in the parasitophorous vacuole of this cell (Fig. 5). 5-F-L-histidine has no effect on the number of knobs or their appearance at 1.0 mM concentration.

The quantitative analyses of surface area of EDM associated with the parasitophorous vacuole membrane, percentage of infected erythrocytes bearing knobs and number of knobs per infected cell at different concentrations of 2-F-L-histidine are plotted for comparison in Figure 6.

DISCUSSION

We have used transmission electron microscopy to establish that the histidine analogue, 2-F-L-histidine, arrests knob formation in P. falciparum asexual blood stage parasites and perturbs the trafficking of EDM from the parasite to the erythrocyte membrane. Our previous studies had shown that in untreated cells EDM accumulates in the parasitophorous vacuole, is transported via Maurer's clefts across the erythrocyte cytoplasm and is finally deposited under knobs (Aikawa et al. 1986). Cryo-thin-section immuno-electron microscopy has shown that the histidine-rich-protein Pf HRP 1 is a constituent of EDM under knobs (unpublished data) and presumably, also a constituent of EDM in transit from parasite to knobs. We had also established that compared with a panel of other histidine analogues, 2-F-L-histidine exerted the most inhibitory effect on in vitro morphological development of K+ P. falciparum and inhibition of Pf HRP 1 synthesis (Howard et al. 1986). The current study set out to answer whether this histidine analogue would affect knob formation and EDM accumulation due to its effects on histidine metabolism and incorporation into protein. Since the structurally related analogue, 5-F-L-histidine, had no effect at 1.0 mM concentration on morphological development or protein synthesis (Howard et al. 1986A), we compared the electron microscopic results with the 5-fluoro- and 2-fluoro-compounds.

Quantitative analysis of EDM in trophozoite-infected erythrocytes (Table 1) showed that as the concentration of 2-F-L-histidine increased (0.025-0.25 mM), so did the amount and frequency of occurence of this material in the parasitophorous vacuole. Conversely, increasing levels of 2-F-L-histidine led to a decrease in the proportion of infected cells bearing knobs and decreased number of knobs per cell. 2-F-L-histidine also increased the amount of EDM seen on Maurer's clefts in the host erythrocyte cytoplasm. At the highest levels of 2-F-L-histidine tested many bizarre accumulations of EDM were noted in the erythrocyte cytoplasm.

It should be noted that movement of EDM from the parasitophorous vacuole through the erythrocyte cytoplasm is a corollary of knob formation. Although K- P. falciparum parasites accumulate EDM in this vacuole, and do have Maurer's clefts in the erythrocyte cytoplasm, transmission electron microscopy failed to identify EDM beyond the vacuolar membrane in these parasites (Aikawa et al. 1986). Taking this observation and the opposite effects of 2-F-L-histidine on knob formation and accumulation of EDM in the parasitophorous vacuole, we suggest the following points of action of this compound:
2-F-L-histidine blocks the trafficking of EDM beyond the parasitophorous vacuole membrane (leading to an increase in its size and increased electron density) and blocks the incorporation of EDM into the erythrocyte membrane (leading to an increase in the amount of EDM associatied with Maurer's clefts and an increase in the density and size

of this material). These effects combine to reduce the number of knobs. As anticipated from the effect documented here of inhibition of knob formation by 2-F-L-histidine, this compound also decreased in vitro cytoadherence of K+ P. falciparum such that after treatment with $\geq$ 0.1 mM concentrations cytoadherence is abolished (unpublished data). In contrast to 2-F-L-histidine, 5-F-L-histidine had no effects on the appearance or amount of EDM or its trafficking to the erythrocyte membrane. The mechanistic difference between effects of these fluoro-analogues has not been explored.

The effects of 2-F-L-histidine lend further support to the concept that EDM in the parasitophorous vacuole is the precursor of EDM under knobs. Since 2-F-L-histidine inhibits knob formation it will be of interest to test whether this compound can be used in vivo to prevent cytoadherence.

ACKNOWLEDGEMENTS

This work was supported in part by the UNDP/World Bank/WHO Special Programme for Research and Training in Tropical Diseases, a USPHS research grant (AI-10645), the Agency for International Development (DPE-0453-A-004027) and the US Army Research and Development Command (DAMD 17-79-C-9029). This is contribution No.1802 to the Army Research Program on Antiparasitic Drugs.

REFERENCES

Aikawa M, Uni Y, Andrutis AT & Howard RJ (1986) Membrane-associated electron-dense material of the asexual stages of Plasmodium falciparum: Evidence for movement from the intracellular parasite to the erythrocyte membrane. Am J Trop Med Hyg 35:30-36

Aikawa M, Rabbege JR & Wellde BT (1972) Junctional apparatus in erythrocytes infected with malarial parasite. Z Zellforsch 124:72-75

Barnwell JW, Howard RJ & Miller LH (1983) Influence of the spleen on the expression of surface antigens on parasitized erythrocytes. In: Malaria & the red cell. CIBA Foundation Symposium, 94, pp 117-136

De Clercq E, Billian A, Edy VA, Kirk KL & Cohen LA (1978) Antimetabolic activities of 2-fluoro-L-histidine. Biochem Biophys Res Commun 82:840-846

Hadley TJ, Leech JH, Green TJ, Daniel WA, Wahlgren M, Miller, LH & Howard RJ (1983) Plasmodium falciparum: A comparison of knobby (K+) & knobless (K-) parasites of two strains. Mol Biochem Parasitol 9:271-278

Howard RJ, Andrutis JH, Ellis WY, Cohen LA & Kirk KL (1986A) Inhibitory effects of histidine analogues on growth & protein synthesis by Plasmodium falciparum in vitro. Biochem Pharm 35:1589-1496

Howard RJ, Uni S, Aikawa M, Aley SB, Leech JH, Lew AM, Wellems TE, Rener

J & Taylor DW (1986B) Secretion of a malarial histidine-rich protein (PF HRP 2) from Plasmodium falciparum-infected erythrocytes. J Cell Biol (In press)

Kilejian A (1979) Characterization of a protein correlated with the production of knob-like protrusions on membranes of erythrocytes infected with Plasmodium falciparum. Proc Natl Acad Sci 76:4650-4653

Leech JH, Barnwell JW, Aikawa M, Miller, LH & Howard, RJ (1984) Plasmodium falciparum malaria: Association of knobs on the surface of infected erythrocytes with a histidine-rich protein & the erythrocyte skeleton. J Cell Biol 98:1256-1264

Luse SA & Miller LH (1971) Plasmodium falciparum malaria: Ultrastructure of parasitized erythrocytes in cardiac vessels. Am J Trop Med Hyg 20:655-660

Oo MM, Aikawa M, Than T, Aye TM, Myint PI, Igarashi I & Schoene W (1986) Cerebral malaria: A pathological study. J Neuropath Exptl Neurol (In press)

Stahl HD, Kemp DJ, Crewther PE, Scanlon DB, Woodrow G, Brown GV, Bianco AE, Anders RF & Coppel RL, (1985) Sequence of a cDNA encoding a small polymorphic histidine- and alamine-rich protein from Plasmodium falciparum. Nucl Acid Res 13:7837-7846

Udeinya IJ, Schmidt JA, Aikawa M, Miller LH & Green I (1981) Falciparum malaria-infected erythrocytes specifically bind to cultured human endothelial cells. Science 213:555-557

Vernot-Hernandez JP & Heidrich HG (1984) Time-course of synthesis, transport & incorporation of a protein identified in purified membranes of host-erythrocytes infected with a knob-forming strain of Plasmodium falciparum. Mol Biochem Parasitol 12:337-350.

Wellems TE & Howard RJ (1986) Homologous genes encode two distinct histidine-rich proteins in a cloned isolate of Plasmodium falciparum. Proc Natl Acad Sci USA (In press)

THE KNOB PROTEIN GENE OF PLASMODIUM FALCIPARUM

A. Kilejian, Y.D. Sharma and Y.F. Yang

The Public Health Research Institute, New York, NY 10016, USA

INTRODUCTION

Membranes of erythrocytes infected with *Plasmodium falciparum* develop protrusions called knobs (Trager *et al*. 1966). These structures mediate the adhesion of parasitized cells to venous endothelium (Luse & Miller 1971) and are essential for the survival of parasites in natural infection; laboratory derived variants which lose their ability to produce knobs (K^-) also lose their pathogenecity in susceptible hosts (Langreth & Peterson, 1985; Green *et al*. 1985).

Current literature suggests that physiologically functional knobs are the result of complex interactions between the host cell membrane and parasite constituents. The best characterized parasite polypeptide which is known to be essential for knob formation is the knob protein (KP). KP has been identified from several geographical isolates as a membrane-associated polypeptide which incorporates relatively large amounts of exogenous histidine (Kilejian 1979,1980; Hadley *et al*. 1983; Leech *et al*. 1984; Gritzmacher & Reese 1984; Vernot-Hernandez & Heidrich 1984). Although KP is not the only histidine-rich polypeptide of *P. falciparum*, it is the only one which is expressed in knob forming (K^+) but not in K^- parasites. Sequence analysis of the KP gene of the Gambian isolate FCR-3 revealed an interesting primary structure (Kilejian *et al*. 1986; Sharma & Kilejian, unpublished); the N-terminal of the polypeptide has a histidine-rich domain while the C-terminal half is lysine-rich. Similar to the histidine-rich protein of the avian parasite *Plasmodium lophurae*, the histidine-rich domain of KP has polyhistidine sequences which most likely form the basis for the serological cross-reactivity of the two proteins (Kilejian 1983).

NATO ASI Series, Vol. H11
Host-Parasite Cellular and Molecular Interactions
in Protozoal Infections. Edited by K.-P. Chang and D. Snary

Although some polypeptides of P. falciparum may be polymorphic and antigenically variant, it is highly likely that functional domains of structural polypeptides, such as KP, which are vital for successful host-parasite interactions, have been conserved through the evolution of different geographical isolates. Here we describe the close similarity of the genomic organization of the KP gene of a Honduran isolate, clone HB-3 (Bhasin & Trager 1985), to the KP gene of the Gambian isolate FCR-3.

MATERIALS AND METHODS

The preparation of K^+ and K^- FCR-3 (Kilejian 1979), FCR-3 K^- clone D-4 (Trager et al. 1981) and Honduran K^+ clone HB-3 (Bhasin & Trager 1985) have been described. DNA was prepared from schizonts of different isolates as described (Kilejian et al. 1986). DNA isolated from HB-3 was digested with Dra I and size-fractionated on an agarose gel. A fraction which gave a strong hybridization signal with a cDNA clone encoding the KP gene of FCR-3 was cloned into the plasmid pUC8 and a selected recombinant clone, A66, was used for further studies. Hybridization studies were carried out with fractionated DNA in dried agarose gels (Purrello & Balazs 1983).

RESULTS AND DISCUSSION

Our previous studies showed strong hybridization of a cDNA encoding the KP gene of the Gambian isolate FCR-3 with DNA fragments of the Honduran K^+ clone HB-3 (Kilejian et al. 1986). For a more detailed comparison of the two geographical isolates, a cross-hybridizing Dra I fragment of HB-3 was cloned and the recombinant plasmid A66 was isolated. The restriction enzyme map of the insert DNA of A66 proved to be identical to the comparable region of the KP gene of FCR-3 (Fig. 1). The presence of internal Dra I sites suggests that clone A66 was derived from partially digested DNA. Also, the

3' end of the insert does not correspond to a Dra I site in FCR-3 clones; there may have been a deletion during cloning or A66 may have an additional Dra I site at the 3' end.

As shown in Fig. 1, the KP gene of FCR-3 is interrupted by an intron which separates the signal peptide sequence from the rest of the coding region. A66 apparently includes the signal peptide sequence, the intron and most of the coding sequence but it is lacking the 3' end of the gene. For additional comparative studies, three probes were prepared using the central coding region, the 3' and 5' flanking fragments of the KP gene of FCR-3; these fragments are shown in Figure 1 as A, B and C, respectively.

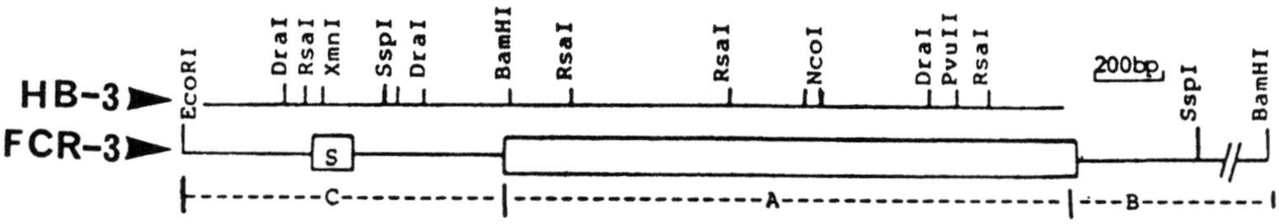

FIGURE 1. Comparison of the restriction enzyme maps of A66 and the KP gene of FCR-3. Boxed regions represent coding sequences. Fragments A, B and C are various regions of FCR-3 genomic DNA clones used as hybridization probes. S, signal peptide sequence. The presence of two SspI and NcoI sites are deduced from the sequence analysis of FCR-3.

A66 and an available recombinant clone (EP9) of FCR-3 were digested with Rsa I and genomic DNAs of HB-3 and FCR-3 were digested with Ssp I. The digested samples were analyzed on an agarose gel and the dried gel was hybridized with probe A. The results (Fig. 2A) did not reveal any apparent differences between the coding regions of the two isolates. Our previous studies had shown that a 2.6 Kb Ssp I fragment includes and extends beyond the second exon of the KP gene of FCR-3 (Fig. 1).

Genomic DNA of HB-3 and K^+ as well as K^- variants of FCR-3 were digested with EcoRI or Xmn I and hybridized with the probes made from the 3' and 5' flaking regions of probe A, respectively. The K^- variants of FCR-3 were included to confirm our previous observations which indicated that while the lack of expression of KP in K^- clone D-4 was due to a deletion, the KP gene was apparently present in uncloned K^- FCR-3. When the hybridized gels were washed with 4XSSC (0.15 M NaCl, 0.015 M Na citrate) at 42°C, multiple DNA fragments showed a hybridization signal in all samples. After washing at 60°C only a 14-15 Kb EcoRI fragment (Fig. 2B) and an 8-8.5 Kb Xmn I fragment (Fig. 2C) retained a strong hybridization signal. While clone D-4 did not hybridize with either probe, there were no detectable differences between the DNA of uncloned K^+ and DNAs prepared from two independently cultured K^- parasites (Fig. 2C, lanes 2 and 6). Also, a DNA sample prepared from FCR-3 parasites which has been previously passaged through <u>Aotus</u> monkeys showed no obvious change (Fig. 2C, lane 1).

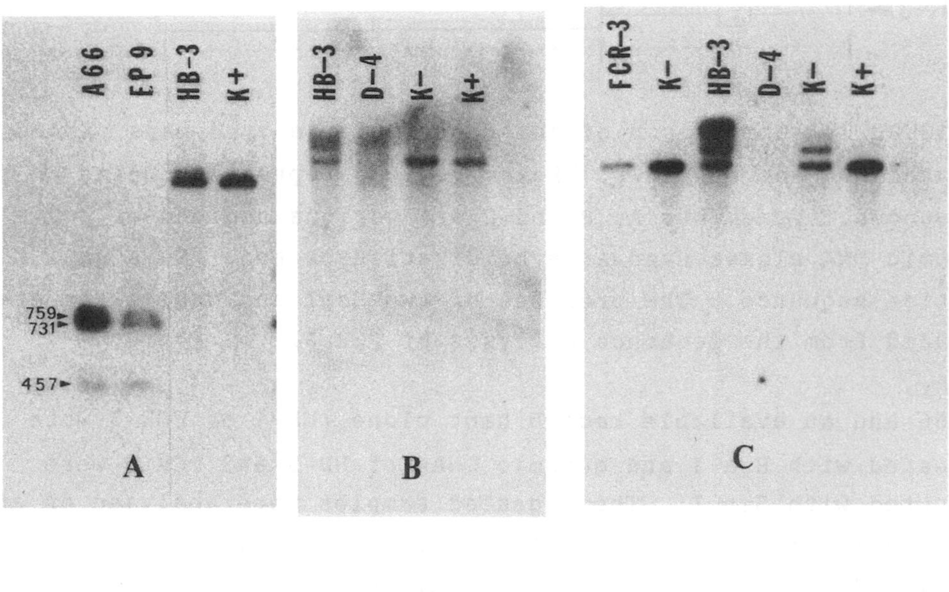

FIGURE 2. Hybridization of probes made from cloned genomic DNA fragments of FCR-3 to plasmids A66 and EP9 and to genomic DNA isolated from HB-3 and from variants of FCR-3. Panel A, B and C were hybridized with the respective probes shown in Fig. 1. Other details are described in text.

In a recent study Pologe and Ravetch (1986) reported differences in the extent of deletions from the 3' end of the KP gene of FCR-3 clone D-4 and a cloned K^{-} variant of FVO, a Vietnamese isolate; the 200 bp 5' end fragment which they used as a probe for these studies is lacking in our genomic and cDNA clones. They made the interesting suggestion that in these cloned K^{-} isolates the truncated KP genes are relocated to the end of chromosomes. Their observation of heterogeneity in the length of rearranged DNAs suggests that after the isolation of the parent clones there has been continued rearrangements. Environmental factors obviously play a role in the expression and probably on genomic reorganization of malarial genes. It would be of interest to examine K^{-} variants which have been derived by passage of parasites in splenectomized monkeys (Barnwell et al. 1983).

One of the interesting and unique structural features of the KP gene of FCR-3 that we had noted previously was a tandemly repeated TTTTA sequence in the intron. To confirm that this is also conserved in the Honduran isolate, we sequenced an Ssp I/Bam HI fragment which had been labeled at the Bam HI site. The sequence proved to be identical to that noted in FCR-3 (Fig. 3). It is of interest that although there are several reports of signal peptide sequences of malarial genes, the only reports of an intron between the signal peptide and the rest of the coding region appear to be limited to histidine-rich proteins: the histidine-rich protein of *P. lophurae* and two small histidine-rich proteins of *P. falciparum* (Ravetch et al. 1984; Wellems and Howard 1986). However, the unique sequence of the intron of the KP gene is quite different from that of the histidine-rich protein of *P. lophurae*.

5'
TTTTATTTTTATTTTATTTTATTTTATTTTAATTTTATTTTATTTTATTTTAATTTTATTTTA
ATTTTATTTTAATTTTATTTTAATTTTATTTTATTTTAATTTTATTTTATTTTATTTTAATTT
TATTTTATTTTATTTTAATTTTATTTTATTTTATTTTAATTTAATTTTTATTTTTTTTTT

FIGURE 3. Partial sequence of the intron of the KP gene of HB-3.

The regulation of expression of the KP gene poses some interesting questions. While asexual stages of K^- variants thrive in laboratory cultures, under natural selection conditions such parasites would be apparently eliminated. However, gametocytes of P. falciparum do not form knobs, yet they obviously survive long enough in circulation to transmit the infection. There is no evidence to suggest that lack of expression of the KP gene in gametocytes is due to genetic rearrangements. Observations on chromosomal and antigenic polymorphism as well as changes in repetitive DNA of P. falciparum have suggested extensive genomic fluidity. Some of the gross changes which are most apparent in cloned parasite isolates may reflect abberant mutations induced by selective pressures of laboratory cultures. In view of the high virulence and prevalence of P. falciparum infections, there have to be extensive regulatory mechanisms to conserve the structural integrity of functionally important domains of malarial polypeptides. The apparent similarity of the coding sequence of the KP gene of two geographical isolates, FCR-3 and HB-3, support the above conclusions.

ACKNOWLEDGEMENTS

Supported by NIH grant AI-19845 and a grant from the WHO/ World Bank/UNDP Program on the Immunology of Malaria. We thank Dr. John Barnwell for the sample of FCR-3 from monkeys.

REFERENCES

Trager W, Rudzinska MA, Bradbury PC (1966) The fine structure of Plasmodium falciparum and its host erythrocytes in natura malarial infection in man, Bull W.H.O. 35:883-236

Luse SA, Miller LH (1971) Plasmodium falciparum malaria: ultra structure of parasitized erythrocytes in cardiac vessels. Am J Trop Med Hyg 20:655-660

Langreth SG, Peterson E (1985) Pathogenicity, stability and immunogenecity of a knobless clone of Plasmodium falciparum in Columbian owl monkeys. Infect Immun 47:760-766

Green TJ, Gadsden G, Seed T, Jacobs R, Morhardt M, Brackett R (1985) Cloning and characterization of Plasmodium falciparum FCR-3/FMG strain. Am J Trop Med Hyg 34:24-30

Kilejian A (1979) Characterization of a protein correlated with the production of knob-like protrusions on membranes of erythrocytes infected with Plasmodium falciparum. Proc Natl Acad Sci USA 76:4650-4653

Kilejian A (1980) Homology between a histidine-rich protein from Plasmodium lophurae and a protein associated with knob-like protrusions on membranes of erythrocytes infected with Plasmodium falciparum. J Exp Med 151:1534-1538

Hadley TJ, Leech JH, Green TJ, Daniel WA, Wahlgren M, Miller LH, Howard RJ (1983) A comparison of knobly (K^+) and knobless (K^-) parasites from two strains of Plasmodium falciparum. Mol Biochem Parasitol 9:271-278

Leech JH, Barnwell JW, Aikawa M, Miller LH, Howard RJ (1984) Plasmodium falciparum malaria: association of knobs on the surface of infected erythrocytes with a histidine-rich protein and the erythrocyte skeleton. J Cell Biol 98:1256-1264

Vernot-Hernandez JP, Heidrich HG (1984) Time-course of synthesis, transport and incorporation of a protein identified in purified membranes of host erythrocytes infected with a knob-forming strain of Plasmodium falciparum. Mol Biochem Parasitol 12:337-350

Gritzmacher CA, Reese RT (1984) Reversal of knob formation on Plasmodium falciparum-infected erythrocytes. Science 226:65-67

Kilejian, A, Sharma YD, Karoui H, Naslund L (1986) Proc Natl Acad Sci USA 83:7938-7941

Bhasin UK, Trager W (1984) Gametocyte-forming and non-gametocyte-forming clones of Plasmodium falciparum. Am J Trop Med Hyg 33:534-537

Trager W, Tershakovec M, Lyandvert L, Stanley H, Lanners N, Gubert E (1981) Clones of the malaria parasite Plasmodium falciparum obtained by microscopic selection: their characterization with regard to knobs, chloroquine sensitivity and formation of gametocytes. Proc Natl Acad Sci USA 78:6527-6530

Purrello M, Balazs I (1983) Direct hybridization of labeled DNA to DNA in agarose gels. Anal Biochem 128:393-397

Pologe LG, Ravetch JV (1986) A chromosomal rearrangement in a P. falciparum histidine rich protein gene is associated with the knobless phenotype. Nature 322:474-477

Barnwell JW, Howard RJ, Miller LH (1983) Influence of the spleen on the expression of surface antigens on parasitized erythrocytes. Ciba Foundation Symposium 94:117-136

Ravetch JV, Feder R, Pavlovec A, Blobel G (1984) Primary structure and genomic organization of the histidine-rich protein of the malaria parasite Plasmodium lophurae. Nature 312:616-620

Wellems TE, Howard RJ (1986) Homologous genes encode two distinct histidine-rich proteins in a cloned isolate of Plasmodium falciparum. Proc Natl Acad Sci USA 6065-6069

PHOSPHOPROTEINS ASSOCIATED WITH THE HOST ERYTHROCYTE MEMBRANE DURING PLASMODIUM CHABAUDI INFECTION

Mark F. Wiser

Max-Planck-Institut für Zellbiologie, Rosenhof, D - 6802 Ladenburg bei Heidelberg, F.R.G.

INTRODUCTION

The malaria parasite alters the structure and function of the host erythrocyte membrane. These alterations include changes in metabolite and cation homeostatis, appearance of new proteins and lipids, increased membrane fluidity and gross structural alterations (Sherman, 1985).

Alteration of the host erythrocyte membrane phosphoproteins during Plasmodium berghei infection has been previously described (Wiser et al., 1983). Two phosphoproteins of 68 and 45 kDa, not found on normal erythrocyte membranes, are associated with the host erythrocyte membrane after infection. These phosphoproteins are of parasite origin and based on differential solubility properties they are peripheral membrane proteins that appear to be tenaciously associated with the submembrane cytoskeleton of the host erythrocyte.

In this report, phosphoproteins of 93 and 76 kDa, associated with the host erythrocyte membrane during P. chabaudi infection, are described.

MATERIALS AND METHODS

Erythrocytes, normal and P. chabaudi infected, were collected, passed over CF-11 cellulose (Whatman) columns, washed and lysed in 0.01 % saponin as described (Wiser and Schweiger, 1985). The lysate was centrifuged at 1000 x g for 5 min and the parasites were purified by subjecting the pellet to discontinuous Percoll gradients as described (Wiser and Schweiger, 1985).

The host erythrocyte membranes were isolated from the 1000 x g supernatant fraction by centrifugation at 27,000 x g for 20 min and washed twice in 20 mM phosphate, pH 7.6. During the washing steps the pigment containing "button" was separated from the membranes and discarded. The purity of the isolated host erythrocyte membranes and parasites was evaluated by phase contrast microscopy and assaying for acetylcholinesterase (AChE) and glutamate dehydrogenase (GDH) as described (Vander Jagt et al., 1982).

NATO ASI Series, Vol. H11
Host-Parasite Cellular and Molecular Interactions
in Protozoal Infections. Edited by K.-P. Chang and D. Snary

Membrane proteins of interest were defined as integral or peripheral by extraction with 1 % Nonidet P-40 in 20 mM phosphate, pH 7.6, or 0.1 N NaOH as described (Wiser et al., 1983).

Phosphorylation of isolated host erythrocyte membranes and parasites and analysis by SDS gel electrophoresis and autoradiography were carried out as described (Wiser and Schweiger, 1985).

RESULTS

P. chabaudi infection results in the appearance of two new phosphoproteins, with molecular masses of approximately 93 and 76 kDa, on the host erythrocyte membrane (Figure 1). These phosphoproteins are not likely the result of parasite contamination since they preferentially fractionate with the host erythrocyte membrane and not the parasite. Furthermore, the parasite specific phosphoproteins are not found in the erythrocyte membrane fraction.

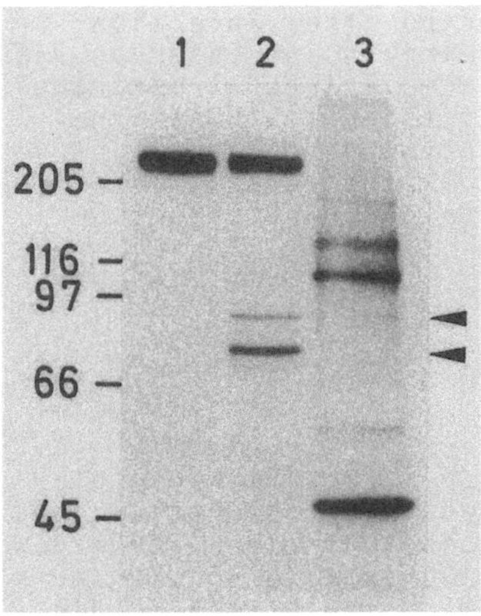

FIGURE 1. Phosphoproteins of the host eryhtrocyte membrane and parasite. Host erythrocyte membranes and parasites were isolated and phosphorylated as described in the Materials and Methods. Shown is the autoradiograph after SDS gel electrophoresis on 10 % polyacrylamide gels of erythrocyte membranes from uninfected cells (lane 1), host erythrocyte membranes from P. chabaudi infected cells (lane 2), and the isolated parasites (lane 3). The molecular mass standards in kDa are as indicated and the arrows indicate Pc(em)-93 and Pc(em)-76 in lane 2.

The purity of the isolated host erythrocyte membranes and parasites was further assessed by measuring ACHE and GDH activities, marker enzymes for the host erythrocyte membrane and parasite, respectively (Table 1). These results confirm the integrity of the subcellular fractionation and demonstrate that the new phosphoproteins are indeed associated with the host erythrocyte membrane. These new phosphoproteins will be designated Pc(em)-93 and Pc(em)-76 to indicate the species of origin, the host erythrocyte localization and the relative molecular masses.

TABLE 1
MARKER ENZYME ANALYSIS[a]

	ACHE	GDH
Host membranes, uninfected	0.122	0.009
Host membranes, infected	0.081	0.073
Isolated parasites	0.006	0.0372

[a] Host erythrocyte membranes and parasites correspond to samples in Figure 1. Activities are expressed as µmoles min^{-1} mg $protein^{-1}$.

Pc(em)-93 can also be labeled metabolically with radioactive proline, but not with methionine (data not shown). This indicates that Pc(em)-93 is of parasite origin as are the P. berghei membrane associated phosphoproteins. The origin of Pc(em)-76 cannot be definitely determined at this time but is also likely of parasite origin.

The solubility properties of the phosphoproteins were also investigated by extracting membranes with either 1 % Nonidet P-40 or 0.1 N NaOH (Figure 2). Pc(em)-93 is partially soluble in 1 % Nonidet P-40 and completely soluble in 0.1 N NaOH, whereas, Pc(em)-76 is not soluble at all in 1 % Nonidet P-40 and cannot be completely solubilized by 0.1 N NaOH. Although the partial detergent solubility of Pc(em)-93 and the incomplete alkali solubility of Pc(em)-76 cannot be completely explained, these proteins are likely to be peripheral membrane proteins as opposed to integral membrane proteins.

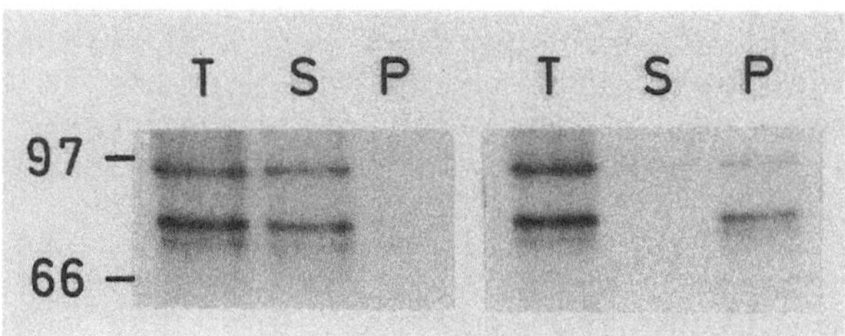

FIGURE 2. Differential solubility of Pc(em)-93 and Pc(em)-76. Host erythrocyte membranes were isolated, phosphorylated and then extracted with either 0.1 N NaOH (left) or 1 % Nonidet P-40 (right) as described in Materials and Methods. Shown is the autoradiograph from the region of interest after electrophoresis on 7.5 % polyacrylamide gels. Lane T indicates the total extract before centrifugation and lanes S and P indicate the supernatant (or soluble) and the pellet (or insoluble) fractions, respectively. Molecular mass standars are as indicated.

DISCUSSION

The results presented clearly show that new phosphoproteins are associated with the host erythrocyte membrane during P. chabaudi infection. The possibility that these phosphoproteins are the result of parasite contamination has been ruled out in that the proteins are only found in the host erythrocyte membrane fraction after subcellular fractionation. The integrity of the subcellular fractionation has also been confirmed by microscopic examination and marker enzyme analysis.

The membrane associated phosphoproteins are insoluble in 1 % Nonidet P-40, but soluble in 0.1 N NaOH indicating that they are peripheral membrane proteins and not integral membrane proteins. Similar solubility properties are also exhibited by the proteins making up the submembrane cytoskeleton of the erythrocyte (spectrin, ankyrin, 4.1 and actin). This suggests that the new parasite derived phosphoproteins may interact with the submembrane cytoskeleton of the erythrocyte, but other interpretations are also possible.

Phosphoproteins associated with the host erythrocyte membrane during P. berghei infection with similar characteristics as the proteins described herein have been previously characterized (Wiser et al., 1983). The 93 kDa P. chabaudi protein and 68 kDa and 45 kDa proteins of P. berghei are also similar in terms of isoelectric points, amino acid compositions and synthesis (M.F. Wiser, in preparation). The similar characteristics may indicate a similar function for these proteins. The differences in molecular masses exhibited by these proteins does not contradict this in that other malarial proteins with similar functions are known to display size heterogenicity between species or even strains (eg. circumsporozoite protein).

An approximately 90 kDa protein which correlates with the expression of "knobs" on the surface of P. falciparum infected erythrocytes has also been described. The "knob associated protein" (KAP), like the rodent malarial phosphoproteins, is insoluble in non-ionic detergents and appears to interact with the submembrane cytoskeleton of the erythrocyte (Leech et al., 1984). The phosphorylation status of KAP is not known.

Despite the similar solubility properties, the KAP is different from the rodent malarial phosphoproteins in that it is histidine rich (Pologe and Ravetch, 1986) and therefore presumably somewhat basic. The rodent malarial phosphoproteins are known to be quite acidic (Wiser et al., 1983) and are relatively poor in histidine (M.F. Wiser, in preparation).

In summary, a class of acidic phosphoproteins synthesized by the malaria parasite, but localized to the host erythrocyte membrane, are described. Furthermore these proteins appear to be associated with the host erythrocyte submembrane cytoskeleton suggesting that alteration in the host erythrocyte membrane during malaria infection may be mediated through perturbations of the membrane skeleton.

ACKNOWLEDGEMENTS

The excellent technical assistance of Brigitte Plitt and secretarial assistance of Christiane Schardt is gratefully acknowledged.

REFERENCES

Leech JH, Barnwell JW, Aikawa M, Miller LH, Howard RJ (1984) Plasmodium falciparum malaria: association of knobs on the surface of infected erythrocyte with a histidine rich protein and the erythrocyte skeleton. J Cell Biol 98:1256-1264

Pologe LG, Ravetch JV (1986) A chromosomal rearrangement in a P. falciparum histidine-rich protein gene is associated with the knobless phenotype. Nature 322:474-477

Sherman IW (1985) Membrane structure and function of malaria parasites and the infected erythrocytes. Parasitol 91:609-645

Vander Jagt DL, Intress C, Heidrich JE, Mrema JEK, Rieckmann KH, Heidrich HG (1982) Marker enzymes of Plasmodium falciparum and human erythrocytes as indicator of parasite purity. J Parasitol 68:1068-1071

Wiser MF, Schweiger HG (1985) Cytosolic protein kinase activity associated with the maturation of the malaria parasite Plasmodium berghei. Mol Biochem Parasitol 17:179-189

Wiser MF, Wood PA, Eaton JW, Sheppard JR (1983) Membrane associated phosphoproteins in Plasmodium berghei infected murine erythrocytes. J Cell Biol 97:196-201

MALARIAL SPOROZOITE - HEPATOCYTE INTERACTIONS MEDIATING INVASION AND EXOERYTHROCYTIC DEVELOPMENT

Michael R. Hollingdale
Biomedical Research Institute
Rockville, Md 20852

Malarial sporozoites rapidly and specifically invade hepatocytes and develop into exoerythrocytic (EE) parasites. Although perfusion experiments have demonstrated sporozoite localization in other organs, notably the spleen and lung, presumably within macrophages, EE development occurs only within hepatocytes. This suggests that a specific interaction(s) between the sporozoite and hepatocyte mediates invasion. Investigation of this event has been greatly assisted by the recent cloning into Escherichia coli of the gene encoding the circumsporozoite (CS) protein, which covers the surface of the sporozoite (Godsen et al. 1983; Enea et al. 1984; Arnot et al. 1985; McCutchan et al. 1985). The importance of the CS protein in mediating invasion was demonstrated by the findings that monoclonal antibodies (Mab's) to the CS protein abolished or reduced sporozoite infectivity to susceptible hosts (Yoshida et al. 1980; Nardin et al. 1982). Further, these Mab's inhibited sporozoite invasion (ISI) of cultured cells in vitro (Hollingdale et al. 1982; 1984). Sera from human volunteers immunized with irradiated sporozoites and immune to sporozoite challenge also blocked ISI in vitro, suggesting that ISI is an in vitro assay of protective antibodies.

CS proteins constitute a major biosynthetic product of sporozoites and are synthesized late in the developmental cycle within mosquitoes, just prior to and mostly after sporozoite migration to the salivary glands. CS proteins contain an immunodominant tandemly repeated epitope to which all Mab's are directed. Nucleic acid sequencing of the CS gene demonstrated that for Plasmodium knowelsi the sequence of 12 amino acids GPPQAQGDGANA is repeated 12 times (Godsen et al. 1983), and for P. falciparum the sequence NANP is repeated 37 times with 4 repeats of the variant sequence NDVP (Dame et al. 1984). Flanking the species-specific repeat region are regions that are highly conserved between P. knowelsi and P. falciparum, maybe exposed on the surface of the sporozoite, and which are poorly immunogenic. These regions include NI and N2, and C2, immediately adjacent to the repeat region towards the 5' and C' ends respectively (Vergara et al. 1985) and Regions I and II situately more towards the 5' and 3' ends respectively (Dame et al. (1984). Sequence homology may indicate conservation for important functions of the protein such as recognition of hepatic receptors.

NATO ASI Series, Vol. H11
Host-Parasite Cellular and Molecular Interactions
in Protozoal Infections. Edited by K.-P. Chang and D. Snary

Recombinant proteins and synthetic peptides containing sequences common to the repeat epitope sequence have been tested for immunogenicity in mice and rabbits (Young et al. 1985; Zavala et al. 1985). Each elicited antibodies, particularly using adjuvants, that recognized intact sporozoites and blocked sporozoite invasion *in vitro* (Young et al. 1985; Zavala et al. 1985). Thus, both the species-specific repeat regions and the conserved region might be involved directly in sporozoite invasion.

The presence of receptors on the surface of the hepatocytes is well documented. The asialoglycoprotein receptor binds serum glycoproteins from which the terminal sialic acid has been removed (Ashwell and Harford 1982). This receptor is present both in cultured hepatocytes (Ashwell and Morell 1974) and the human hepatoma cell line HepG2-A16 (Schwartz et al. 1981). Since the HepG2-A16 line is metabolically and ultrastructurally closely similar to hepatocytes, contains a similar distribution of receptors, and supports the invasion and EE development of rodent and human malarias (Hollingdale et al. 1983a; 1983b; 1985) it is well suited for studies on the molecular biology of the sporozoite-binding receptor.

Sporozoites attach to the surface of HepG2-A16 cells, and CS protein appears to be shed from the sporozoite surface onto the target cell surface (Hollingdale et al. 1983a). As the sporozoite enters, the HepG2-A16 cell membrane invaginates opposite the apical end of the sporozoite, finally enclosing the sporozoite within a vacuole bounded by a parasitophorous vacuole membrane (PVM) derived from the host cell plasmamembrane (Aikawa et al. 1984). This is closely similar to invasion *in vivo* (Meis et al. 1981). Both the sporozoite and PVM strongly stain by the immunoperoxidase antibody technique for the presence of CS protein. Thus binding of CS protein to the hepatocyte surface is a necessary step in mediating sporozoite invasion, and the availability of recombinant proteins and synthetic peptides, particularly of *P. knowelsi* and *P. falciparum* of various regions of the CS protein have been investigated as potential binding sites, or ligands, recognizing hepatocyte receptors.

Both recombinant proteins and synthetic peptides common to the repeat region of *P. falciparum* have been tested for specific binding to HepG2-A16 cells (Aley et al; in press a;b) . R32tet$_{32}$ contains the sequence $(NANP)_{30}$ $(NVDP)_2$ with 32 additional amino acids on their carboxyl terminal end derived from the host plasmid (Young et al. 1985). R32tet$_{32}$ bound to HepG2-A16 cell cultures, measured using Mab to *P. falciparum* CS protein and ^{125}I-protein A, and the level of binding increased until all available sites were saturated at about 1 µg/ml. However, similar binding was observed on blank plastic or glass. The plasmid-derived 32 amino acid tail contained 12 highly positive arginine residues, 6 were histidine, none were acid, resulting in a shift in charge distribution and PI of $R32tet_{32}$ compared to the repeat region alone. When $R32tet_{32}$ was treated with trypsin to cleave at arginine residues to produce R32-leucine-arginine ($R32tet_2$), no specific binding was observed. The recombinant construct $R32tet_2$ was also produced, labelled with ^{125}I, and bound to HepG2-A16 cells in the presence of increasing amounts of unlabelled $R32tet_2$. Again, no specific binding was observed (Aley et al, in press a). In similar experiments (Aley et al. in press b), no specific (saturable) accumulation of ^{125}I-labelled $(NANP)_5$ by HepG2-A16 cells was observed.

However, Aley has shown that peptides containing about 10 amino acids immediately adjacent to the repeat region specifically bound to HepG2-A16 cells (Aley et al. in press b). These contained the highly conserved region N1, and HepG2-A16 incubated with ^{125}I-labelled N1 peptides from both P. falciparum (PfN1;LKQPGDGNPDP) and P. knowelsi (PkN1:KLKQPNEGOP) rapidly accumulated radiolabel. Binding was inhibited by the simultaneous addition of excess unlabelled peptide. Scatchard analysis demonstrated similar curves for either PfN1 or PkN1, and binding was inhibited equally well by unlabelled peptide of either 13 species. In contrast, however, a longer peptide, PfN2, that contained the PfN1 sequence plus the next five amino acids towards the 5' end, KPKHKKLKQPGDGNPOP, did not specifically bind to HepG2-A16 cells. Aley speculated that since the malarial CS protein is processed into at least 3 different forms, the multiple lysine residues preceding the N1 region may act as a target for proteolytic enzymes, either in the mosquito salivary glands, sporozoite or the host, that activate the protein for binding to hepatocytes.

Aley also found that antibodies to the N1 region were active in the ISI assay with P. falciparum sporozoites. Antibodies to Region I and Region II, however, did not block sporozoite invasion suggesting that these regions may not be involved in sporozoite invasion, especially since Region I lies beyond the multiple lysine residues suggested as a site of CS processing. Aley suggested that the ISI activity of antibodies to both the CS repeat region and N1 region may be best explained by the possibility that both regions are involved in sporozoite invasion. The repeat region may act through non-specific interactions, and the N1 region through specific interactions.

The nature of the hepatocyte sporozoite-binding receptor was determined by chemical and enzymic treatment of HepG2-A16 cells (Pancake and Hollingdale 1986). Preliminary experiments in other cells had determined the receptor was a surface protein susceptible to chymotrypsin (Hollingdale, unpublished results). However, chymotrypsin or other proteases had no effect on the HepG2-A16 receptor unless surface structures were first removed by solvent extraction and air drying. Protease treatment of the exposed hydrophobic regions then abolished sporozoite binding (Pancake and Hollingdale 1986). The existence of a possible hydrophobic hepatocyte protein acting as a sporozoite-binding receptor may be analogous to the binding of P. falciparum merozoites to the hydrophobic core of glycophorin A of erythrocytes (Perkins 1984). Other similarities between sporozoite invasion of hepatocytes and merozoite invasion of erythrocytes exist, including inhibition of invasion but not attachment by cytochalasin B (Hollingdale et al. 1983a).

Aley has identified the HepG2-A16 cell components to which the PfN1 and PKN1 peptides bind (Aley et al. in press b). HepG2-A16 cells were incubated with ^{125}I-labelled N1 peptide, and then treated with the bifunctional reagent disuccinyl suberimidate to cross-link the N1 peptide with its putative receptor. After solubilization and SDS-PAGE, labelled N1 was covalently linked to two polypeptides of M_{rs} of 55,000 and 35,000. The relationship of these components with the protease-sensitive components (Pancake and Hollingdale, 1986) is yet to be determined.

Cell surface receptors often recycle into the cell in the presence or absence of their ligands, and once within the cell an acidification step dissociates ligand-receptor interactions. The receptor is then passaged to the surface to re-initiate the cycle. This receptor-mediated endocytosis is the means of entry into cells of asialoglycoprotein, hormones, viruses and toxins (Steinman et al. 1983). The acidification step can be prevented by treatment of the cells with lysomotropic amines, weak bases that concentrate within intracellular compartments, and disrupt normal cycling of receptors by causing their intracellular sequestration. Binding and invasion of P. berghei sporozoites to HepG2-A16 cells was inhibited in a dose-dependent manner by such lysomotropic amines as primaquine, chloroquine or ammonium chloride (Schwartz and Hollingdale, 1985). Under identical conditions each amine also inhibited the receptor-mediated binding and endocytosis of asialoglycoproteins. Thus, it appears likely that at least part of the sequence of events mediating sporozoite invasion of hepatocytes involves recognition of a continuously recycling surface receptor. Since it is unlikely that the hydrophobic protein regions are involved in such a recycling process, sporozoite invasion of hepatocytes may be a multistep process involving cell surface recycling receptors and inner hydrophobic protein regions.

REFERENCES

Aley SB, Bates MD, Hockmeyer WT, et al. Repeat region of P. falciparum circumsporozoite protein does not recognize or bind to human hepatoma target cell. Mem Inst Oswaldo Cruz Rio de Janeiro in press.

Aley SB, Bates MD, Tam JP, et al. Synthetic peptides from the circumsporozoite proteins of Plasmodium falciparum and P. knowelsi recognize the human hepatoma cell line HepG2-A16 in vitro. J Exp Med in press.

Aikawa M, Schwartz AL, Uni S, et al. 1984. Ultrastructure of in vitro cultured exoerythrocytic stage of Plasmodium berghei in a hepatoma line. Am J Trop Med Hyg 33: 792-799.

Arnot DE, Barnwell JW, Tam JP, et al. 1985. Circumsporozoite protein of Plasmodium vivax: gene cloning and characterization of the immunodominant epitope. Science 230:815-818.

Ashwell G, Harford J. 1982. Carbohydrate recognition systems of liver. Ann Rev Biochem 51:531-554.

Ashwell G, Morell AG. 1974. The role of surface carbohydrates in the hepatic recognition and transport of circulating glycoproteins. Adv Enzymol Relat Areas Mol Biol 41:99-129.

Dame JB, Williams JL, McCutchan TF, et al. 1984. Structure of the gene encoding the immunodominant surface antigen of the sporozoite of the human malaria Plasmodium falciparum. Science 225:593-599.

Enea V, Ellis J, Zavala F, et al. 1984. DNA cloning of Plasmodium falciparum circumsporozoite gene: amino acid sequence of repetitive epitope. Science 225:628-629.

Godsen GN, Ellis J, Svec P, et al. 1983. Identification and chemical synthesis of a tandemly repeated immunogenic region of Plasmodium knowelsi circumsporozoite protein. Nature 305: 29-33.

Hollingdale MR, Ballou RW, Aley SB, et al. Administration of peptides and recombinant CS proteins of Plasmodium falciparum to mice elicits circulating antibodies that inhibit sporozoite invasion (ISI) of cultured cells. Am J Trop Med Hyg in press.

Hollingdale MR, Collins WE, Campbell CC, et al. 1985. In vitro culture of two populations (dividing and non-dividing) of exoerythrocytic parasites of Plasmodium vivax. Am J Trop Med Hyg 34:216-222.

Hollingdale MR, Leland P, Leef JL, et al. 1983a. Entry of Plasmodium berghei sporozoites into cultured cells and their transformation into trophozoites. Am J Trop Med Hyg 32:685-690.

Hollingdale MR, Leland P, Schwartz AL. 1983b. In vitro cultivation of the exoerythrocyte stage of Plasmodium berghei in a hepatoma cell line. Am J Trop Med Hyg 32:682-684.

Hollingdale MR, Nardin EH, Tharavanij S, et al. 1984. Inhibition of entry of Plasmodium falciparum and P. vivax sporozoites into cultured cells; an in vitro assay of protective antibodies. J Immunol 132:909-913.

Hollingdale MR, Zavala F, Nussenzweig RS, et al. 1982. Antibodies to the protective antigen of Plasmodium berghei sporozoite prevent entry into cultivated cells. J Immunol 128:1929-1930.

McCutchan TF, Lal AA, de la Cruz VF, et al. 1985. Sequence of the immunodominant epitope for the surface protein on sporozoite of Plasmodium vivax. Science 230:1381-1383.

Meis JFGM, Verhave JP, Jap PHK et al. 1981. An ultrastructural study of the developing stages of exoerythrocytic Plasmodium berghei in rat hepatocytes. Parasitol 82:195-204.

Nardin EH, Nussenzweig V, Nussenzweig RS et al. 1982. Circumsporozoite proteins of human malaria parasites Plasmodium falciparum and Plasmodium vivax. J Exp Med 156:20-30.

Pancake SJ, Hollingdale MR. 1986. Plasmodium berghei: reaction of sporozoites with chemically and enzymatically modified hepatoma cells. Exp Parasitol 62:120-126.

Perkins ME. 1984. Binding of glycophorins to Plasmodium falciparum merozoites. Mol Biochem Parasitol 10:67-78.

Schwartz AL, Fridovich SE, Knowles BB et al. 1981. Characterization of the asialoglycoprotein receptor in a continuous hepatoma line. J Biol Chem 256:8878-8881.

Schwartz AL, Hollingdale MR. 1985. Primaquine and lysomotropic amines inhibit malaria sporozoite entry into human liver cells. Mol Biochem Parasitol 14:305-311.

Steinman RM, Mellman IS, Muller W, et al. 1983. Endocytosis and the recycling of plasma membrane. J Cell Biol 96:1-27.

Vergara V, Gwadz R, Schlesinger D, et al. 1985. Multiple non-repeated epitopes on the circumsporozoite protein of Plasmodium knowelsi. Mol Biochem Parasit 14:283-292.
Yoshida N, Nussenzweig R, Potcnjak V, et al. 1980. Hybridoma produces protective antibodies directed against the sporozoite stage of malaria parasite. Science 207:71-73.

Young JF, Hockmeyer WT, Gross M, et al. 1985. Expression of P. falciparum circumsporozoite protein derivatives in E. coli for development of a human malaria vaccine. Science 228:958-962.

Zavala F, Tam JP, Hollingdale MR, et al. 1985. Rationale for development of a synthetic vaccine against Plasmodium falciparum malaria. Science 228:1436-1440.

IMMUNO-ELECTRON MICROSCOPIC OBSERVATIONS ON PLASMODIUM FALCIPARUM SPOROZOITES USING ULTRACRYOMICROTOMY.

G. Posthuma[1], J.F.G.M. Meis[2], P.J.M. Rijntjes[3], J.P. Verhave[2], M.R. Hollingdale[4], J.H.E.T. Meuwissen[2], H.J. Geuze[1], S.H.Yap[3].

1. Department of Cell Biology, Medical Faculty, State University of Utrecht, the Netherlands.
2. Department of Medical Parasitology, Catholic Universty of Nijmegen, The Netherlands.
3. Department of Gastroenterology, Catholic University of Nijmegen, The Netherlands.
4. Biomedical Research Institute, Rockville, USA.

INTRODUCTION

The surface membrane of Plasmodium sporozoites contains stage specific antigens, the circumsporozoite (CS) protein. This CS protein is the major target molecule for the immune response of the host against the sporozoite infection (Nussenzweig & Nussenzweig, 1985). Monoclonal antibodies (MoAb) against different malaria parasites have been used to localize parasites antigens (Fine et al, 1984; Oka et al, 1984). Here we demonstrate the immuno-electron microscopic localization of immunogenic determinants on P. falciparum sporozoites using cryo-ultramycrotomy, a MoAb and protein-A-gold.

MATERIALS AND METHODS

Parasites. Sporozoites and infected salivary glands were obtained from Anopheles stephensi mosquitoes, membrane-fed 14 days previously on cultured P. falciparum gametocytes (isolate NF54).

Electronmicroscopy. Parasites were fixed for 24 h in 1%

NATO ASI Series, Vol. H11
Host-Parasite Cellular and Molecular Interactions
in Protozoal Infections. Edited by K.-P. Chang and D. Snary

glutaraldehyde in 0.1M phosphate buffer and embedded in gelatin.

Ultracryotomy and immunolabelling were performed as previously described (Posthuma et al, 1984) using MoAb 3SP2 (Verhave et al, 1986) and protein-A-gold Ø 6 nm (Slot and Geuze, 1985). As a control, sporozoites were simultaneously incubated with an unrelated MoAb (directed against Semliki Forest Virus) or only protein A-gold.

RESULTS AND DISCUSSION

MoAb 3SP2 raised against the repetitive epitope of the CS protein of *P. falciparum* reacts with sporozoites and mature oocystes in ELISA (Verhave et al, 1986). Localization of immunogenic determinants on the surface of *P. falciparum* sporozoites was possible with 3SP2 and immuno-electron microscopy of ultra thin cryosections. The sporozoite ultrastructure was comparable to that found in plastic sections, although the membranes were negatively stained. The pellicular membranes were clearly visible and gold particles could be observed all along the surface including the apical complex (Fig. 1,2,4). Both inner and outer pellicular membranes were labelled (Fig. 1,2,4). In contrast to results obtained with *P. knowlesi* (Fine et al, 1984), neither the micronemes, rhoptries nor any other cytoplasmic components of *P. falciparum* were labelled (Fig.1,2,4) using Moab 3SP2. Since we used ultrathin cryosections the accessability of the CS protein in the sections cannot be the reason for the absence of gold particles over these organelles. It is more likely that the MoAb failed to detect precursor CS protein because relevant antigenic sites were masked by some amino acids that are removed during maturation to a membranous form (Vergara et al, 1984).

Polyclonal human anti-sporozoite antibodies will be used next to detect the precursor CS protein. Evidence of shedding of protein from the sporozoite surface was found inside the salivary gland cells. Gold particles were observed in 3 different localizations in the gland cells representing the stages of invasion.

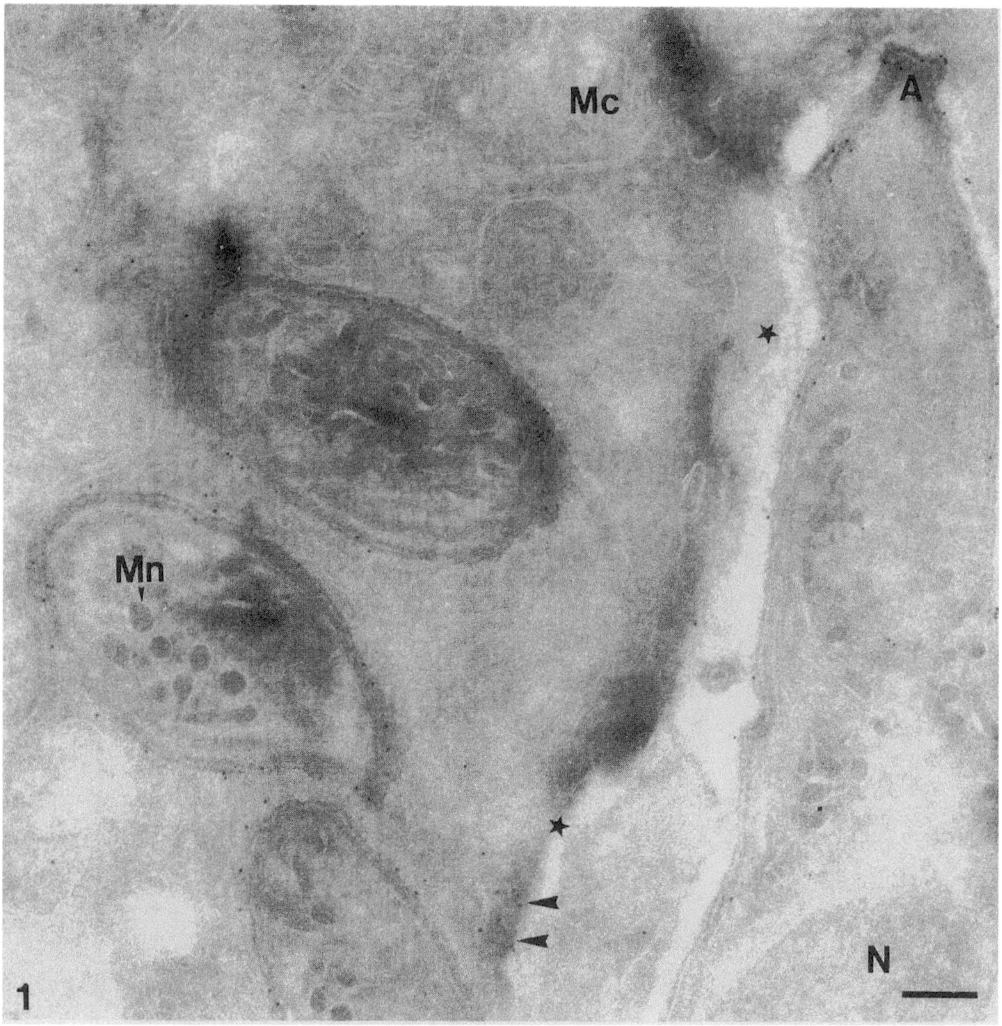
Mc
A
Mn
N
1

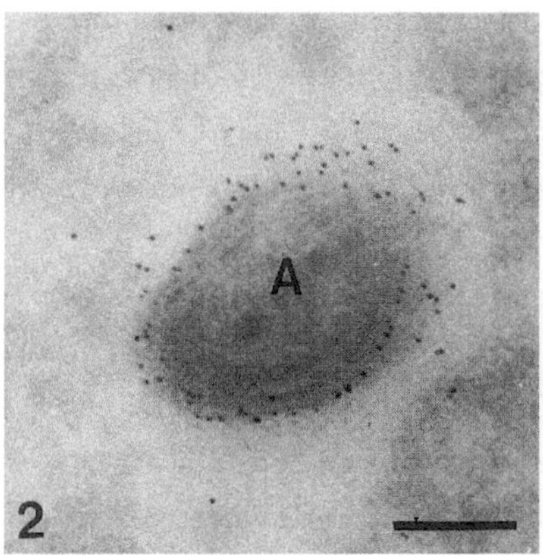
A
2

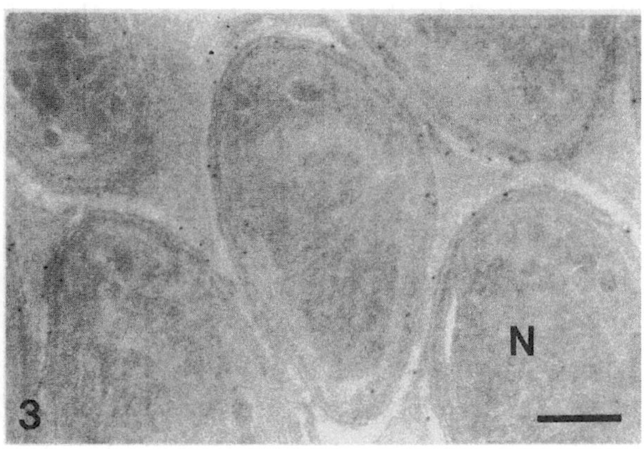
N
3

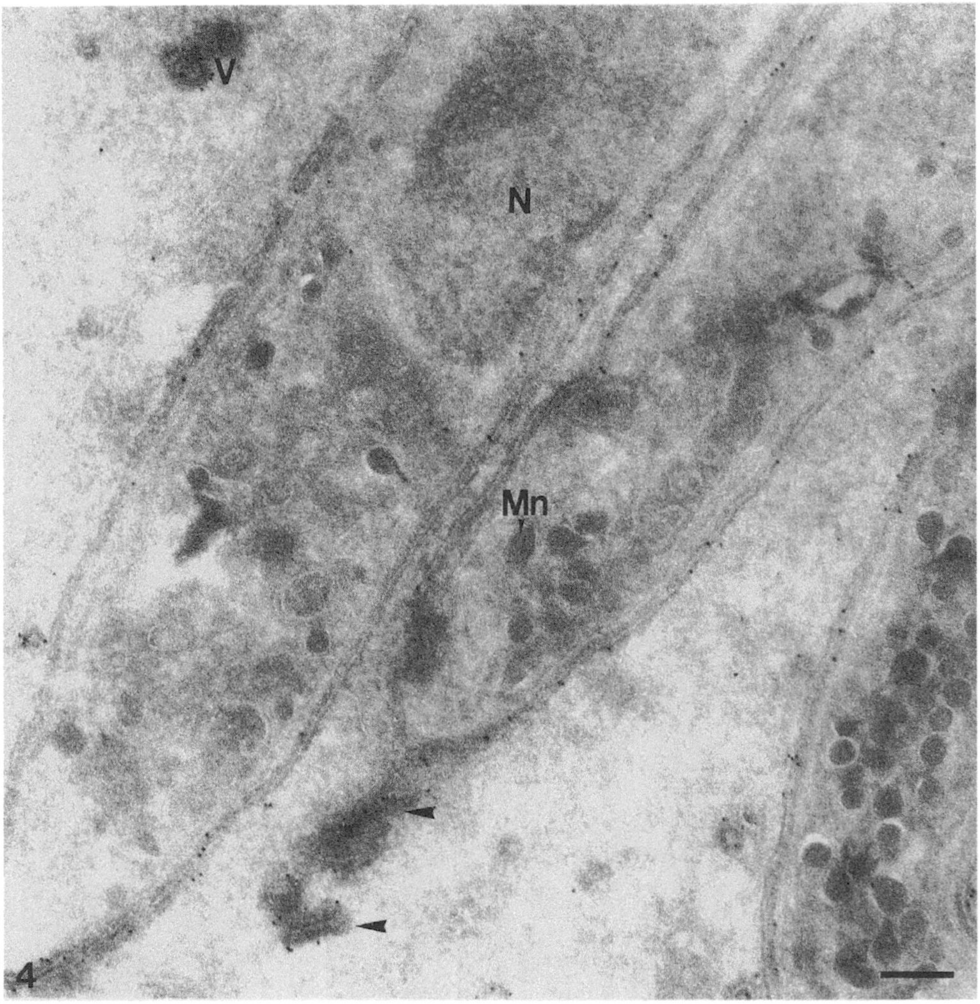

Figure 1,2,4. Electron micrographs of ultrathin cryosections from salivary glands from *Anopheles stephensi* infected with *Plasmodium falciparum.* The circum sporozoite (CS) protein was visualized using a monoclonal antibody (3SP2) and protein A-gold (6 nm).

Figure 3 Electron micrograph of an ulthathin cryosection from free *P.falciparum* sporozoites labeled with 3SP2 and protein A-gold. Fig. 1,3,4 X 63000; fig. 2 X 80000; bar = 0.15 um N = Nucleus, Mc = Mitochondrion, V = endocytotic vesicle, A = Apical complex, Mn = Microneme, * = Basal membrane of a salivary gland cell. Arrows indicate shedded CS protein.

1. Labeling was observed at the basal membrane of the salivary gland cell.This might be the result of shedding during entrance of the sporozoite.

2. Gold particles were seen over small vesicles with an electron-dense content. These structures were similar to endocytotic vesicles described in rat liver cells after contact with a ligand (Geuze et al, 1983). It is thus presumed, that the CS protein is internalized via an endocytotic mechanism.

3. After internalization, the sporozoite seemed to continue the shedding of CS protein, visualized by electron-dense labelled material along the sporozoite's surface.

These preliminary results with CS protein distribution in the mosquito salivary gland will be extended to the occurence of CS protein throughout the hepatic cycle and oocyst development.

ACKNOWLEDGEMENT

This work has been supported by the US Agency for International Development (Contract DPE 0453-C-3051-00).

REFERENCES

Fine E, Aikawa M, Cochrane AH, Nussenzweig RS (1984). Immuno-electron microscopic observations on Plasmodium knowlesi sporozoites: localization of protective antigen and its precursors. Am J Torp Med Hyg 33: 220-226.

Geuze HJ, Slot JW, Strous GJAM, Lodish HF, Schwartz AL (1983). Intracellular site of asialoglycoprotein receptor-ligand uncoupling: double-label immunoelectron microscopy during receptor mediated endocytosis. Cell 32: 277-287.

Nussenzweig RS, Nussenzweig V (1985). Circumsporozoite proteins of malaria parasites. Cell 42:401-403.

Oka M, Aikawa M, Freeman RR, Holder AA, Fine E (1984). Ultrastructural localization of protective antigens of Plasmodium yoelii merozoites by the use of monoclonal antibodies and ultrathin cryomicrotomy.
Am J Trop Med Hyg 33:342-346.

Posthuma G, Slot JW, Geuze HJ (1984). Immunocytochemical assays of amylase and chymotrypsinogen in rat pancreas secretory granules.J Histochem Cytochem 32:1028-1034.

Slot JW, Geuze HJ. (1985). A new method of preparing gold probes for multiple-labeling cytochemistry. Eur J Cell Biol 38:87-89.

Vergara U, Ruiz A, Ferreira A, Nussenzweig RS, and Nussenzweig V (1985).Conserved group specific epitopes of the circumsporozoite proteins revealed by antibodies to synthetic peptides. J Immunol 134: 3445-3448.

Verhave JP, Leeuwenberg A.D.E.M., Ponnudurai T, Meuwissen JHETh, Van Druten J (1986). The biotin-streptavidin system in a two-site ELISA for the detection of sporozoite antigen in mosquitoes (submitted)

STRUCTURAL HOMOLOGY OF MEMBRANE PROTEINS OF BABESIA RODHAINI

David Snary

Department of Molecular Biology, Wellcome Biotechnology, Langley Court, Beckenham, Kent BR3 3BS, England

INTRODUCTION

Babesia species are tick-borne parasitic protozoa which multiply within red blood cells of infected mammalian hosts inducing clinical symptoms such as pyrexia, haemoglobinuria and haemoglobinaemia. Although primarily considered to be a parasite of veterinary importance, under some circumstances man can become infected and if the patient has been immune compromised, for example by splenectomy, then the outcome is usually death (Hoare 1980). A vaccine which protects against *Babesia bovis* infection has been developed and is used very effectively in Australia, the vaccine is live infected blood where the parasite has been attenuated by repeated passage through splenectomised calves (Timms *et al*, 1981). The molecular basis for the attenuation is not known but antigens which will provide a degree of protection against challenge in cattle have been isolated by monoclonal antibody affinity chromatography (Wright *et al*, 1983 and 1985).

Relatively little work on the molecular biology of *Babesia* species has been undertaken, but an unusual genetic locus defined as Bab-R has been identified in *Babesia bovis* (Cowman *et al*, 1984a). This locus contains a group of tandemly linked genes which have considerable lengths of common DNA sequence, unique sequence being found only at the

NATO ASI Series, Vol. H11
Host-Parasite Cellular and Molecular Interactions in Protozoal Infections. Edited by K.-P. Chang and D. Snary

extreme 3' end of the putative open reading frames. Southern and Northern blotting analyses of a number of different strains have shown that there were extensive variations not only in and around this locus, but also in the number and size of the transcripts from this region (Cowman et al, 1984b). Subsequently, a series of monoclonal antibodies raised against Babesia rodhaini have been shown to bind to a group of membrane proteins which share some common epitopes amongst themselves and with a red blood cell protein (Snary & Smith 1986). It is not known at this stage whether the monoclonal antibodies to B.rodhaini recognise products of genes analogous to the Bab-R locus, but work on the rodent parasite has provided further evidence for the existance of unusual genetic events such as gene rearrangement and duplication in Babesia species.

MATERIALS AND METHODS

The Antwerp strain of parasite was used throughout these studies. Methods used are essentially described elsewhere (Snary & Smith 1986, Odink et al, 1984, Holder et al, 1985).

RESULTS AND DISCUSSION

Monoclonal antibodies to B.rodhaini have been raised and several selected for further study. These antibodies were specific for a group of proteins of molecular weights 37 000, 35 000, 33 000 and 30 000 which have been shown by charge shift electrophoresis and immunoprecipitation from microsomal membrane preparations of infected red cells to be membrane bound (Snary & Smith 1986). Mice recovered from infection or drug cured during the course of an infection were resistant to reinfection, and sera from these mice reacted strongly to a small number of proteins amongst which were the group of membrane proteins defined by the monoclonal antibodies. These observations suggested that the membrane proteins may be expressed on the parasite cell surface and could be potential candidates for protective antigens. The proteins were purified by two cycles of immunoaffinity chromatography from detergent solubilised infected mouse blood (Fig.1) and then used to immunise mice. Saponin, used as an adjuvant, together with 10ug of protein was given subcutaneously on two occasions two weeks apart, challenge with infected blood was two weeks after the final immunisation. Adjuvant alone and untreated mice had

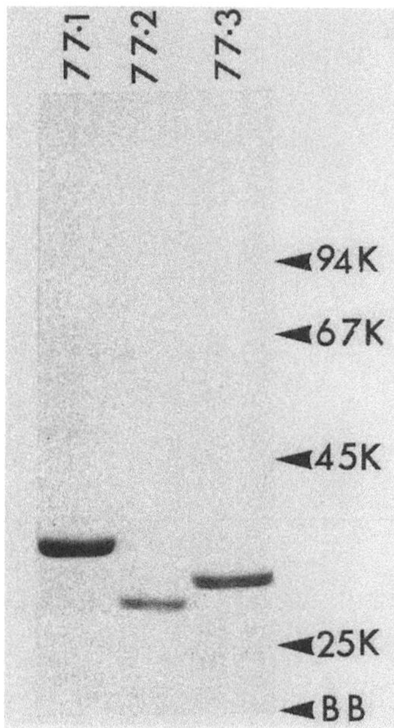

FIGURE 1. SDS PAGE of proteins purified by monoclonal antibody affinity chromatography and stained with coomasie blue. 77.1, 77.2 and 77.3 are the monoclonal antibodies used.

high parasitaemias and in the majority of cases the outcome was death. Mice immunised with the 37 000 mol. wt. protein had equally high parasitaemias but a significantly better survival rate, whereas mice immunised with either the 33 000 or the 30 000 mol. wt. protein had not only improved survival rates but also reduced parasitaemias (Table 1). The level of protection achieved with whole lysed infected blood

TABLE 1

PROTECTION OF MICE WITH PURIFIED PARASITE PROTEINS

Immunogen	Survivors/No in group	Peak blood parasitaemia
No antigen with or without saponin	2/25	75%
37 000 mol.wt. antigen & saponin	6/15	70%
33 000 mol.wt antigen & saponin	10/15	27%
30 000 mol.wt. antigen & saponin	11/15	28%

Mice were immunised with 10ug of protein s.c. using 50ug of saponin as an adjuvant, immunisation was on day 0 and day 14, and challenge with infected blood was on day 28. Peak parasitaemia was on day 6 for all groups.

using this immunisation protocol was not significantly different from that given by the 33 000 and the 30 000 proteins demonstrating that these antigens were capable of giving a reasonable degree of protective immunity.

Although peptide mapping experiments (Snary & Smith 1986) and the monoclonal antibodies used to purify the proteins clearly demonstrated differences between the proteins, a further series of monoclonal antibodies revealed that these proteins contained epitopes which were shared between one or more of the proteins (Fig.2). Furthermore, the ability of the crossreacting monoclonal antibodies to immunoprecipitate the expected multiple proteins from in vitro translation products demonstrated that these antibodies recognised common amino acid sequences and not post-translation modifications such as carbohydrate side chains. To understand the structural basis for the common epitopes it was necessary to clone and sequence the genes coding for these proteins, a study of this type would have the added advantage of enabling the proteins to be expressed, possibly in E.coli, when sufficient material would become available to extend the protection and immunological studies. Preliminary studies revealed that the mRNA species coding for these proteins were polyadenylated and that the proteins were synthesised as larger precursors of molecular weights 43 000, 39 000, 35 000 and 34 000, presumably because of the presence of signal peptides. The cloning strategy used to isolate the genes was to prepare a cDNA library from polyadenylated RNA, size enriched by sucrose gradient centrifugation, and cloned into pUC 9 by homopolymer tailing, the library was screened by hybrid selection and in vitro translation, followed by immunoprecipitation with the monoclonal antibodies. Three clones were identified, one (p17) selected the mRNA species coding for the 33 000 mol. wt.protein, the other two (p25 and p26) selected mRNA species coding for all four proteins (Fig.3), suggesting that whereas p17 contained sequence unique to the 33 000 mol.wt.gene product, p25 and p26 contained common sequence. However, these latter two plasmids hybridised only weakly if at all to each other and therefore were not overlapping and could well have been derived from different genes.

The availability of cDNA clones allowed preliminary studies to be undertaken on the genomic organisation of the

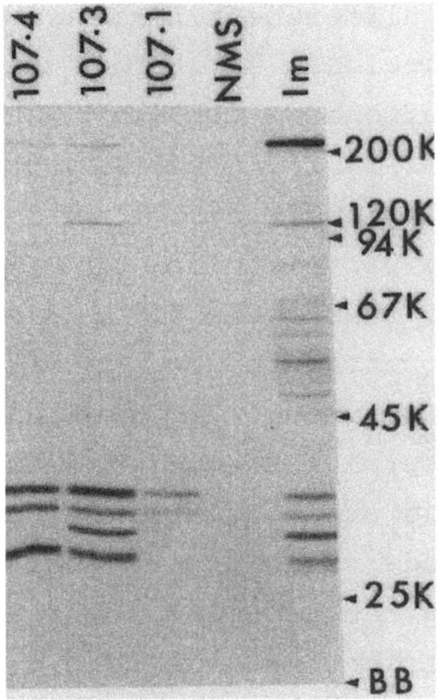

FIGURE 2. SDS PAGE of immunoprecipitates from ^{35}S methionine labelled infected blood. Im.immune serum; NMS, normal mouse serum; 107.4, 107.3 and 107.1 are monoclonal antibodies.

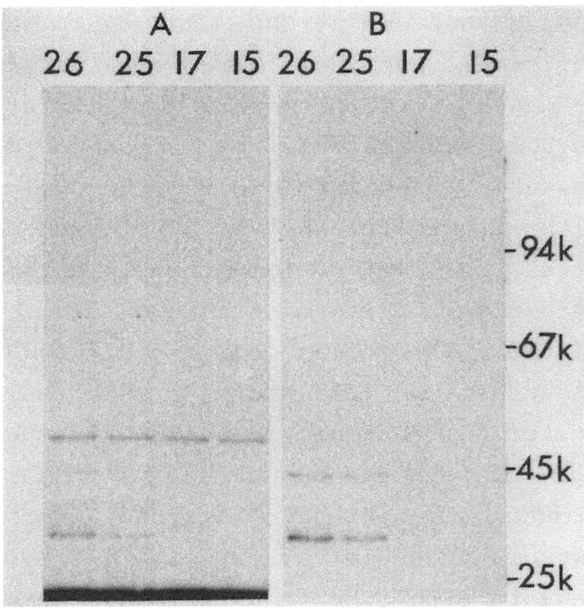

FIGURE 3. SDS PAGE of hybrid select translations. A, translates of selected mRNA; B, immunoprecipitates of products in panel A with a pool of monoclonal antibodies. 15,17,25 and 26 are the plasmids used for the selection.

genes coding for the four proteins. Significantly, in Southern blot analyses, all three cDNA probes hybridised to a single large Pst 1 fragment of a similar size suggesting that the four genes may be tandemly arranged along a single length of chromosome. To test this postulate, Pst 1 digested DNA was fractionated on agarose gels and DNA of a size equivalent to that seen by the cDNA probes eluted and cloned into Pst 1 cut pUC 8, a positive clone was detected by probing with the cDNA clones. The cloned genomic fragment was extremely large (over 15kb) and very unstable, however the pattern of bands seen on Southern blot analysis of the large genomic clone, using individual cDNA clones as probes, was closely similar to the pattern given with total DNA (Fig.4). The similarity of the pattern of fragments seen is further evidence to support the suggestion that the four genes were tandemly arranged along a chromosome. The multiple bands seen may represent some partial or incomplete digestion, but the fact that clones p25 and p26 both contained sequence common to all four genes would suggest that in these instances multiple bands should have been the expected result.

The epitope defined by one of the monoclonal antibodies and found to be present on all four of the parasite membrane proteins was also shown to be present on a red blood cell protein. Western blot analysis revealed that of the numerous monoclonal antibodies prepared to <u>B.rodhaini</u>, only one bound to a protein which was also present in normal uninfected red blood cells, this protein had a molecular weight of 22 000 and existed in the red cell as a disulphide bridged dimer of 44 000 mol.wt.(Snary & Smith 1986). The same monoclonal antibody was the only antibody which bound to all four of the membrane proteins. The significance of this observation is not clear, especially since an epitope can constitute only three or four amino acids and it may be no more than a coincidence that similar sequences were found in the group of parasite surface membrane proteins and a host cell protein. Alternatively, the fact that there was

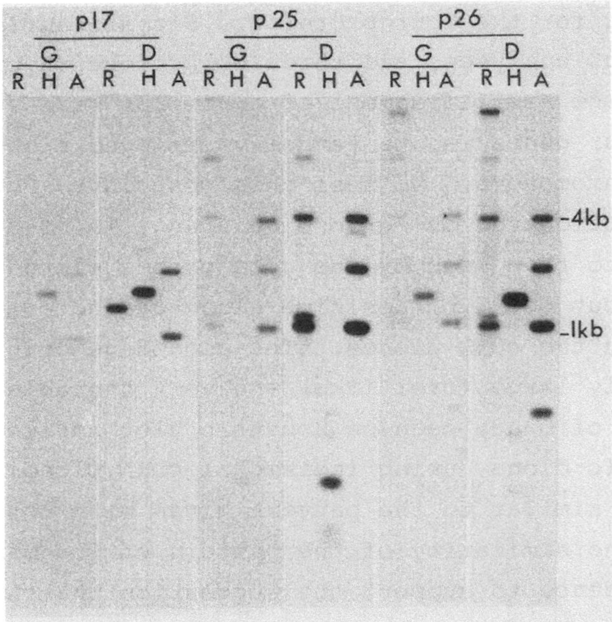

FIGURE 4. Southern blot of D, parasite DNA or G, genomic clone; R, EcoR1; H, Hind III; or A, Hha 1 digested; and probed with p17, p25 or p26.

sequence homology between host cell and parasite proteins may be evidence of some functionality of these molecules in the host-parasite inter-relationship. The basis for the homology and some indication of its possible functional significance may become clearer when the extent of the sequence homology between host and parasite is determined and further studies on the biochemistry and immunology of the red cell protein are complete.

The tandem arrangement of four genes, which have a degree of sequence homology with each other, along a single length of chromosome suggests that either gene rearrangement or gene duplication has taken place. Although the true extent of sequence homology between the genes is not yet known it would seem from peptide mapping and serological studies that the regions of homology may not be extensive. This is in contrast to the Bab-R gene complex identified in B.bovis where over 80% of open reading frame sequences were common, differences being found only at the 3'end of the gene. This could imply that the B.rodhaini gene complex is not necessarily analogous to the Bab-R locus and may suggest that gene rearrangement or gene duplication in Babesia species is a relatively common occurrence. However, it is intriguing that not only is there an antigenic epitope and presumably at least amino acid sequence homology between host and parasite, but also that the parasite genes coding for this common epitope have evolved by duplication and or rearrangement, coincidences which suggest that there may indeed be a functional basis for the host-parasite commonality. If this is the case then has the parasite sequence arisen by mimicry or by acquisition and integration of host nucleic acid, elucidation of the sequences of the parasite and host genes may well go some way towards explaining these phenomena.

B.rodhaini has been reported to be capable of undergoing antigenic shift under immune pressure in rats (Phillips 1971), although it is not clear whether the immune response in the rats was selecting antigenically altered subpopulations of the parasite or stimulating antigenic change, but it is possible that genetic mechanisms of the type implied by the studies described here could have been utilised to accommodate these changes. Clearly there are many interesting biological phenomena associated with Babesia species which require the application of molecular techniques to explain their relevance to the host-parasite interaction.

REFERENCES

Cowman AF, Bernard O, Steward N, Kemp DJ (1984a) Genes of the protozoan parasite Babesia bovis that rearrange to produce RNA species with different sequences. Cell, 37: 653 - 660.

Cowman AF, Timms P, Kemp DJ (1984b) DNA polymorphisms and subpopulations in _Babesia bovis_. Mol Biochem Parasitol 11: 91-103.

Hoare CA (1980) Comparative aspects of human babesiosis. Trans R Soc Trop Med Hyg 74: 143-148.

Holder AA, Lockyer MJ, Odink KG, Sandhu JS, Riveros-Moreno V, Nicholls SC, Hillman Y, Davey LS, Tizard MLV, Schwarz RT, Freeman RR (1985) Primary structure of the precursor to the three major surface antigens of Plasmodium falciparum merozoites. Nature 317: 270-273.

Odink KG, Nicholls SC, Lockyer MJ, Hillman Y (1984) Cloning of malarial genes coding for high molecular weight antigens: isolation of fragments of _Plasmodium falciparum_ genes coding for proteins of 145 000 molecular weight. Mol Biochem Parasitol 10: 55-66.

Snary D, Smith MA (1986) An antigenic determinant common to both mouse red blood cells and several membrane proteins of the parasitic protozoa Babesia rodhaini. Mol Biochem Parasitol 20: 101-109.

Timms P, McGregor W, Dagleish J (1981) Tick fever vaccines: how they are made and how to use them. Queensland Agric J Nov/Dec: 311-316.

Wright IG, White M, Tracey-Patte PD, Donaldson RA, Goodger BV, Waltisbuhl DJ, Mahoney DF (1983) _Babesia bovis_: isolation of a protective antigen using monoclonal antibodies. Infect Immun 41: 244-250.

Wright IG, Mirre GB, Rode-Braminis K, Chamberlain M, Goodger BV, Waltisbuhl DJ (1985) Protective vaccination against virulent _Babesia bovis_ with a low molecular weight antigen. Infect Immun 48: 109-113.

THE MECHANISM BY WHICH INTERFERON GAMMA BLOCKS THE GROWTH OF *TOXOPLASMA GONDII* IN CULTURED FIBROBLASTS

E. R. Pfefferkorn

Department of Microbiology, Dartmouth Medical School, Hanover, NH 03756 USA

INTRODUCTION

Toxoplasma gondii is an obligate intracellular protozoan parasite. Overt disease, except in the context of transplacental infection or of immunosuppression, is rare. The immunological basis for the prevention of symptomatic infection is not completely understood. Experiments with passive transfer of immunity in mice suggest that the cellular component is more important than the humoral. To explore the role of T-cells, we concentrated on the activity of interferon gamma (IFN-γ).

MATERIALS AND METHODS

All experiments were carried out with a cloned line of the RH strain of *T. gondii* growing in human fibroblasts. The growth of the parasite was measured by the parasite-specific incorporation of [^{3}H]uracil into acid-precipitable form (Pfefferkorn & Guyre 1984). Human recombinant IFN-α, -β, and -γ were supplied by Genentech, Inc. All IFN-γ titers are reported in units/ml as determined by antiviral activity measured in parallel with an international standard (Pfefferkorn, Rebhun & Eckel 1986).

RESULTS AND DISCUSSION

Preliminary experiments showed that as little as 4 units/ml IFN-γ exerted a potent antitoxoplasma effect in human fibroblasts. (Previous publications (Pfefferkorn 1984; Pfefferkorn & Guyre 1984) that reported a minimal antitoxoplasma concentration of 8 units/ml had not been corrected to conform to an international IFN-γ standard.) Significantly, human recombinant IFN-α and -β had no antitoxoplasma effect at 1000 units/ml, the highest

NATO ASI Series, Vol. H11
Host-Parasite Cellular and Molecular Interactions
in Protozoal Infections. Edited by K.-P. Chang and D. Snary

concentration tested.

Assay of viable parasites by plaque formation showed that the activity of IFN-γ was predominantly parasitostatic. In an effort to determine the mechanism of this effect, we initially followed the antiviral paradigm. In many ways, the antitoxoplasma and antiviral actions of IFN-γ appeared similar (Pfefferkorn & Guyre 1984). IFN-γ had no effect on extracellular *T. gondii*. Instead, pretreatment of uninfected fibroblast cultures was required to establish an "antitoxoplasma state". Once established, this "antitoxoplasma state" did not require the continued presence of IFN-γ, which could be eliminated by neutralization with a specific monoclonal antibody. The establishment of the "antitoxoplasma state" in response to IFN-γ required messenger RNA synthesis and, by implication, protein synthesis by the host cell (Pfefferkorn 1986).

We found that much or all of the *in vitro* antitoxoplasma activity of IFN-γ could be explained by the induction of a potent indoleamine 2,3-dioxygenase (IDO), which degrades tryptophan to N-formylkynurenine. The first clue to the importance of tryptophan metabolism came from the observation that when the tryptophan content of the medium was increased, more IFN-γ was required to block the growth of the parasite (Pfefferkorn 1984). Conversely, when media with reduced tryptophan were used, less IFN-γ was required (Fig. 1). These results suggested that tryptophan might be destroyed in cultures treated with IFN-γ. This suggestion was confirmed by analysis of the medium of human fibroblasts treated with IFN-γ and [^{14}C]tryptophan. When medium samples taken one day after treatment with IFN-γ were analyzed by two dimensional chromatography-electrophoresis, most of the labeled amino acid was found to have been converted to two metabolites, N-formylkynurenine and kynurenine (Pfefferkorn 1984). The metabolic pathway leading to these tryptophan metabolites is shown in Fig. 2.

We measured the two enzymes of the pathway from tryptophan to kynurenine. The first enzyme, IDO, proved to be efficiently induced by IFN-γ in a time- and dose-dependent manner. The induction of this enzyme was inhibited by low concentrations of actinomycin D or cycloheximide and thus presumably required *de*

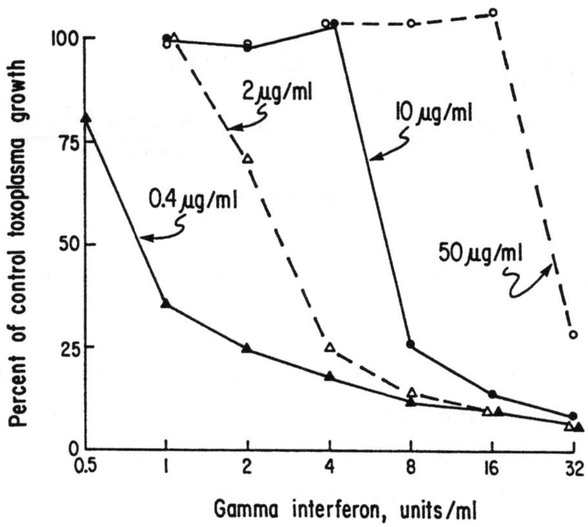

Fig. 1. Titration of the antitoxoplasma activity of IFN-γ in media with four tryptophan concentrations. Redrawn from Pfefferkorn (1984) with the IFN-γ concentrations converted to international units/ml.

novo synthesis of both messenger RNA and protein. Conversely, the second enzyme, N-formylkynurenine formamidase was a constitutive enzyme (Pfefferkorn, Rebhun & Eckel 1986).

Further characterization of the IDO induced by IFN-γ showed that it was different from either of the two well-characterized mammalian tryptophan degrading enzymes. The inducible hepatic tryptophan dioxygenase has a high K_m for tryptophan and does not attack D-tryptophan. Conversely, the constitutive intestinal IDO has an intermediate K_m for tryptophan and attacks a variety of other indoleamines including D-tryptophan. The IDO induced by IFN-γ had a very low K_m (3 µM) and failed to attack D-tryptophan (Pfefferkorn, Rebhun & Eckel 1986).

At least two mechanisms could explain how the induction of a potent IDO suppressed the growth of *T. gondii*. Assuming that tryptophan is an essential amino acid for *T. gondii*, extensive destruction of this amino acid could simply starve the parasite and block its protein synthesis. Alternatively, the degradation of tryptophan could yield toxic products that adversely affect the multiplication of *T. gondii*. Since the principal products of tryptophan degradation appeared in the medium, we assumed that

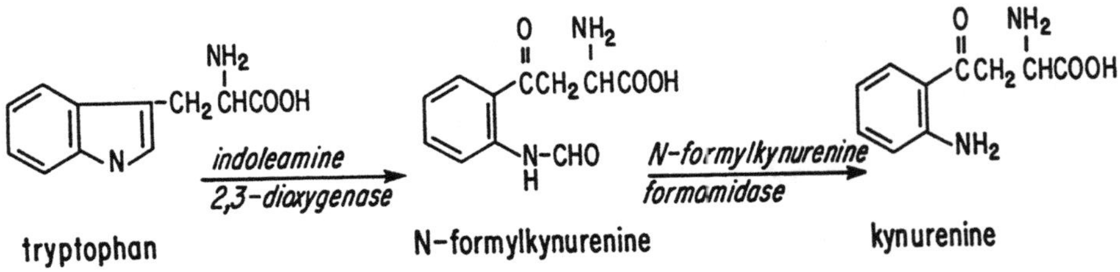

Fig. 2. The pathway from tryptophan to kynurenine

the putative toxic products might also be found there. A simple test for their presence would be to treat uninfected cultures with IFN-γ for one day, sufficient time to establish an "antitoxoplasma state", and then transfer the medium to fresh cultures and determine if the growth of *T. gondii* was blocked in these cultures. This protocol was complicated by the fact that much of the IFN-γ was not taken up by the cells during the first incubation and remained in biologically active form in the medium. When this medium was transferred to fresh cultures, the IFN-γ that it contained could suppress the growth of *T. gondii* even though the medium contained no toxic products. This artifact was circumvented by neutralizing the residual IFN-γ with a specific monoclonal antibody. When the active IFN-γ was eliminated in this way, the medium of cultures treated with IFN-γ had no effect on the growth of *T. gondii* (Table 1).

An alternative search for toxic products examined the ability of N-formylkynurenine and kynurenine to inhibit the growth of *T. gondii*. Neither of these substances had any effect on the growth of the parasite at concentrations equimolar to the tryptophan in the medium. Higher concentrations of N-formylkynurenine (1.6 mM) and of kynurenine (3.1 mM), did suppress the growth of the parasite (Pfefferkorn, Eckel & Rebhun 1986). However these elevated concentrations could not have been achieved *in the medium* since the metabolites were derived from tryptophan and could not exceed its concentration.

Neither of the above approaches revealed the presence of any toxic metabolites that might account for the effect of IFN-γ. However our experiments were confined to a search for stable

TABLE 1

Absence of toxic metabolites in the medium of cultures treated with IFN-γ provided that the residual IFN-γ was neutralized with antibody.

IFN-γ in used medium: units/ml	Anti-IFN-γ treatment of used medium: anti-IFN units/ml	Percent of control growth of *T. gondii* in cultures incubated with used medium
0	0	100
32	0	9
0	64	105
32	64	96

toxic metabolites in the medium. If a toxic metabolite were labile or failed to escape from human fibroblasts after its production, it would have eluded detection. Nonetheless, in the absence of evidence for toxic metabolites, we turned to the alternative hypothesis that *T. gondii* was starved for tryptophan in human fibroblast cultures treated with IFN-γ. Experiments in which cultures incubated in media with various concentrations of tryptophan were infected with *T. gondii* showed that the growth of the parasite was suppressed at concentrations below 0.8 μM. Since concentrations of L-tryptophan in this range were achieved in the medium of uninfected cultures treated with IFN-γ, starvation for this amino acid was a reasonable explanation for the antitoxoplasma effect (Pfefferkorn, Eckel & Rebhun 1986).

Since *T. gondii* is an intracellular parasite, the tryptophan concentration to which it is exposed is not that of the medium but rather that of the intracellular pool of its host cell. We therefore measured tryptophan and its metabolites in the acid soluble pools of control human fibroblasts and of human fibroblasts treated with 32 units/ml IFN-γ for 24 hr and then incubated for a few minutes in medium that contained [^{3}H]tryptophan. The tryptophan pool of control cells showed the expected kinetics of labeling. Equilibrium was reached within 90 sec and thereafter the radioactivity per cell remained constant. In the cultures pretreated with IFN-γ for one day and then incubated in medium with [^{3}H]tryptophan, the intracellular [^{3}H]tryptophan remained at a barely detectable level, less than 3% of the concentration seen in the control cells (Pfefferkorn,

Eckel & Rebhun 1986). Apparently a steady state was achieved in which the [^{3}H]tryptophan was degraded as rapidly as it entered the cell.

Although the [^{3}H]tryptophan content of the IFN-γ treated cells was much lower than that of the controls, the total radioactivity was notably higher. Nearly all of this increased radioactivity was associated with the two principal tryptophan metabolites, N-formylkynurenine and kynurenine. Two rate-limiting steps caused the build-up of these substances within cells treated with IFN-γ. First, the activity of the induced IDO exceeded that of the constitutive formamidase. Thus N-formylkynurenine was a prominent intracellular metabolite of tryptophan. Second, neither N-formylkynurenine nor kynurenine were able to diffuse (or be transported) out of the cells as rapidly as tryptophan entered. The intracellular parasites were therefore exposed to concentrations of N-formylkynurenine and kynurenine that were significantly higher than the molar concentration of tryptophan in the medium. Thus we were forced to re-examine the conclusion that these metabolites played no role in the antitoxoplasma action of IFN-γ. We measured the amount of radioactivity associated with tryptophan and its two metabolites over the entire course of a two-day treatment of human fibroblasts with IFN-γ. [^{3}H]Tryptophan and IFN-γ were added at the same time and the soluble pool was analyzed at intervals.

Figure 3 shows that the tryptophan content of the pool gradually fell to a low level. The slow decline in radioactivity of the tryptophan pool reflects the time required for IFN-γ to induce a sufficient amount of IDO. As the tryptophan concentration of the pool fell, the concentrations of radioactive N-formylkynurenine and kynurenine rose dramatically to greatly exceed that of the radioactive tryptophan in the control cells. Then, just as rapidly, the concentrations of these metabolites fell to low levels. The explanation for the rise in concentration has already been considered. The decline of intracellular N-formylkynurenine and kynurenine began when the tryptophan content of the medium was nearly exhausted. At this point no more [^{3}H]tryptophan was available for destruction and the tryptophan metabolites gradually leaked out, down the

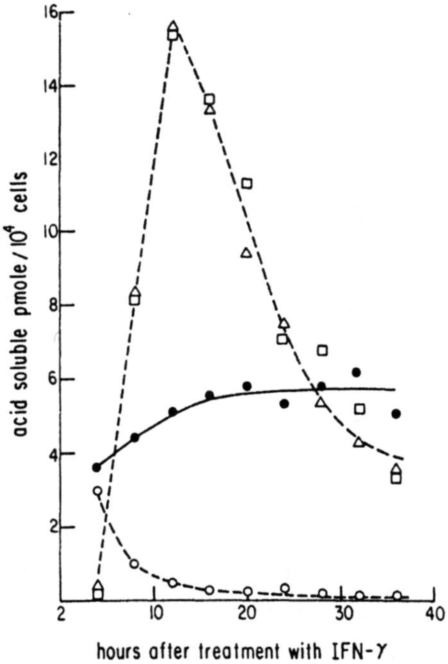

Fig. 3. The fate of [^{3}H]tryptophan in the acid soluble pools of control human fibroblasts and of fibroblasts treated with 32 units/ml IFN-γ in medium that contained [^{3}H]tryptophan. Tryptophan in control cells (●). Tryptophan (○), N-formylkynurenine (△), and kynurenine (□) in cells treated with IFN- γ. From Pfefferkorn, Eckel and Rebhun (1986).

concentration gradient between the intracellular pool and the medium. The critical question is: Did the tryptophan metabolites ever achieve intracellular concentrations sufficiently high to compromise the growth of *T. gondii*? In interpreting these data, it should be noted that the tryptophan transport system (the N system) of mammalian cells does not depend upon the sodium gradient across the plasma membrane and does not substantially concentrate this amino acid within the cell (Oxender *et al* 1977). Rather, it merely facilitates its transport across the plasma membrane. Thus in the control cells the intracellular concentration of radioactive tryptophan probably reflects the tryptophan content of the medium. If we assume that the intracellular water volume of the fibroblasts was unaffected by treatment with IFN-γ, the ratios of the tryptophan radioactivity in the pool of the control cells and the radioactivities of the N-formylkynurenine and kynurenine in the IFN-γ treated cells should provide an approximation of the intracellular concentrations of these metabolites relative to the concentration of the tryptophan in the medium. Calculated upon this assumption, the peak intracellular concentration of both N-formylkynurenine and kynurenine was only 0.2 mM, well below the concentrations required to inhibit the parasite when these metabolites were added to the medium. If we assume that these metabolites leak into cells as efficiently as they leak out, these data give no support to the hypothesis that *T. gondii* was inhibited by toxic concentrations of either of the principal metabolites of typtophan.

Two additional arguments make it unlikely that intracellular toxicity of the principal metabolites of tryptophan could account for the antitoxoplasma effect of IFN-γ. The intracellular concentrations of these metabolites reached a peak 12 hr after treatment with IFN-γ and then rapidly fell to lower levels. We routinely infect cultures 24 hr after treatment with IFN-γ, when the peak intracellular concentrations of tryptophan metabolites are long past. Indeed, IFN-γ has much less antitoxoplasma activity in the first 24 hr of treatment than in the second (Pfefferkorn & Guyre 1984). Furthermore, if toxic metabolites played a major role, we would expect that reducing the tryptophan content of the medium would reduce the potency of IFN-γ because lower concentrations of toxic metabolites could be generated. However, lowering the tryptophan content of the medium actually increased the apparent activity of the IFN-γ (Fig. 1).

Having found no evidence to support the theory that a toxic intracellular metabolite mediated the antitoxoplasma activity of IFN-γ, we returned to the alternative hypothesis that the parasitostatic effect is simply a consequence of starvation for an essential amino acid. Treatment of human fibroblasts with IFN-γ for 24 hr reduced the L-tryptophan content of the medium to a level known to compromise the growth of *T. gondii*. Furthermore, the IDO probably further reduced the intracellular tryptophan concentration. Starvation for an essential amino acid is thus a reasonable explanation for the parasitostatic effect of

IFN-γ. Starvation of the host cell must occur at the same time. The host cell remains viable in the absence of net protein synthesis through rapid protein turnover. The tryptophan transiently produced by this protein turnover cannot be readily available to the intracellular *T. gondii* since their growth is severely restricted when the tryptophan content of the medium was reduced, even in the absence of IFN-γ.

In summary, the most economical explanation for the *in vitro* antitoxoplasma activity of IFN-γ is that the induction of a potent IDO depletes the tryptophan and starves the parasite for an essential amino acid. The fact that this enzyme has a low K_m for tryptophan makes it particularly efficient in reducing intracellular concentrations of this amino acid. Although this mechanism probably prevails *in vitro*, its significance *in vivo* remains to be established. McCabe *et al* (1984) have shown that recombinant murine IFN-γ will protect mice against an otherwise fatal challenge with *T. gondii*. IFN-γ, in addition to its *in vitro* antiviral and antitoxoplasma activity, is a potent immunoregulatory lymphokine responsible for many other effects (Trinchieri & Perussia 1985). In the context of experimental toxoplasmosis, IFN-γ is a potent macrophage activating factor (Nathan *et al* 1983). Activation dramatically alters the interaction of macrophages with *T. gondii*. Nonactivated macrophages are readily infected and eventually killed by the rapidly multiplying parasites. In contrast, activated macrophages readily phagocytose and kill *T. gondii*. Macrophage activation undoubtedly plays a major role in the *in vivo* antitoxoplasma activity of IFN-γ. However, *Toxoplasma gondii* infects not only macrophages but probably nearly every cell type in the body. Any mechanism that serves to suppress the multiplication of the parasite in these other cells may offer an opportunity for other defense mechanisms, including macrophage activation, to play their roles. It should be noted that the only *in vitro* property of *T. gondii* that correlates with virulence measured *in vivo* is the rate of growth (Kaufman *et al* 1959). Any mechanism that slows growth may thus be advantageous. At present nothing is known of the *in vivo* effects of IFN-γ on tryptophan metabolism in humans. It is possible that cancer chemotherapy trials now underway will afford the opportunity to determine if there is an effect on tryptophan metabolism.

ACKNOWLEDGMENTS

Genentech Inc., generously supplied the human recombinant IFNs used in these experiments. I am indebted to Matthew Eckel and Scott Rebhun for both ideas and skillful assistance. This investigation was supported by Grant AI-14151 from the National Institutes of Health.

REFERENCES

Kaufman HE, Melton ML, Remington JS, Jacobs L (1959) Strain differences of *Toxoplasma gondii*. J Parasitol 45:189-190

McCabe RE, Luft BJ, Remington JS (1984) Effect of murine interferon gamma on murine toxoplasmosis. J Infect Dis 150:961-962

Nathan CF, Murray HW, Wiebe ME, Rubin BY (1983) Identification of interferon-γ as the lymphokine that activates human macrophage oxidative metabolism and antimicrobial activity. J Exp Med 158:670-689

Oxender DL, Lee M, Moore PA, Cecchini G (1977) Neutral amino acid transport systems of tissue culture cells. J Biol Chem 252: 2675-2679

Pfefferkorn ER (1984) Interferon γ blocks the growth of *Toxoplasma gondii* in human fibroblasts by inducing the host cells to degrade tryptophan. Proc Nat Acad Sci USA 81:908-912

Pfefferkorn ER (1986) Interferon-γ and the growth of *Toxoplasma gondii* in fibroblasts. Ann Inst Pasteur/Microbiol 137A:348-352

Pfefferkorn ER, Eckel M, Rebhun S (1986) Interferon-γ suppresses the growth of *Toxoplasma gondii* in human fibroblasts through starvation for tryptophan. Molec Biochem Parasitol, in press

Pfefferkorn ER, Guyre PM (1984) Inhibition of growth of *Toxoplasma gondii* in cultured fibroblasts by human recombinant gamma interferon. Infect and Immun 44:211-216

Pfefferkorn ER, Rebhun S, Eckel M (1986) Characterization of an indoleamine 2,3-dioxygenase induced by IFN-γ in cultured human fibroblasts. J Interferon Res 6:267-279

Trinchieri G, Perussia B (1985) Immune interferon: a pleiotropic lymphokine with multiple effects. Immunol Today 6:131-136

MODIFICATION OF HOST CELL PHAGOCYTIC COMPARTMENTS BY INTRACELLULAR *TOXOPLASMA GONDII*.

L. David Sibley

Department of Zoology/Physiology, Louisiana State University, Baton Rouge, LA 70803, and Immunology Research Department, National Hansen's Disease Center, Carville, LA 70721, USA

INTRODUCTION

The success of *Toxoplasma* as an intracellular parasite can be attributed to a capacity to infect a wide range of host cell types, both phagocytic and nonphagocytic, and to extensive modification of host cell endocytic vacuoles. Within macrophages, *Toxoplasma* occupy membrane bound vacuoles that are uniquely modified to resist host cell killing and digestion. Consistent with this, *Toxoplasma* cells survive the microbicidal capacity of normal macrophages where they occupy vacuoles which fail to stimulate production of oxygen intermediates (Wilson *et al*. 1980) and resist fusion with lysosomes (Jones and Hirsch 1972). In activated macrophages, which produce elevated levels of oxygen intermediates, *Toxoplasma* are killed within vacuoles that readily undergo fusion with lysosomes and are digested (Wilson *et al*. 1980).

We have recently expanded the physiological study of macrophage phagosomes occupied by *Toxoplasma* to examine the role of organelle acidification. Macrophage phagosomes containing yeast cells undergo acidification that can occur independently of lysosome fusion (Geisow *et al*. 1982). In normal macrophages, *Toxoplasma* enter vacuoles that remain at near neutral pH; however, in activated macrophages *Toxoplasma* occupy vacuoles that rapidily acidify after formation (Sibley *et al*. 1985). Given the susceptibility of *Toxoplasma* to low pH (Sibley *et al*. 1985), it seems evident that increased capacity for acidification by activated macrophages may

NATO ASI Series, Vol. H11
Host-Parasite Cellular and Molecular Interactions
in Protozoal Infections. Edited by K.-P. Chang and D. Snary

contribute to killing of intracellular microorganisms such as Toxoplasma.

Macrophage endocytic vacuoles occupied by Toxoplasma are characteristically surrounded tightly by both host rough endoplasmic reticulum and mitochondria (Jones et al. 1972, Nichols and O'Connor 1981). In addition, an extensive array of membrane-like tubules are found within vacuoles occupied by viable Toxoplasma (Jones et al. 1972). This intraphagosomal membrane (IPM) network is produced by live Toxoplasma entering macrophages but not when heat killed or specific antibody coated parasites are engulfed and digested by macrophages (Jones et al. 1972, Nichols, and O'Connor 1981). The origin and function of the IPM network are unknown; however, similar profiles appear in host cell phagosomes containing other protozoa that evade endocytic digestion (see Sibley et al. 1986b). The IPM network forms an interface that is ideally situated to evade microbicidal events at the level of the phagosome membrane and to provide nutrient transport. We have recently developed a method to purify membrane vesicles from this interface and have described the Toxoplasma proteins elaborated within the IPM network (Sibley et al. 1986b). The results discussed below provide evidence that Toxoplasma resides in vacuoles that contain protein-rich vesicles derived from the parasite surface membrane.

RESULTS AND DISCUSSION

Ca^{++} and Purification of IMP Networks

Toxoplasma released from ruptured host cells are connected with aggregates of IPM network tubules. In the presence of Ca^{++}, the networks adhere to Toxoplasma cells and can be pelleted by low speed centrifugation. Filtration followed by repeated washing removes host cells and yields a pellet containing intact Toxoplasma cells and the associated IMP networks. When placed in Ca^{++}-free buffer, the networks disperse allowing intact Toxoplasma cells to be separated by differential centrifugation. The remaining supernatant contains membrane vesicles derived from the IMP networks, that are recovered in the 100,000g pellet. IMP vesicles contain epitopes recognized by chronically infected mouse sera, as demonstrated by specific absorbtion onto antibody-coated formvar-carbon EM grids. In addition to the structural matrix, Toxoplasma secretes soluble excretory-secretory (E-S) products during incubation in Ca^{++}-free buffer. Toxoplasma cells remain morphologically intact and viable folowing purification of the IMP network vesicles.

Protein Characterization

Components of the IMP networks and E-S products show characteristic profiles when analyzed by SDS-PAGE. IMP network vesicles contain primarily a 30 kD protein but also contain significant amounts of 60, 58, 43, 35, and 20 kD proteins. The E-S soluble fraction contains a discrete set of proteins that differs from the characteristic pattern of whole Toxoplasma cell lysate. Ca^{++}-binding activity of proteins separated by SDS-PAGE and transferred to nitrocellulose was examined by $^{45}Ca^{++}$-overlay and autoradiography (Maruyama et al. 1984). Ca^{++}-binding activity is not found in the IMP proteins, but is readily detected in a 22 kD protein secreted by Toxoplasma cells and found in the soluble E-S fraction. The 22 kD soluble Ca^{++}-binding protein may function in mediating the observed Ca^{++}-dependent cell adhesion and stabilization of the IMP network.

Calcium levels may also be involved in formation of the IMP network as suggested by prominent membrane protrusions formed from the plasma membrane of Toxoplasma cells incubated in Ca^{++}-free buffer. When stained with tannic acid, these protrusions appear as lamellar whorls with 5-7 nm spacing. Similar profiles have been observed in malaria cell rhoptries and membraneous profiles within host cell vacuoles (Bannister et al. 1986, Stewart et al. 1985). Morphological evidence suggests these membrane components may contribute to formation of the vacuole surrounding merozoites entering red cells (Bannister et al. 1986). A similar relationship between

Toxoplasma rhoptry contents and the extensive production of membraneous tubules during host cell entry has been observed (Nichols _et al_. 1983).

Localization of IMP Network Proteins in Infected Macrophages

Polyclonal antibody prepared against the IMP network proteins, primarily 30 kD, was used to localize these proteins by immunoperoxidase EM. This polyclonal antibody recognizes primarily a 30 kD protein in total _Toxoplasma_ extracts analyzed by immunoblot. In _Toxoplasma_ infected macrophages, IMP proteins are located on the _Toxoplasma_ cell surface as well as the IMP network. This result indicates that _Toxoplasma_ derived proteins released from the cell surface form an elaborate protein-rich network within the vacuole. Although no labeling is detected on the macrophage cell surface, the _Toxoplasma_-containing phagosome membrane is strongly labeled by the anti-IMP antibody. The association of the IMP proteins with the phagosome membrane is supported by transmission EM examination of thick (300nm) sections showing numerous continuities between the IMP tubules and the vacuole membrane. Connections with the _Toxoplasma_ cell surface are also apparent; thus, the elaborate structural matrix within the vacuole connects the parasite cell surface with the membrane of the vacuole.

Localization of Toxoplasma Surface Proteins in IMP Networks

The antigenic similarity and morphologic connections between the _Toxoplasma_ cell surface, IPM networks and the vacuole membrane suggest that similar proteins occur in these membranes. To test this hypothesis we have used monoclonal antibodies (McAb) of defined specificity to _Toxoplasma_ surface proteins as probes for immuno-EM, radioimmune precipitation (RIP), and immunoblot analysis. The surface of _Toxoplasma_ is relatively simple, containing primarily proteins of 43, 35, 27-30, and 14 kD as shown by surface iodination and SDS-PAGE (Handman _et al_. 1980, Johnson _et al_. 1983). McAb's to these surface determinants react positively in the Sabin-Feldman dye test and by indirect immunofluorescence assay, confirming the exposure of the recognized epitopes at the surface (Handman and Remington 1980).

McAb's 1E11 to a 27 kD surface protein and 2G11C to 35 and 14 kD surface proteins (Handman et al. 1980) were used to identify components of purified IPM network vesicles iodinated with ^{125}I using chloramine T. McAb 1E11 immunoprecipitated the predominant 30 kD protein found in the IPM vesicles with similar efficiency to the polyclonal anti-IPM antibody. McAb 2G11C immunoprecipitated proteins of 35, 20, and 14 kD from the labeled IPM networks. A similar result has previously been noted by Johnson et al. (1983) who reported McAb 2G11C precipitated proteins of 21 and 14 kD, suggesting these two subunits may combine to form the 35 kD form. McAb's 1E11 and 2G11C both gave strong colloidal gold labeling on whole Toxoplasma cells and on the IPM vesicles formed during in vitro incubation. These results confirm that the predominant surface proteins of Toxoplasma are secreted in vesicles that form the extensive network within the vacuole (Table 1).

Table 1. Identification of Toxoplasma membrane proteins in IPM networks with monoclonal antibodies

Ag-MW	Surface	SDS-PAGE		EM	
		Blot	R.I.P.#	C-gold*	Peroxidase
30 [1]	+	@	+	+	+
35/20/14 [2]	+	@	+	+	+
58/28 [3]	-	+		-	+(T-X-100)

#Radioimmune precipitation; *Colloidal gold; [1]McAb 1E11 and [2]2G11C;(Handman et al.,1980); [3]McAb F3G3 (Sharma et al.,1984); @McAbs which do not recognize epitopes after SDS-PAGE

McAb F3G3 was also used to investigate the content of IPM vesicles. This McAb recognizes a 58 kD cytoplasmic protein that protects mice from lethal toxoplasmosis (Sharma et al. 1984). The epitopes recognized by both McAb F3G3 and a polyclonal antibody prepared against purified 58 kD protein are not normally exposed on the surface of *Toxoplasma* cells, as confirmed by lack of immunogold labeling. To examine the location of this antigen, lightly fixed *Toxoplasma* cells were extracted with Triton X-100, labeled by immunoperoxidase, and examined by EM. Cells treated with Triton X-100 showed peroxidase labeling by F3G3 beneath the outer membrane and in the IMP vesicles produced *in vitro*. The localization of proteins recognized by F3G3 was confirmed by purification of IMP vesicles and immunoblot analsyis. Whereas McAb F3G3 and polyclonal antibody to F3G3-purified protein labeled both a 58 and a 28 kD protein in extracts of whole *Toxoplasma*, the 28 kD protein was the predominant form recognized in the purified IMP vesicles. The IMP vesicles contain proteins not normally exposed on the cell surface, amoung them a 58 kD protein that may undergo processing to a 28 kD form at the membrane (Table 1), (Sibley and Sharma in preparation).

The composition of the IPM networks was also analyzed by immunoblot and RIP using normal mouse sera, chronically infected mouse sera, and chronically infected human sera. Both mouse and human chronic sera recognized the 30 kD protein comprising the major component of the IPM network, thus confirming previous observations that this protein is the immunodominant surface epitope of *Toxoplasma* (Handman *et al*. 1980, Kasper *et al*. 1983). In addition, 35, 20, and 14 kD proteins were recognized by RIP, while several higher MW proteins ranging from 58-62 kD were detected by immunoblot analysis. Figure 1 represents a model of the IMP vesicles depicting the predominant proteins of 43, 35, 30, 20, and 14 kD that have been specifically localized to the external cell surface either by results presented here or previous reports identifying *Toxoplasma* surface proteins (Handman *et al*. 1980, Johnson *et al*. 1983, Kasper *et al*. 1985). On the other hand, proteins of 28, 58, 60, and 62 kD are not exposed on the surface and have been represented as occurring on the inside of the membrane. It is not known if these proteins are associated with the membrane bilayer or merely packaged within the vesicles.

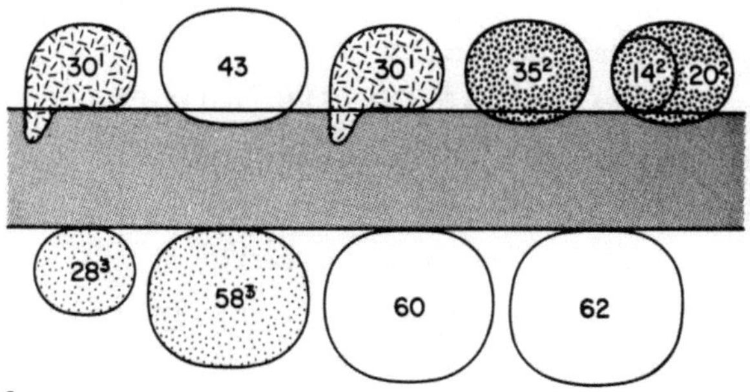

1

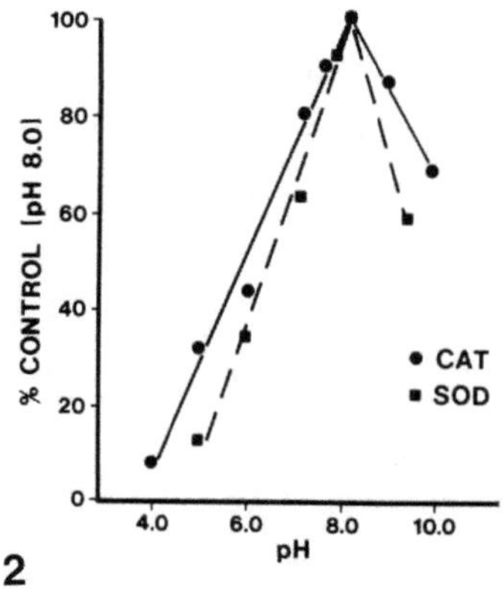

2

Figure 1. Diagramatic representation of Toxoplasma proteins located within the IPM vesicles. Superscripts as in Table 1.
Figure 2. pH dependence of Toxoplasma SOD and CAT activity.

Enzyme Activities in Excretory-Secretory Products

The IPM network represents a considerable extension of the Toxoplasma cell surface which may play a role in neutralizing macrophage metabolites generated at the plasma membrane during entry. We therefore examined both the IPM network vesicles and the E-S soluble fraction for scavenging enzymes superoxide dismutase (SOD) and catalase (CAT). While the IPM vesicles are devoid of activity, considerable enrichment of SOD and CAT activity is detected in the E-S soluble fraction (Table 2). This activity is 8-10 times greater than that found in Toxoplasma lysates when expressed on a U/mg basis, suggesting these enzymes are selectively secreted. The activity of SOD in the E-S fraction is inhibited 50% by azide but not by cyanide, characteristic of the Fe-containing enzyme previously described from Toxoplasma (Sibley et al. 1986b). If SOD and CAT are also secreted by Toxoplasma into the modified vacuole, the acidification capacity of this compartment is of considerable significance. Toxoplasma SOD and CAT have narrow pH optima and are rapidly neutralized by acidic pH (Fig 2). Thus activated macrophages, which have an increased capacity to acidify Toxoplasma-containing vacuoles, would be effective in neutralizing this defence mechanism.

As an obligate intracellular parasite, Toxoplasma tachizoites have a limited survival capacity outside the host cell. Presumedly the elaborate matrix of the IPM network within Toxoplasma-containing vacuoles, in both phagocytic and nonphagocytic host cells, contributes to a favorable micro-environment. To support their prolific growth, Toxoplasma cells may rely on active transport at the interface formed by this extensive surface area within the modified vacuole. One likely transport system would involve purine nucleotides which intracellular Toxoplasma do not synthesize *de novo*, but acquire from the host cell (Schwartzman and Pfefferkorn 1982). Consistent with this idea, we were able to detect ATPase activity in the soluble portion of whole cell lysates and in the E-S fraction but not in the membrane fractions (Table 2). We were not able to detect the extremely high levels of this enzyme previously reported (Asai *et al.* 1983) and found that activity was similar with Ca^{++}, Mg^{++}, or Na^{+} ATP and with n-nitrophenylphosphate as substrates. The secretion of ATPase activity by Toxoplasma may allow processing of nucleotides within the modified vacuole prior to incorporation within the Toxoplasma cell.

Table 2. Enzyme activities in soluble lysate, excretory-secretory and membrane fractions of Toxoplasma.

Fraction	Catalase U/mg	SOD U/mg	ATPase U/mg#
Lysate sup.	16.9 ± 0.9	16.3 ± 2.2	6.7 ± 1.2
Lysate pel.	ND	ND	ND
E-S sup.	162.3 ± 19.2	126.3 ± 8.7	8.6 ± 1.7
IPM network	ND	ND	ND

$\overline{X}$ ± SE; N = 3; ND = not detected, #30 min

ACKNOWLEDGEMENTS

Support by the LSU Alumni Federation Fellowship program is gratefully acknowledged. Monoclonal antibodies to Toxoplasma were generously provided by Dr. Jack Remington and Dr. Somesch Sharma, Palo Alto Medical Research Foundation, Palo Alto, CA.

REFERENCES

Aikawa M, Komata Y, Asai T, Midorikawa O (1977) Transmission and scanning electron microscopy of host cell entry by Toxoplasma gondii. Am J Pathol 87:285-296.

Asai T, O'Sullivan W, Tatibana M (1983) A potent nucleoside triphosphate hydrolase from the parasitic protozoan Toxoplasma gondii. J Biol Chem 258:6816-6822.

Bannister LH, Mitchell GH, Butcher GA, Dennis ED (1986) Lamellar membranes associated with rhoptries in erythrocytic merozoites of Plasmodium knowlesi: a clue to the mechanism of invasion. Parasitol 92:291-303.

Geisow MJ, Hart PD, Young MR (1982) Temporal changes of lysosome and phagosome pH during phagolysosome formation in macrophages: studies by fluorescence spectroscopy. J Cell Biol 89:645-652.

Handman E, Remington JS (1980) Serological and immunochemical characterization of monoclonal antibodies to Toxoplasma gondii. Immunol 40:579-588.

Handman E, Goding JW, Remington JS (1980) Detection and characterization of membrane antigens of Toxoplasma gondii. J Immunol 124:2578-2583.

Johnson AM, McDonald PJ, Neoh SH (1983) Monoclonal antibodies to Toxoplasma gondii cell membrane surface antigens protect mice from toxoplasmosis. J Protozool 30:351-356.

Jones TC, Yeh S, Hirsch JG (1972) The interaction between Toxoplasma gondii and mammalian cells. I. Mechanism of entry and intracellular fate of the parasite. J Exp Med 136:1157-1172.

Jones TC, Hirsch JG (1972) The interaction between Toxoplasma gondii and mammalian cells. II. The absence of lysosomal fusion with phagocytic vacuoles containing living parasites. J Exp Med 136:1173-1194.

Kasper LH, Crabb JH, Pfefferkorn ER (1983) Purification of a major membrane protein of Toxoplasma gondii by immunoabsorbtion with a monoclonal antibody. J Immunol 130:2407-2412.

Maruyama K, Mikawa T, Ebashi S (1984) Detection of calcium binding proteins by ^{45}Ca autoradiography on nitrocellulaose membrane after sodium dodecyl sulfate gel electrophoresis. J Biochem 95:511-519.

Nichols BA, O'Connor GR (1981) Penetration of mouse peritoneal macrophages by the protozoan Toxoplasma gondii. Lab Invest 44:324-335.

Nichols BA, Chiappino ML, O'Connor GR (1983) Secretion from the rhoptries of Toxoplasma gondii during host-cell invasion. J Ultrstr Res 83:85-98.

Scwartzman JD, Pfefferkorn ER (1982) Toxoplasma gondii: purine synthesis and salvage in mutant host cells and parasites. Exp Parasit 53:77-86.

Sharma SD, Araujo FG, Remington JS (1984) Toxoplasma antigen isolated by affinity chromatography with monoclonal antibody protects mice against lethal infection with Toxoplasma gondii. J Immunol 133:2818-2820.

Sibley LD, Weidner E, Krahenbuhl JL (1985) Phagosome acidification blocked by intracellular Toxoplasma gondii. Nature 315:416-419.

Sibley LD, Lawson R, Weidner E (1986a) Superoxide dismutase and catalase in Toxoplasma gondii. M B Parasit 19:83-87.

Sibley LD, Krahenbuhl JL, Adams GM, Weidner E (1986b) Toxoplasma modifies macrophage phagosomes by secretion of a vesiclular network rich in surface proteins. J Cell Biol 103:867-874.

Stewart MJ, Schulman S, Vandeberbg JP (1985) Rhoptry secretion of membraneous whorls by Plasmodium berghei sporozoites. J Protozool 32:280-283.

Wilson CB, Tsai V, Remington JS (1980) Failure to trigger the oxidative burst of normal macrophages. Possible mechanism for survival of intracellular pathogens. J Exp Med 151:328-346.

CHARACTERIZATION OF A 42 KDA RHOPTRY ANTIGEN OF <u>TOXOPLASMA GONDII</u>

Jean François DUBREMETZ, Abderrahim SADAK, Zoubida TAGHY and Bernard FORTIER

Unité de Biologie et Biochimie Parasitaires et Fongiques, INSERM U 42, 369, Rue Jules Guesde, 59650 Villeneuve d'Ascq, France

INTRODUCTION

The rhoptries of Sporozoan zoites are a set of intracellular organelles supposed to be secretory and to play a role during the invasion of the host cell by these parasites. A prerequisite to the determination of their function is the identification of the nature and properties of their contents. Little data have been obtained so far in this field except in the genus *Plasmodium* were a 235 kDa protein in *P. yoelii* (Holder & Freeman 1981) and 2 proteins of 140 and 80 kDa in *P. falciparum* (Holder, Freeman, Uni & Aikawa 1985; Campbell, Miller, Hudson, Franco & Andriziak 1984) have been identified in the rhoptries of erythrocytic merozoites by monoclonal antibodies (McAb). Schwartzman (1986) has obtained McAbs specific for the rhoptries of *Toxoplasma gondii* tachyzoites but has not identified the corresponding antigens. We report here on the McAbs characterization of a 42 kDa protein located in the rhoptries of tachyzoites and bradyzoites of *T. gondii* .

MATERIALS AND METHODS

A rhoptry enriched preparation was prepared by subcellular fractionation of *T. gondii* tachyzoites (modified from Dubremetz & Dissous 1981) and used to immunize Balb/c mice. Hybridomas were produced by fusion with SP2/O cells according to Galfre *et al* (1977). Screening of secreting hybridomas was performed by immunofluorescence assay (IFA) and characterization of protein antigens by McAbs was achieved by western immunoblotting.

For immunoelectronmicroscopy, *T. gondii* RH tachyzoites invading lymphosarcoma cultivated cells (TG180) and cysts of *T. gondii* 76K from infested mouse brain were fixed and embedded in Lowicryl K4M according to Roth *et al* (1981). Immunodetection on thin sections was done with Protein A Gold prepared according to Slot & Geuze (1985).

For immunosorption, *T. gondii* RH tachyzoites were grown *in vitro* in Hela cells and the culture was labeled for 4 hours either with 35S methionine (40 uCi/ml) or 3H isoleucine (80 uCi/ml). Infested labeled monolayers were then lysed with 1% NP 40 and the lysate used for immunosorption.

NATO ASI Series, Vol. H11
Host-Parasite Cellular and Molecular Interactions
in Protozoal Infections. Edited by K.-P. Chang and D. Snary

Monoclonal immunoglobulin purification, immunosorbent preparation, SDS PAGE analysis and fluorography were as described elsewere (Delplace, Dubremetz, Fortier & Vernes 1985).

RESULTS

One of the monoclonal antibodies obtained (1E5) gave a spotted intracellular fluorescence in the anterior half of zoites; in some cases elongated fluorescent organelles could be observed. By immunoelectronmicroscopy on tachyzoites, this antibody bound exclusively to the contents of dense bodies which could be easily identified as the rhoptries (Fig.1); the label was essentially found on the posterior, inflated part of the organelles whereas little gold was observed on peduncles. Identical results were obtained on bradyzoites (Fig.2).

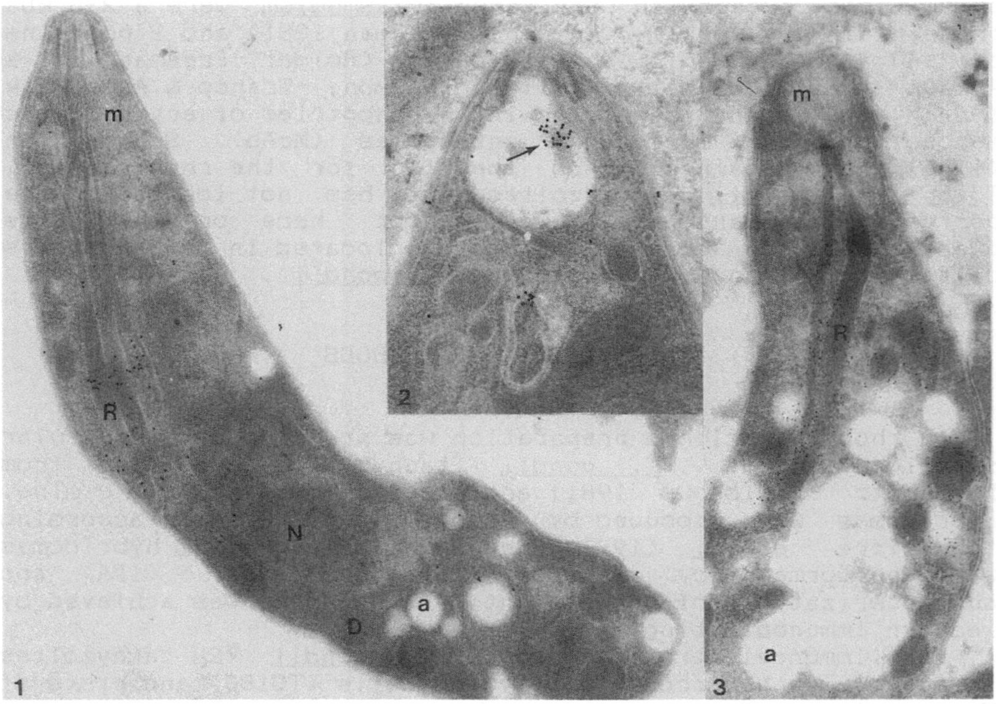

Figure 1,2,3. Immunogold labeling with McAb 1E5 on Lowicryl K4M sections (R: rhoptry, m: micronemes, D: dense granule, N: nucleus, a: amylopectin)

Figure 1. Extracellular tachyzoite (x 25,000 ; 8nm gold)

Figure 2. Tachyzoite fixed after host cell invasion: the label is found on an "empty rhoptry" (x 36,000 ; 8nm gold)

Figure 3. Mouse brain cyst bradyzoite (x 28,000 ; 5nm gold)

The micronemes, which were more abundants in bradyzoites, were never labeled. Some dense organelles similar to cross sectioned rhoptries were not labeled: they were interpreted as dense granules, the third group of zoite electron dense organelles. On tachyzoites fixed during host cell invasion, some label was found on an anterior vacuole corresponding to the structure described by Nichols et al (1983) as an empty rhoptry (Fig 3), but no gold was found outside of the zoite.

Probing western blots of tachyzoites with McAb 1E5 led to the characterization of a 42 kDa antigen (Fig 4), the molecular weight of which was not altered upon reduction of the sample by dithiothreitol.

Immunosorption of labeled parasitized cells lysates by McAb 1E5 allowed the purification of 35S methionine or 3H isoleucine labeled 42 kDa antigen (Fig 5).

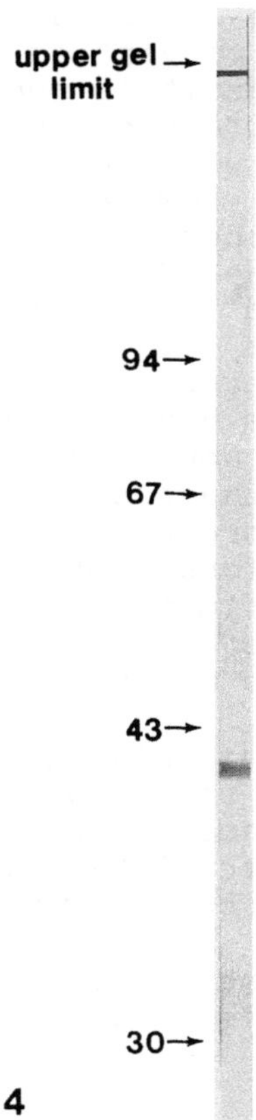

4

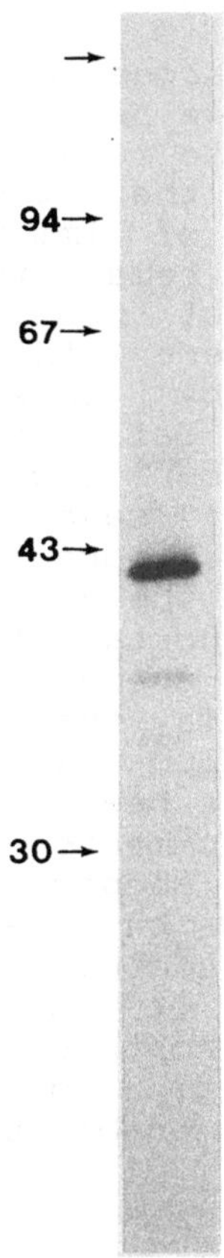

5

Figure 4. Immunoblotting of T. gondii tachyzoites with McAb 1E5. Molecular weights (kDa) of standard proteins are given.

Figure 5. SDS PAGE and Fluorography of 35S methionine labeled T. gondii 42 kDa rhoptry antigen purified on immunosorbent 1E5. Molecular weights (kDa) of standard proteins are given.

DISCUSSION

The present report is the first characterization of a T. gondii rhoptry antigen. Electronmicroscopy enabled us to demonstrate that this antigen is rhoptry specific and cannot be detected in the other electron dense organelles of the zoite. This give further support to the assumption of complete distinction between rhoptries, micronemes and dense granules in Sporozoan zoites suggested by previous works (Dubremetz & Dissous 1981, Entzeroth, Dubremetz, Hodick & Ferreira 1986). The detection of the rhoptry antigen in structures described previously as emptying rhoptries strongly enforces this morphological identification and is a new step towards the demonstration of the exocytosis of rhoptry contents during host cell invasion. Our inability to detect the 42 kDa antigen outside of the zoite during invasion calls for some comments since this later result tends to argue against the exocytic properties of the organelle; we believe however that this result is more likely to be due to technical failures, our antibody having too low affinity or the antigen being too diluted with respect to the sensitivity of immunogold detection on Lowicryl sections. These points remain to be investigated further.

It is not possible to correlate the 42 kDa protein described here with the antigens localized in the T. gondii rhoptries by Schwartzman (1986) due to the lack of biochemical data on these antigens. Moreover, preliminary results in our laboratory with other McAbs have suggested the existence of several distinct proteinic rhoptry antigens in T. gondii ; more investigations are therefore needed to characterize the entire contents of these organelles.

It is important to notice that two different stages of the life cycle share the same rhoptry antigen which is thus likely to play the same function. This has not been described in Plasmodium where only erythrocytic merozoites rhoptry antigens have been identified so far.

Little comments can be made comparing Plasmodium and Toxoplasma rhoptry antigens since we still lack too many informations on these molecules. As underlined by Sinden (1985) and because a commun function is very likely for these organelles, it would be important to search for similarities between rhoptries antigens among Apicomplexa. The characterization of this first T. gondii 42 kDa rhoptry protein will enable us to go further in this way.

ACKNOWLEDGEMENTS

The authors are indebted to P. Delplace and A. Vernes for stimulating discussion during the course of this study and to C. Ansel, I. Duprez, A. Loyens, M. Mortuaire for expert assistence. This work was supported by INSERM and RHONE MERIEUX.

REFERENCES

Campbell GH, Miller LH, Hudson D, Franco EL, Andryziak PM (1984) Monoclonal antibody characterization of Plasmodium falciparum antigens. Am J Trop Med Hyg 33:1051-1054

Delplace P, Dubremetz JF, Fortier B, Vernes A (1985) A 50 kilodaltons exoantigen specific to the merozoite release-reinvasion stage of Plasmodium falciparum . Mol Biochem Parasitol 17:239-251

Dubremetz JF, Dissous C (1980) Characteristic proteins of micronemes and dense granules from Sarcocystis tenella zoites (Protozoa, Coccidia). Mol Biochem Parasitol 1:279-289

Entzeroth R, Dubremetz JF, Hodick D, Ferreira E (1986) Immunoelectronmicroscopic demonstration of the exocytosis of dense granule contents into the secondary parasitophorous vacuole of Sarcocystis muris (Protozoa,Apicomplexa). Eur J Cell Biol 41:182-188

Galfre G, Howe SC, Milstein C, Butcher GW, Howard JC (1977) Antibodies to major histocompatibility antigens produced by hybrid cell lines. Nature 266:550-552

Holder AA, Freeman RR (1981) Immunization against bloodstage rodent malaria using purified parasite antigens. Nature 294:361-364

Holder AA, Freeman RR, Uni S, Aikawa M (1985) Isolation of a Plasmodium falciparum rhoptry protein. Mol Biochem Parasitol 14:293-303

Nichols BA, Chiappino ML, O'Connor GR (1983) Secretion from the rhoptries of Toxoplasma gondii during Host-Cell invasion. J Ultrastruct Res 83:85-98

Roth J, Bendayan M, Carlemalm E, Villinger W, Garavito M (1981) Enhancement of structural preservation and immunocytochemical staining in low temperature embedded pancreatic tissue. J Histochem Cytochem 29:663-671

Schwartzman JD (1986) Inhibition of a penetration enhancing factor of Toxoplasma gondii by monoclonal antibodies specific for rhoptries. Infect Immun 51:760-764

Sinden RE (1985) A cell biologist's view of host cell recognition and invasion by malarial parasites. Trans Roy Soc Trop Med Hyg 79:598-605

Slot JW, Geuze HJ (1985) A new method for preparing gold probes for multiple labeling cytochemistry. Eur J Cell Biol 38:87-93

Entamoeba histolytica: ACTIVATION AND RELEASE OF MEMBRANE DENSE BODIES

A. Martínez-Palomo, I. Meza, B. Chávez, J.L. Rosales-Encina, M.L. Muñoz, A. González-Robles, and M. Rojkind

Center for Research and Advanced Studies, National Polytechnical Institute, Apartado Postal 14-740, 07000 Mexico, D.F., Mexico

INTRODUCTION

The protozoan parasite *Entamoeba histolytica* is the etiological agent of invasive amebiasis, a common human disease in certain developing countries. The requirement of close contact between the parasite and target cells for cytolysis to occur (Ravdin & Guerrant 1981; Martínez-Palomo, 1982) and the isolation of amebic proteins capable of forming aqueous pores upon their insertion into the plasma membrane of target eukaryotic cells (Lynch et al. 1982; Young et al. 1982) suggest that membrane interactions have an important role in the powerful cytolytic activity of pathogenic amebas. Tissue invasion may be facilitated by the collagenolytic and proteolytic activities described in this protozoan (Muñoz et al. 1982; Gadasi & Kessler 1983; Keene et al. 1986).

The present report studies cellular changes occurring during contact and degradation of collagen by pathogenic amebas cultured under axenic conditions. The process involves the formation and liberation of membrane dense bodies simultaneous with the intracellular accumulation and the extracellular liberation of collagenase.

MATERIALS AND METHODS

Cells

In order to study the cellular changes occurring during contact of a highly virulent amebic strain with collagen, axenic cultures of *E. histolytica* HM1:IMSS maintained in BI-S-33 medium (Diamond et al. 1978) were plated on collagen-coated culture flasks, incubated for 16 hrs, and harvested after cooling and low-speed centrifugation. Following resuspension in 2 ml of serum-free culture medium, aliquots were used for membrane fractionation, collagenase determination, and ultrastructural examination. Controls were carried out in parallel experiments with amebas cultured in regular plastic culture flasks.

Membrane Fractions

Membranes were purified according to Aley et al. (1980) with some modifications. Intact cells (10^8) were washed with 19 mM potassium phosphate buffer (pH 7.2) containing 0.27 M NaCl. Cells were resuspended in cold phosphate buffer

NATO ASI Series, Vol. H11
Host-Parasite Cellular and Molecular Interactions
in Protozoal Infections. Edited by K.-P. Chang and D. Snary

containing 10 mM $MgCl_2$, rapidly mixed with an equal volume of cold concanvalin A (Con A, 1 mg/ml), and incubated at 4°C for 5 minutes. The cells were gently centrifuged and the pellet was resuspended in Tris buffer 10 mM Tris-HCl (pH 7.5), 1 mM $MgCl_2$ containing 2 µM leupeptin, 2 mM diisopropyl fluorophosphate, and 5 mM N-ethylmaleimide. After 10 min in this hypotonic solution, cells were ruptured by homogenization with a glass Dounce homogenizer, layered on a two-step gradient (8 ml of 0.5 M mannitol over 4 ml of 0.58 M sucrose in Tris buffer), and centrifuged at 250 g for 30 min. Large membrane fragments and various cellular debris formed the pellet at the bottom of the gradient, as seen by phase contrast microscopy. The pellet was resuspended in 1 ml of Tris buffer containing 1 M alpha-methylmannoside and incubated at 4° C for 40 to 60 min. The plasma membranes, now free of Con A were diluted with three volumes of Tris buffer, and centrifuged at 250 g for 30 min. Vesiculated plasma membranes floating on top of the sucrose were collected and treated again with alpha-methylmannoside, floated on a 20% sucrose gradient, and concentrated by centrifugation at 40,000 g for one hr. The pellet containing the enriched membranes was resuspended in Tris buffer at protein concentrations of 3 to 4 mg/ml and stored at -20° C. Collagenase determination was carried out as indicated elsewhere (Muñoz et al. 1982). The collagenolytic activity of control and collagen-activated trophozoites and their isolated plasma membrane fractions was measured in duplicate experiments. The activities of the intact trophozoites were estimated from the slopes of time curves, because both the kinetics and the total amount of collagen degraded were dissimilar in the two conditions.

Electron Microscopy.

After 16 hrs of interaction with the specific substrate, amebas were fixed with 1.5% glutaraldehyde in 0.1M sodium cacodylate for 30 min at 36°C. After fixation with osmium tetroxide, dehydration in increasing concentrations of ethanol, and embedding in Epon 812, thin sections were cut with diamond knives. Sections were observed using a Jeol 100SX transmission electron microscope. Quantitation of the number of dense bodies per length of plasma membrane profiles was carried out in photographs of thin sections, studied at a total magnification of X 50,000. The total length of cell profiles was measured with a GTCO digitizer, using an IMB-PC XT microcomputer and a digitizer program kindly provided by Dr. M. Kurt of Baylor College of Medicine, Houston, Texas. For each experimental procedure, a total of 25 cell profiles selected at random were measured at X 1,000. Data represent the average number of dense bodies per 10 µm of membrane profile.

RESULTS AND DISCUSSION

Electron microscopy of thin sections of control amebas revealed less than 0.01 dense bodies/10 µm of plasma membrane. In contrast, collagen-activated trophozoites

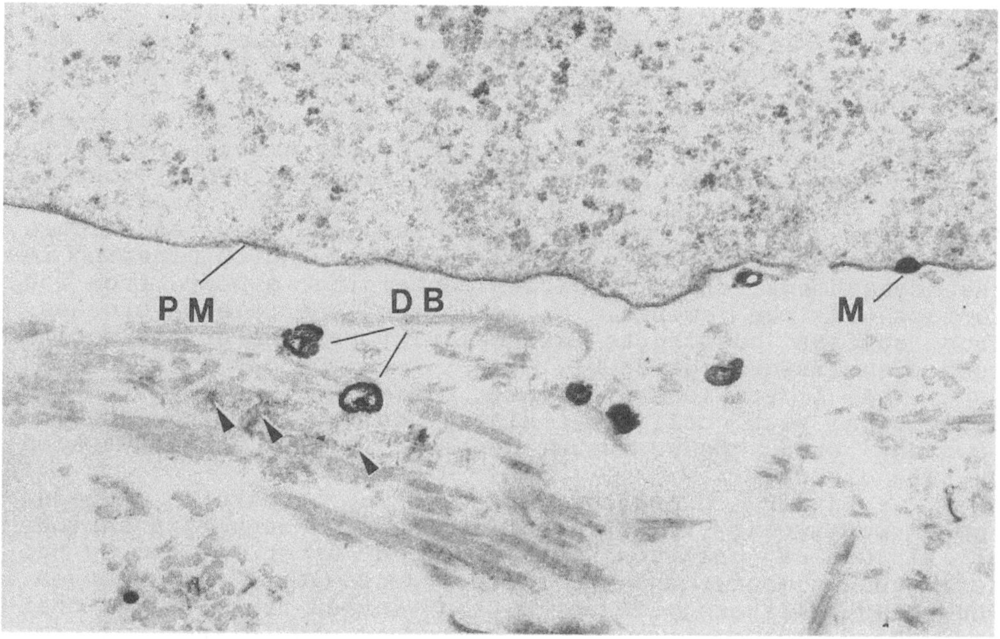

FIGURE 1. Axenic culture of the highly virulent amebic strain <u>Entamoeba histolytica</u> HM1:IMSS were plated on collagen-coated culture flasks. This electron micrograph shows the surface of a trophozoite (PM) near partially degraded collagen fibers (arrowheads). Several dense bodies (DB) can be observed in the extracellular space in contact with collagen. A dense body is seen beneath the plasma membrane (M) (X40,000).

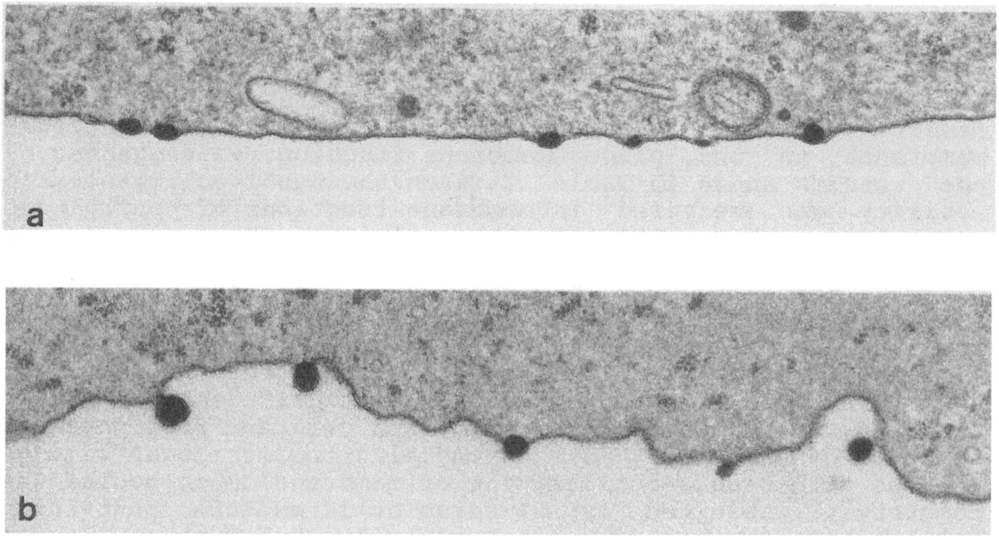

FIGURE 2. Various stages of the release of plasma membrane-associated dense bodies from activated trophozoites of <u>Entamoeba histolytica</u> are demonstrated in these electron micrographs (a-b) (X 32,000).

revealed an average of up to 2.11 dense bodies/10 µm of plasma membrane. Dense bodies were irregular in size and shape, averaging 30 to 50 nm in diameter both in control and in activated amebas.

In addition to the increase in the number of dense bodies, micrographs of collagen-activated amebas frequently showed the release of dense bodies (Fig. 1). This occurred through a budding process involving the progressive separation of the bodies which remained covered with a portion of the parasite plasma membrane, and culminating in the detachment of a vesicle containing a body, from the surface of the amebas. All stages of the release process, from surface budding to extracellular release of the bodies were visualized (Fig. 2a-b).

The liberation of dense bodies was not specific for activation with collagen, since the phenomenon was also apparent when amebas were incubated in 0.1% albumin-coated culture dishes.

Intact trophozoites used as controls degraded approximately 12 to 13 µg of collagen. However, a maximum of 21 µg of collagen was degraded after 24 hrs by the activated amebas, and no further degradation was observed. The mean collagenolytic activity of intact trophozoites that had been in contact with collagen was 0.156 units/mg protein, whereas in controls, it was only 0.068 units/mg protein. Although some variation was seen among different cultures of amebas, for each set of experiments, the differences between control and activated trophozoites was always greater than two fold. The difference was even more striking when the distribution of collagenolytic activity in membrane fractions of control and activated parasites was compared (Table 1). The specific activity of the enzyme in the plasma membrane fraction of activated amebas was 1.378 units/mg protein, nine fold greater than in whole activated cells (0.156 units/mg protein), suggesting a membrane localization of the enzyme. In the plasma membrane fraction of control cells the enzymatic activity was 0.497. The activation in collagenolytic activity is apparently the result of translocation of the enzyme from internal membranes to the plasma membrane fraction, as suggested by the results shown in Table 1. When the total collagenolytic activity was measured in membrane fractions of control and activates amebas, 72% of the activity in the controls was found in the internal membrane fraction and 20% in the plasma membrane. In contrast, in the activated trophozoites, 67% of the activity was found in the plasma membrane fraction, and only 9% was left in the internal membrane fraction. Fractions of isolated membranes of amebas activated through contact with collagen were mainly composed of large sheets of plasma membrane fragments and numerous dense bodies, as shown by electron microscopy.

The formation and liberation of membrane dense bodies was completely inhibited by 5% serum or 10 mM EDTA, both known to inhibit the collagenolytic activity (Muñoz et al. 1982). Treatment of amebas with 0.5 M NaCl results in elution of the collagenase (Gadasi & Kessler, 1983); in our samples,

TABLE 1
DISTRIBUTION OF COLLAGENOLYTIC ACTIVITY IN MEMBRANE FRACTIONS IN E.HISTOLYTICA

Fraction	Activity			
	Control		Activated	
	Units/mg protein	(%)*	Units/mg protein	(%)
Internal membranes	0.784±0.01	72.32	0.126±0.01	8.59
Nonvesiculated membranes	1.012±0.03	7.81	0.410±0.01	23.98
Plasma membrane	0.464±0.01	19.86	1.436±0.01	67.06

* % of total activity in submembranal fractions.

such treatment reduced the number of dense bodies to 0.17/10 µm.

In summary, our observations demonstrate that following contact with collagen, E. histolytica trophozoites show a considerable increase in the number of membrane-associated dense bodies, the release of these bodies from the parasite through a process of budding, and a ninefold increase in collagenase activity of purified plasma membrane fractions containing dense bodies, compared to whole cells. The results suggest that exposure to collagen substrates triggers the formation and liberation of membrane dense bodies, and that this process is related to increased collagenolytic activity of the trophozoites. Purification and chemical analysis of the bodies will prove whether they contain collagenase, and possibly other enzymes. The dense bodies here described have been observed since early electron microscopical observations of amebas (Rondanelli et al. 1965; Sobota 1985), and although it was speculated that they could be related to the pathogenicity of these parasites, their presence in pathogenic as well as in nonpathogenic amebas discouraged further studies. Our results now suggest that the intercellular formation and the extracellular liberation of dense bodies could be related to the lytic activity of the parasite. The observations described here may be of significance for the understanding of the cellular mechanisms involved in the degradation of mammalian extracellular components by pathogenic amebas isolated from man.

ACKNOWLEDGEMENTS

This work was supported in part by grants from the National Council for Science and Technology, Mexico, (CONACYT), the Rockefeller Foundation, the MacArthur Foundation, the McConnell Clark Foundation, and the N.I.H. (AI 21304).

REFERENCES

Aley SB, Scott WA, Cohn ZJ (1980) Plasma membrane of Entamoeba histolytica. J Exp Med 152:391-404

Diamond LS, Harlow DR, Cunnick CC (1978) A new medium for the axenic cultivation of Entamoeba histolytica and other Entamoeba. Trans Roy Soc Trop Med Hyg 72:431-432

Gadasi H, Kessler E (1983) Correlation of virulence and collagenolytic activity in Entamoeba histolytica. Infec Immun 39:528-531

Keene WE, Petitt MG, Allen S, McKerrow JH (1986) The major neutral proteinase of Entamoeba histolytica. J Exp Med 163:536-549

Lynch EC, Rosenberg IM, Gitler C (1982) An ion-channel forming protein produce by Entamoeba histolytica. Euro Mol Biol Organ J 1:801-804

Martinez-Palomo A (1982) The Biology of Entamoeba histolytica. Research Studies Press/John Wiley, Chichester

Muñoz ML, Calderón J, Rojkind M (1982) The collagenase of Entamoeba histolytica. J Exp Med 155:42-51

Ravdin JL, Guerrant RL (1981) Role of adherence in cytopathogenic mechanisms of Entamoeba histolytica. J Clin Invest 68:1305-1313

Rondanelli EG, Osculati F, Gerna G, Fiori GP (1965) Ricerche microelettroniche su Entamoeba invadens. Boll Ist Sieroter Milan 44:218-239

Sobota A (1985) Subplasmalemmal calcium-binding microregions in Acanthamoeba. J Cell Sci 79:217-235

Young JD, Young TM, Lu LP, Unkless JC, Cohn Z (1982) Characterization of a membrane pore-forming protein from Entamoeba histolytica. J Exp Med 156:1677-1690

ENTAMOEBA HISTOLYTICA: EFFECT OF BACTERIAL ASSOCIATES ON ISOENZYME PATTERNS AND VIRULENCE

David Mirelman,[1] Rivka Bracha,[1] Ann Chayen,[1] and Louis S. Diamond,[2]

[1]Dept. of Biophysics and Unit for Molecular Biology of Parasitic Diseases, Weizmann Institute of Science, Rehovot 76100, Israel and [2]Laboratory of Parasitic Diseases, National Institutes of Health (NIAID), Bethesda, MD 20892, USA.

INTRODUCTION

Electrophoresis patterns of isoenzymes (zymodemes) is widely used as a tool for taxonomic and genetic identification of parasitic protozoa and helminths (Flockhart and Denham 1984). Studies of isoenzymes of *E. histolytica* trophozoites isolated from thousands of asymptomatic and symptomatic human cases around the world have resulted in the characterization of at least 20 distinct zymodemes (Sargeaunt 1985; Sargeaunt *et al.* 1982; Sargeaunt and Williams 1978, Sargeaunt *et al.* 1982a). Based on the clinical picture of the source of the isolate, these zymodemes are placed in one or two categories: pathogenic or non pathogenic. The more valuable isoenzyme pattern for determining *E. histolytica* pathogenicity has been the hexokinase, whose double bands migrate differently in the two cases. The non pathogenic zymodemes, which are the majority of cases, are in isolates from asymptomatic carriers with negative serology, while the pathogenic zymodemes are derived from individuals with positive serology and presence or history of some clinical symptoms. Sargeaunt and his colleagues, who studied the isoenzyme patterns of numerous amebal isolates, have never reported the finding of a non pathogenic zymodeme in conjunction with a pathogenic one in any single host (Sargeaunt 1985; Sargeaunt *et al.* 1982a). Nor were they able to demonstrate evidence of alteration of isoenzyme patterns, i.e. alterations from non pathogenic to pathogenic or vice versa, in any of their longitudinal culture studies (Sargeaunt 1985; Sargeaunt *et al.* 1982a), conducted in the presence of viable bacteria in Robinson's medium (Robinson 1968). Moreover, short term interaction of

NATO ASI Series, Vol. H11
Host-Parasite Cellular and Molecular Interactions
in Protozoal Infections. Edited by K.-P. Chang and D. Snary

axenically grown amebae with bacteria (for 1 hour) which markedly enhanced their virulence (Bracha and Mirelman 1984), didn't cause any changes in their pathogenic zymodeme (Mirelman *et al.* 1984). Based on these studies, it has been proposed that zymodemes can be used as biochemical markers to distinguish between the different *E. histolytica* species, and that persons harboring amebae with non pathogenic zymodemes may not require treatment (Editorial 1985; Sargeaunt and Williams 1982).

Axenization of Non Pathogenic *E. histolytica*

The development of axenic culture medium (Diamond *et al.* 1978), in which amebae could grow and divide in the absence of any other living organism or cell, has provided the basis by which studies of various aspects of the ameba and its interrelationship with bacteria and host cells could be undertaken. Little progress, however, has been made towards our understanding of the differences between pathogenic and non pathogenic isolates of *E. histolytica.* This has been mainly due to the unavailability of any axenic strains of the latter, as in contrast to pathogenic isolates, all attempts to cultivate non pathogenic amebae under axenic conditions failed. This feat has recently been accomplished (Mirelman *et al.* 1986). Moreover, propagation of the non pathogenic, cloned strain SAW 1734R clAR (zymodeme group III) was also done in the TYI-S-33 medium for axenic cultivation, provided that it contained a mixture of antibiotics (amikacin, cefotaxime and erythromycin, at 50 μg/ml) to suppress the growth of the associated bacterial flora, and a nutritional supplement consisting of lethally irradiated bacteria (2×10^8/ml). Under these new culture conditions the trophozoites gradually increased their growth rate and 21 days later the supplement of irradiated bacteria was no longer required. Although at this stage the cultures were not yet completely axenic and the presence of a small number of bacteria ($<10^5$/ml) could still be detected in sterility tests, a change was observed in the morphology of the trophozoites from a relatively smooth endoplasm with few vacuoles to one which possessed many small vesicles. At this same time a change was also noted in the virulent properties of the trophozoites, as they became capable of inducing hepatic abscesses in hamsters and of destroying monolayers of tissue cultured cells (Bracha and Mirelman 1984). The most unexpected alteration, however, was in the hexokinase and phosphoglucomutase isoenzyme electrophoretic migration patterns of the amebae which changed from non pathogenic to pathogenic (Mirelman *et al.* 1986) (Fig. 1A,1B).

Conversions from Pathogenic to Non Pathogenic Zymodeme

The question that remained open was whether pathogenic zymodemes could convert or revert back to non pathogenic. This was accomplished (Mirelman *et al.* 1986a) by reassociating and subculturing in xenic conditions (Diamond 1982) the pathogenic amebae of strain *E. histolytica* SAW 1734R clAR with the bacterial flora that accompanied this strain in the original xenic culture. A change to non pathogenic zymodeme was observed after seven days (Fig. 2). No alteration in the pathogenic zymodeme was obtained, however, when the reassociation of the newly converted pathogenic amebae was done with bacterial flora that accompanied a pathogenic strain of amoebae *E. histolytica* SAW 408, isolated from a symptomatic case (Reed *et al.* 1983).

In both cases, and similarly to what was previously shown (Mirelman *et al.* 1984; Sargeaunt *et al.* 1982a), electrophoretic migration of the bands from the bacterial isoenzymes appeared in other regions of the gel and did not interfere or affect the reading of the patterns given by the amebae. Using this same method we have succeeded in converting a number of well known, axenically grown, cloned strains of pathogenic amebae, *E. histolytica* HM-1:IMSS (clone 6), Rhaman (clone 1) and 200:NIH, to non pathogenic zymodeme (D. Mirelman, R. Bracha, A. Wexler and A. Chayen, submitted for publication). In all these cases too, the shift to non pathogenic zymodeme was reproducible and appeared 7 days after the initial association with the bacterial flora taken from the non pathogenic isolate *E. histolytica* SAW 1734R clAR (Fig. 3). No shift in zymodeme was observed if the association was done with the bacterial flora

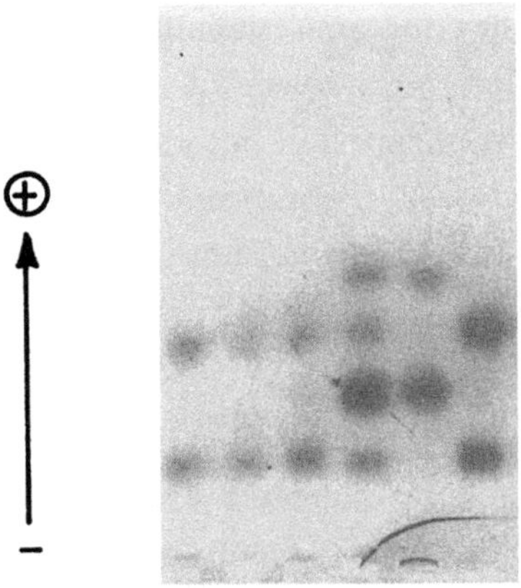

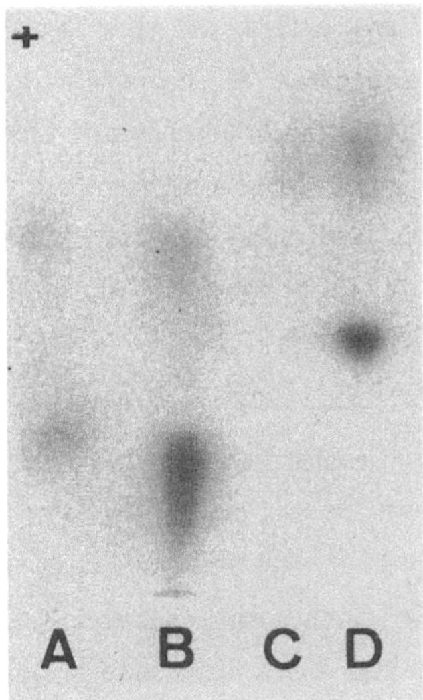

FIGURE 1A (LEFT). Electrophoretic pattern of hexokinase isoenzymes from *E. histolytica* strain SAW 1734R clAR. Trophozoite samples were harvested from each growth condition, lysates prepared and applied to agarose minigels (Mirelman *et al.* 1986). (A) Original pattern of strain SAW 1734R clAR cultured together with their accompanying bacterial flora; (B) SAW 1734R 17 days after transfer to a medium for axenic growth (TYI-S-33) (Diamond *et al.* 1978) supplemented with γ-irradiated bacteria (Mirelman *et al.* 1986); (C) Pattern of cultures 18 days after transfer to TYI-S-33 medium; (D) Pattern taken 20 days after transfer to TYI-S-33 medium; (E) Pattern of SAW 1734R clAR 21 days after transfer to TYI-S-33 medium; (F) Non pathogenic pattern of hexokinase from *E. histolytica* strain CDC:0784:4 (group I) used as non pathogenic marker; (G) Pathogenic pattern from *E. histolytica* strain HM-1:IMSS used as a pathogenic marker. Samples containing lysates of bacterial cells without ameba did not show any hexokinase bands in this region.

FIGURE 1B (RIGHT). Electrophoretic pattern of phosphoglucomutase isoenzymes from *E. histolytica* strain SAW 1734R clAR. Trophozoite samples were harvested from each growth condition, lysates prepared and applied to agarose minigels (Mirelman *et al.* 1986). (A) *E. histolytica* strain SAW 1734R clAR grown in a medium for axenic cultivation for 25 days; (B) *E. histolytica* strain 200:NIH grown axenically and used as pathogenic marker; (C) Lysate containing only the bacterial flora that grew in the xenic cultures of SAW 1734R clAR; (D)SAW 1734R clAR. - original culture grown in TYSGM-9 medium together with its associated bacterial flora.

from the pathogenic isolate *E. histolytica* SAW 408 (Reed *et al.* 1983).

Bacterial contribution

It is generally recognized that pathogenicity of *E. histolytica* is related at least in part to the association of amebae with suitable bacterial species (Bracha

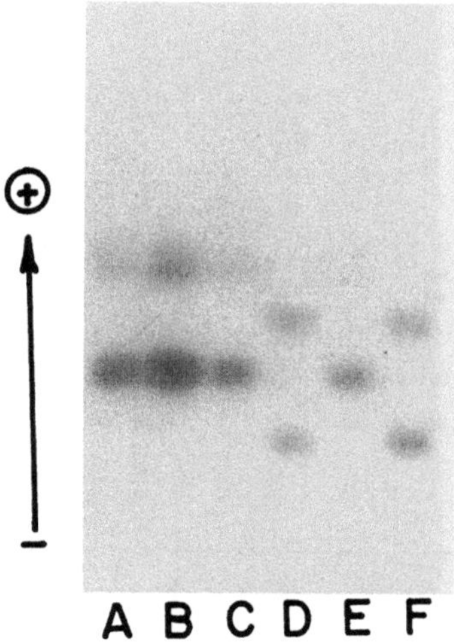

FIGURE 2. Electrophoretic pattern of hexokinase isoenzymnes from *E. histolytica* strain SAW 1734R clAR. Trophozoite samples were harvested from each growth condition, lysates prepared and applied to agarose minigels (Mirelman *et al.* 1986). (A) Samples of *E. histolytica* strain HM-1:IMSS used as a pathogenic marker; (B) Pattern of strain SAW 1734R clAR 30 days after transfer to a medium for axenic growth (TYI-S-33) (see also Fig. 1);(C) Pattern of strain SAW 1734R clAR (as in B) after culturing them for 2 days in TYSGM-9 medium (Diamond 1982), together with the bacterial flora that accompanied this strain in its original culture; (D) Pattern of *E. histolytica* strain 1734R clAR (as in B) after culturing them for 7 days in TYSGM-9 medium together with the bacterial flora that accompanied it in its original culture; (E) Pattern of trophozoites (as in B) after culturing them for 7 days in TYSGM-9 medium together with the bacterial flora that accompanied the pathogenic strain *E. histolytica* SAW 408 (Reed *et al.* 1983); (F) Marker of original culture of *E. histolytica* strain 1734R clAR grown together with its accompanying bacterial flora.

and Mirelman 1984; Martinez Palomo 1982). The definitive influence exerted by certain bacteria for the growth of the amoebae as well as in the production of amoebic dysentery has been observed repeatedly, but deserves a much more careful study.

One of the reasons why zymodeme conversion have not been observed until today in any of the longitudinal culture studies of Sargeaunt and colleagues (Sargeaunt 1985; Sargeaunt *et al.* 1982; Sargeaunt and Williams 1978; Sargeaunt *et al.* 1982a) may be due to the way the amebae cultures are grown in Robinson's xenic medium (Robinson 1968). In this system, stool samples containing *E. histolytica* are inoculated together with erythromycin to suppress the growth of the original fecal bacterial flora, while separately grown viable *Escherichia coli* B cells are added to the cultures. The addition of

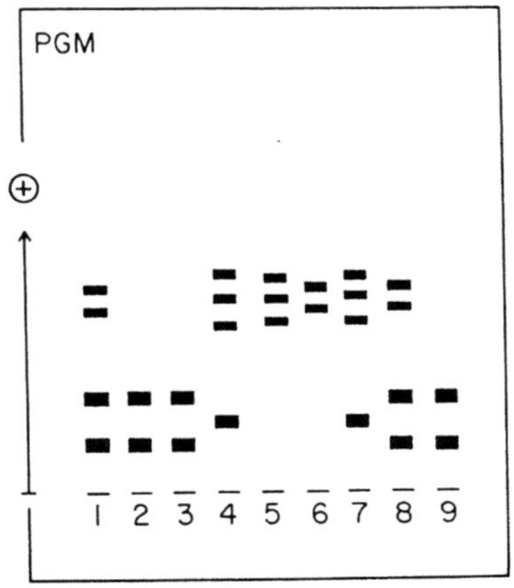

FIGURE 3. Electrophoretic pattern of phosphoglucomutase isoenzymes from various *E. histolytica* strains. Lane 1: *E. histolytica* strain SAW 408 grown together with its bacterial flora; (2) *E. histolytica* strain 200:NIH grown under axenic conditions; (3) *E. histolytica* strain Rhaman grown under axenic conditions; (4) *E. histolytica* non pathogenic strain SAW 1734R (clAR) grown together with its bacterial flora; (5) Lysates containing only the bacterial flora associated with strain SAW 1734R (clAR); (6) Lysate containing the bacterial flora associated with strain SAW 408; (7) Strain HM-1:IMSS grown together with the bacterial flora associated with non pathogenic strain SAW 1734R clAR for 7 days; (8) Strain HM-1:IMSS grown together with the bacterial flora associated with pathogenic strain SAW 408; (9) Strain HM-1:IMSS grown under axenic conditions.

these particular *E. coli* cells to the amebae cultures supports their growth and apparently does not cause any changes in their zymodeme. In our case, the trophozoites are grown in the rich medium for axenic cultivation (Diamond *et al.* 1978). which contains 15% bovine serum, and their dependence on bacterial cells for metabolites is diminished. This change in growth conditions may cause the induction of inactive metabolic systems in the amebae, which in turn could also cause the conversion of their isoenzymes.

The composition of the bacterial flora associated with the non pathogenic amebae isolates, which were successful in inducing the shift from pathogenic to non pathogenic zymodeme, as well as the flora from the pathogenic amebae, which were unable to induce such a shift, are currently being analysed. Preliminary observations indicate that each consists of a mixture of different bacterial species and that *Escherichia coli* is one of the common ones.

Isoenzyme Changes and Pathogenicity

E. histolytica is not the first organism in which zymodemes have been found to change. Somewhat similar changes have been shown to happen in other protozoa (*Tetrahymena, Paramecium*) where variations in isoenzyme patterns are subject to conditions of growth, media components and the presence or absence of bacteria in the medium (Allen 1968; Allen and Nerad 1978; Rowe *et al.* 1970). In *E. histolytica* the change to pathogenic zymodeme appears to coincide with the capability of the ameba to become virulent, and this could be of extreme importance for development of symptomatic amebiasis in humans. Moreover, the reversion from pathogenic to non pathogenic ameba is also of great interest as it may help us understand the multitude of asymptomatic cases. The molecular mechanisms which induce these changes in pathogenicity are not yet well understood, and whether they are of a phenotypic or genotypic nature remains to be determined.

Coming back to the initial question of whether *Entamoeba histolytica* consists of a pathogenic species, *E. dysenteria*, and a non pathogenic one, *E. dispar* (Brumpt 1925), our present results suggest that all *E. histolytica* strains are one species that possess the pathogenic potential. The virulent properties of the amebae may, however, become activated or expressed only under certain conditions and the trigger may very well depend on their growth conditions, which in turn may be affected by microenvironmental factors that the amebae encounter in the human intestine, such as the nature of the bacterial flora, redox potential, and other variables about which at the moment we know very little (Bracha and Mirelman 1984; Martinez Palomo 1982). It appears that *E. histolytica* is an opportunist organism which can exist in two states of activity, non pathogenic and pathogenic. Both states are apparently interconvertible and once they acquire the pathogenic character they may also vary considerably in their degree of virulence (Bracha and Mirelman 1984; Gitler and Mirelman 1986; Martinez Palomo 1982).

CONCLUSIONS

In principle, we support the notion that amebae in the pathogenic state are capable of causing disease in humans. However, contrary to the established hypothesis and previous beliefs (Sargeaunt *et al.* 1982a, Mirelman *et al.* 1984 Editorial 1985), classification of pathogenic and non pathogenic isolates by their isoenzyme electrophoretic patterns could be misleading. Our results (Mirelman *et al.* 1986; Mirelman *et al.* 1986a) clearly demonstrate that isoenzymes o cloned cultures can change from one state to the other, and consequently we must conclude that zymodeme analysis of human isolates may be of little diagnostic value and should not be used to predict a clinical condition or serve as a basis for recommendation of therapy (Editorial 1985; Sargeaunt and Williams 1982). It is possible that asymptomatic carriers may harbour amebae in a non pathogenic state for even long periods, but these could rapidly change and become pathogenic if suitable conditions, such as a change in bacterial flora, develops in the intestine. Learning more about these conversions and what regulates them may perhaps allow us in the future to better control this disease.

ACKNOWLEDGEMENTS

This research was supported in part by grants from the Rockefeller Foundation and the John D. and Catherine T. MacArthur Foundation.

REFERENCES

Allen SL (1968) Genetic and epigenetic control of several isozymic systems in *Tetrahymena.* Ann N Y Acad Sci 151:190-207.

Allen SL, Nerad TA (1978) Effect of acetate on esterase C activity during the growth cycle of *Paramecium.* J Protozool 25:273-279.

Bracha R, Mirelman D (1984) *Entamoeba histolytica* trophozoites: Effects of bacteria, microaerobic conditions and Metronidazole. J Exp Med 160:353-369.

Brumpt E (1925) Etude sommarie de l'Entamoeba dispar amibe a kystes quadrinuclees, parasite de l'homme. Bulletin de l'Acadamie Medicine (Paris) Bull de l'Acad Med (Paris) 94:942-952.

Diamond LS (1982) A new liquid medium for xenic cultivation of *Entamoeba histolytica* and other lumen dwelling protozoa. J Parasitol 68:958-959.

Diamond LS, Harlow DR, Cunnick CC (1978) A new medium for the axenic cultivation of *Entamoeba histolytica* and other *Entamoeba.* Trans R Soc Trop Med Hyg 72:431-432.

Editorial (1985) Is that amoeba harmful or not? Lancet I:732-734.

Flockhart HA, Denham DA (1984) Differentiation of species and life cycle stages of *Bruggia spp.* by isoenzyme analysis. *J. Parasitol.* 70, 378-384.

Gitler C, Mirelman D (1986) Factors contributing to the pathogenic behaviour of *Entamoeba histolytica.* Ann Rev Microbiol 40:237-261.

Martinez Palomo A (1982) The Biology of *Entamoeba histolytica.* In: Brown KN (ed), Tropical medicine research study series 2. Research Studies Press, Chichester.

Mirelman D, Bracha R, Chayen A, Aust-Kettis A, Diamond LS (1986) *Entamoeba histolytica*: Effect of growth conditions and bacterial associates on isoenzyme patterns and virulence. Exp Parasitol 62:142-148.

Mirelman D, Bracha R, Wexler A, Chayen A (1986a) Changes in isoenzyme patterns of a cloned culture of non pathogenic *Entamoeba histolytica* during axenization. Infect Immun (in press). (accepted for publication).

Mirelman E, Bracha R, Sargeaunt PG (1984) *Entamoeba histolytica*: Virulence enhancement of isoenzyme-stable parasites. Exp Parasitol 57:172-177.

Reed SL, Sargeaunt P.G, Braude AI (1983) Resistance to lysis by human serum of pathogenic *Entamoeba histolytica*. Trans R Soc Trop Med Hyg 77:248-253.

Robinson GL (1968) The laboratory diagnosis of human parasitic amoeba. Trans R Soc Trop Med 62:285-293.

Rowe E, Gibson I, Cavill A (1970) Effects of growth conditions on the esterases of *Paramecium aurelia*. Biochem. Gen. Biochem Gen 5:151-159.

Sargeaunt PG (1985) Zymodemes expressing possible genetic exchange in *Entamoeba histolytica*. Trans R Soc Trop Med Hyg 79:86-89.

Sargeaunt PG, Jackson TFHG, Simjee AE (1982) Biochemical homogeneity of *Entamoeba histolytica* isolates, especially those from liver abscess. Lancet 1:1386-1390.

Sargeaunt PG, Williams JE (1978) Electrophoretic isoenzyme patterns of *Entamoeba histolytica* and *Entamoeba coli*. Trans R Soc Trop Med Hyg 72:164-166.

Sargeaunt PG, Williams JE (1982) A study of intestinal protozoa including nonpathogenic *Entamoeba histolytica* from patients in a group of mental hospitals. Am J Pub Health 72:178-179.

Sargeaunt PG, Williams JE, Bhojnani R, Kumate J, Jimenez E (1982a) A review of isoenzyme characterization of *Entamoeba histolytica* with particular reference to pathogenic and non pathogenic stocks isolated in Mexico. Arch Invest Med(Mex) 13(suppl.3):89-94.

Sargeaunt PG, Williams JE, Neal RA (1980) A comparative study of *Entamoeba histolytica* (NIH-200,HK9, etc.), "*Entamoeba histolytica*-like" and other morphologically identical amoebae using an isoenzyme electrophoresis. Trans R Soc Trop Med Hyg 74:469-474.

TRYPANOSOMATID-VECTOR INTERFACES - IN VITRO STUDIES ON PARASITE SUBSTRATE INTERACTIONS

David H Molyneux, Keith R Wallbanks and George A Ingram

Department of Biological Sciences, University of Salford, Salford M5 4WT, England

INTRODUCTION

All trypanosomatids studied to date within their invertebrate host attach to some of the surfaces of the organs which they parasitise, be it gut, salivary gland or Malpighian tubule. Commonly, particularly in Trypanosoma and Leishmania, attachment is to the cuticle-lined ectodermal invaginations forming the foregut and hindgut or in the microvillus-lined midgut; other non-pathogenic monoxenous flagellates of the genera Blastocrithidia, Leptomonas Crithidia and Herpetomonas have a similar association with the gut of their insect hosts (see reviews by Molyneux 1977, 1983; Killick-Kendrick 1979). The flagellum typically becomes an attachment organ, often expanding its membrane to increase the area of cell-substrate contact presumably to facilitate the adhesion mechanism. In all parasite-cuticle associations and in some parasite-microvillus interactions flagellum-host contact is associated with formation of characteristic electron dense plaques on the inner leaflet of the plasmalemma of the flagellum in contact with the insect surface called by Vickerman (1973) hemidesmosomes. Such plaques are probably analogous to rather than homologous to the hemidesmosomes seen in cells from vertebrates; none of the many components of these hemidesmosomes (Suhrbier and Garrod 1986) have been demonstrated in the plaques of trypanosomatids.

Trypanosomatids attach to microvillous surfaces in the midgut of insects or leeches and to the malpighian tubules and salivary glands by interdigitation of unmodified flagella, (Killick-Kendrick et al 1974; Molyneux 1975), by interdigitation of expanded flagella with or without formation of hemidesmosome-like plaques (HDLPs) (Tetley & Vickerman 1985; Lewis and Ball 1979; Tieszen et al 1986) or by causing a loss of microvilli and attaching with HDLPs to remaining membrane (Tieszen et al 1983). In salivarian trypanosomes in Glossina (see Evans and Ellis 1983) and in Leishmania (Killick-Kendrick 1979) flagella are known to penetrate deep into epithelial cells of the midgut usually via a deep invagination of the plasmalemma of these cells.

Flagellates attaching to cuticular surfaces of insect fore and hindgut always form HDLPs and the flagella are usually expanded (Vickerman 1973; Killick-Kendrick et al 1974; Molyneux 1975) although in T.cruzi in Triatomids attachment via HDLPs without any flagellar expansion has been described (Zeledon et al 1977).

The function of the attachment mechanisms of trypanosomatids is usually regarded as a prerequisite to maintain position in a flowing aqueous environment (Livesey et al 1980;

NATO ASI Series, Vol. H11
Host-Parasite Cellular and Molecular Interactions
in Protozoal Infections. Edited by K.-P. Chang and D. Snary

Jefferies et al 1986) where expulsion, except following production of infective forms during the act of transmission, will lead to the death of the parasite. Trypanosoma, Leishmania and monoxenous flagellates may also make use of the attached phase to divide and/or produce infective forms for example the cysts of Blastocrithidia (Tieszen et al 1985) metacyclics of Trypanosoma (Tetley and Vickerman 1985) or infective forms of Leishmania (Sacks et al 1985). The need for attachment as a prelude to metacyclic trypanosome production in T.vivax in vitro has been shown by Hirumi et al (1983).

Attachment of trypanosomatids to substrates in vitro has been demonstrated by Brooker (1971) to millipore filters in Crithidia fasciculata and by Hommel and Robertson (1976) and Gray et al (1981) to plastic in Herpetosoma trypanosomes and T.congolense respectively. In these studies attachment to substrates by flagella resembled flagellum-cuticle attachment in the insect. The ability of at least some parasites to successfully attach in vitro somewhat contradicts the attractive belief that attachment mechanisms may play a role in trypanosomatid/vector specificity. This idea is supported by, for example, the localisation and attachment of Peripylaria Leishmania to the pylorus and ileum of sandflies (Lainson and Shaw 1979) and T.vivax and T.congolense to the labrum and hypopharynx of Glossina and not to the proboscis of other biting flies.

One advantage of in vitro attachment systems however is that it allows a detailed study of the mechanism of attachment as the ultrastructure of attachment is remarkably similar both in vitro and in vivo (Molyneux 1977, 1983; Hommel and Robertson 1975), large numbers of parasites can be obtained and cultured and more readily studied than when in insects, the mechanisms of such interactions are poorly if at all understood. This paper summarises recent results.

RESULTS AND DISCUSSION

Parasite adhesion may be a result of polymer interactions of divalent cation bridging and or electrostatic charge. The adhesion of vertebrate cells in vitro which have been studied in much greater detail involves an ever increasing array of proteins forming and linking with the extra-cellular matrix with or without cation involvement (Hay 1981). This bond is readily broken by proteases or chelators although steric hindrance may protect the bond from these lytic agents (Whur et al 1977). In contrast our studies (Wallbanks & Molyneux in preparation) have revealed that, using T.microti (a Herpetosoma trypanosome isolated from Microtus agrestis) and L.major, parasite-substrate attachment in in vitro cultures displays a remarkable stability. Parasites cannot be detached using a multitude of enzymes singly, in combination with chelators or in any binary combination (Table I). pH changes or changes in the osmotic pressure of the culture medium over a wide range also fails to induce parasite detachment. If parasites are physically scraped from the surface they rapidly reattach if they are viable.

TABLE I
ENZYMES AND CHELATORS USED IN ATTEMPTS TO DETACH
T.MICROTI AND *L.MAJOR*

Test solution	Maximum concentration units/ml	pH
Enzymes		
Protease (Pronase E)	75	7.5
Trypsin	2,500	7.6
Dextranase	75	6.0
Hemicellulase	75	5.5
Hyaluronidase	300	7.0
Chondroitinase ABC	3	7.3
Collagenase	75	7.4
Neuraminidase	1	5.0
Lipase	500	7.7
α-Glucosidase	25	6.8
β-Glucosidase	25	5.0
α-Amylase	300	6.9
α-Mannosidase	4	7.5
β-N-acetylglucosaminidase	1.7	7.0
Phospholipase C	75	7.3
Acid Phosphatase	10	4.8
Chelators		
Ethylenediaminetetraacetic Acid (EDTA)	2mg/ml	7.4
Ethyleneglycol-bis (β-aminoethyl ether) N,N,N ,N -tetra-acetic acid (EGTA)	2mg/ml	7.4

There is a possibility that trypanosomatids attach via complex carbohydrates such as galactomannans, glycans, xylans (Mendoza-Previato *et al* 1979; Gorin *et al* 1979) polysaccharide surface coats (Dwyer *et al* 1974) or excreted factors (El-On *et al* 1979) and that these are sufficiently complex to resist degradation by the enzymes listed. Excreted factors of *Leishmania* produced in culture (El-On *et al* 1979) have only been considered with respect to *Leishmania* survival in macrophages whereas, as they are derived from promastigotes, they may also have a role in parasite-sandfly interaction. *Leishmania* and *Trypanosoma* are known to produce gels and filamentous precipitates within vectors (Killick-Kendrick 1979; Heywood and Molyneux 1986) but their formation and chemical nature is unknown.

T.microti attachment sites *in vitro* stain heavily with colloidal iron at low pH showing the presence of glycoproteins (or other sources of negative charge) (Fig. 1) and vesicles frequently observed at attachment sites both *in vitro* and *in vivo* may represent secretion although they are usually regarded as products of pinocytosis. Interference microscopy

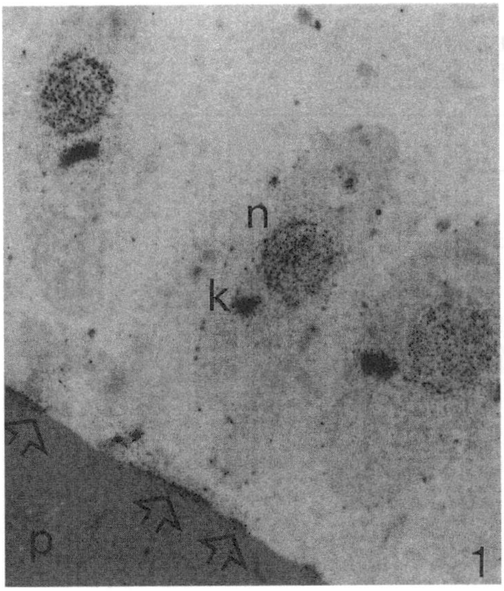

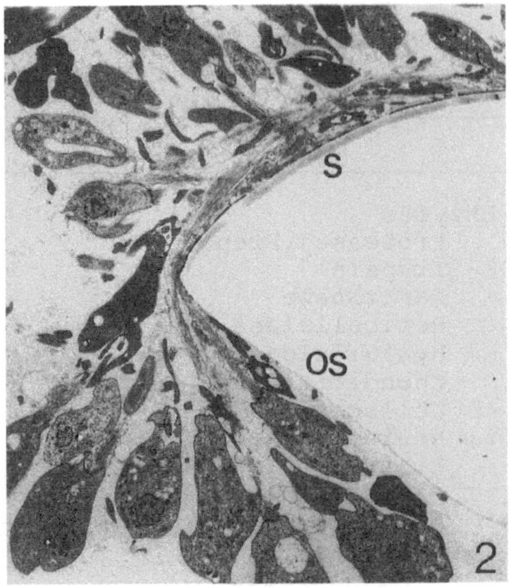

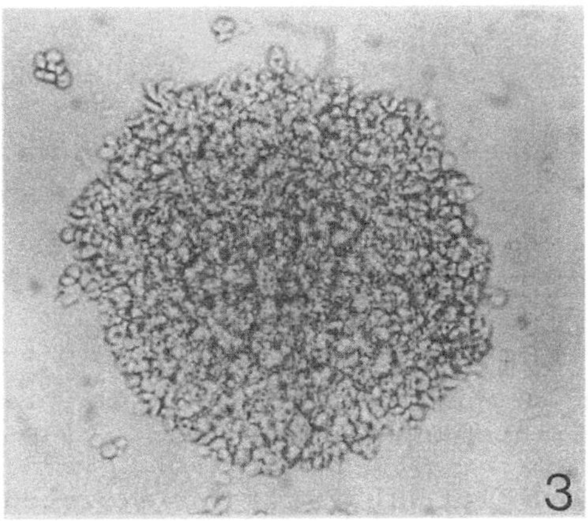

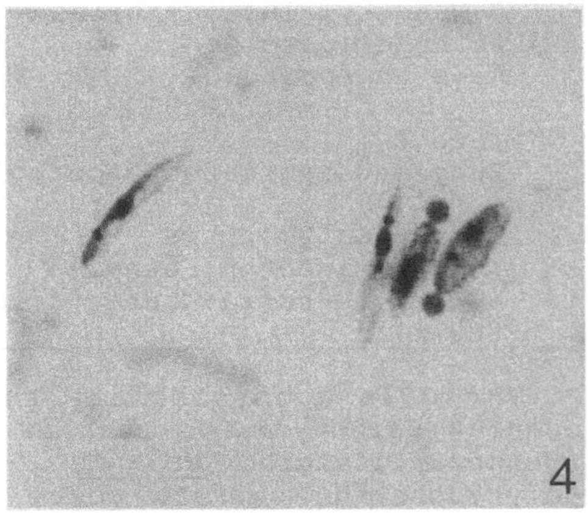

FIGURE 1. *Trypanosoma microti* attached to tissue-culture plastic (p). Attachment sites (arrows), kinetoplasts (K) and nuclei (n) stained with colloidal iron at low pH. Electron-micrograph x 12000.

FIGURE 2. *Leishmania major* promastigotes attached to the side of a scratch in tissue culture plastic (s) and not to the original surface (os). Electron-micrograph x 3000.

FIGURE 3. Circular colony of *L.major* promastigotes attached to a plastic Petri dish. Phase-contrast light-micrograph x 600.

FIGURE 4. *L.major* parasites scraped from the Petri-dish surface (figure 3). The two promastigotes on the right show short, greatly expanded flagella. Bright field light-micrograph, Giemsa stained, x 2000.

has shown that neither trypsin nor EDTA cause detachment of attached T.congolense epimastigotes in vitro (Hendry and Vickerman 1986). Trypsin apparently weakens the bond since parasite substrate distance increases but EDTA causes a surprising reduction in this distance suggesting it enhances the attachment bond. It seems therefore unlikely that divalent cations are involved in attachment of T.microti, or L.major to any significant extent (see Table I).

There is however strong evidence to suggest that electrostatic charge participates in attachment mechanisms in at least some trypanosomatids.

Initial attempts to discover a suitable substrate on which to attach L.major using gelatin, collagen, polyanion and polycation surfaces were unsuccessful. However, the finding that other trypanosomatids vigorously attached to scratches in polystyrene tissue culture flasks (Maraghi et al submitted) led to the production of linear colonies of attached L.major. The surface of these scratches appears to carry a higher negative charge than the surrounding plastic, resulting in an electron dense subsurface layer visible in electron micrographs (Fig. 2). Similar attachment has since been observed in one of six varieties of commercially available sterile petri-dishes and culture flasks. L.major seeded onto these surfaces produced expanding circular colonies of attached promastigotes (Fig. 3). When physically detached and stained some promastigotes revealed an intense patch of deposit at the tip of the flagellum (Fig. 4). Such intensive staining is not usually seen in preparations of parasites obtained from vectors.

Whether or not this attachment is a result of substrate surface change, which is probably negative, remains to be shown. Eukaryotic cells usually carry an overall negative charge and thus might not be expected to attach to like charged substrates. In bacteria which adhere to epithelial surfaces in vertebrates the attraction between hydrophobic molecules on the bacterial surface and the hydrophilic phospholipid molecules in the lipid bilayer membrane of the epithelial cells probably overcomes a similar charge repulsion (Watt and Ward, 1980).

One argument that could be deployed against involvement of electrostatic charge phenomena is the observaion that if in vitro parasites are scraped physically from the surface they only reattach if they are viable; heat killed, fixed or enzyme killed parasites some of which presumably retain their original charge distribution fail to reattach; such an observation suggests that in vivo in the insect a combination of mechanisms are involved. Some host surfaces certainly carry negative charge which can be visualised using cationic ferritin (Perrone et al 1986).

Although our in vitro studies have clearly demonstrated that charge related phenomena may induce attachment and that the attachment bond is a particularly stable one when exposed to a range of enzymes, pH, chelating agents and osmotic stresses the question of the delicacy of the specificity of Trypanosoma and Leishmania in their vectors remains; site specificity in insects thus may result from a combination of interactions such as chemotaxis (Bray 1983) to surfaces which

(a) carry a sufficient degree of -ve charge to allow adherence (b) have sufficient firmness to allow attachment to stabilise with or without cytoskeletal involvement and (c) are in regions of a suitable physiological microenvironment for the parasite.

In addition the suggestion that specificity and establishment in *Trypanosoma* may be dependent on lectin-parasite interaction in the vector (Pereira *et al* 1981; Ibrahim *et al* 1984) is difficult to reconcile with trypanosomatid attachment *in vitro* unless both mechanisms simply bring parasite and substrate into sufficiently close proximity to allow other attachment systems to operate. There is however a considerable body of evidence that lectin-induced adherence of micro-organisms via intestinal receptor recognition systems plays an important role in mammalian gut infections (Farthing 1985). Evidence of the presence of lectins (specific for mannose) on the lumen side of the peritrophic membrane of *Calliphora* but not on the surface of the midgut cells (Peters *et al* 1983) and the differential localisation of lectins on the basal lamina of the lobes of *Aedes aegypti* salivary glands (Perrone *et al* 1986) suggests that different exposed residues may allow penetration of sporozoites of *Plasmodium* at selected sites. These studies complement the work of Pereira *et al* 1981 and Ibrahim *et al* (1984) on the lectins of *Rhodnius prolixus* and *Glossina austeni* gut extracts respectively which clearly demonstrated difference in agglutinogenic properties to RBC species and appropriate parasites and identified the specificity of the molecules involved via carbohydrate inhibition experiments. Our studies have been extended to other species of *Glossina* and a wider battery of RBCs from different human blood groups (Molyneux *et al* 1986) which demonstrate haemagglutinins to both Rh+ and Rh- erythrocytes of the human ABO(H) blood groups in haemolymph of *Glossina* with titres ranging from 2^{-9} to 2^{-16}. Inhibition studies using 45 different carbohydrates and 10 glycoproteins indicated the lectin was directed towards glucose, mannose and galactose and their derivatives, depending on red cell type and species of tsetse examined; in one instance however xylose was the specific sugar which inhibited the reaction. Extracts of hind gut also agglutinated erythrocytes predominantly of BRh+, AB+ and AB- groups where specificity was linked to glycosyl and mannosyl residues.

CONCLUSIONS

Although the specificity of *Trypanosoma* and *Leishmania* for vectors is widely recognised and has an integral role in our understanding of the epidemiology of disease the molecular basis of the interactions are not understood. This specificity not only relates to species (or strain) of vector but to the location within the insect. The determinants of this site specificity are via attachment mechanisms. *In vitro* studies demonstrate parasite substrate bonding is resistant to a variety of enzymes, chelating agents pH and osmotic changes and attachment may be mediated via an electrostatic charge phenomenon as a contributory mechanism. In addition the role of lectins in determining the location of parasite establishment in the insect is discussed following studies on Rhodnius and Glossina where molecules exist within the gut and haemolymph which have specific agglutinogenic properties to trypanosomes and defined red cell types and which are inhibited by various carbohydrates. There is a need to extend such studies to Leishmania and sandflies but also to characterise the molecules responsible for these differences in agglutination in different regions of the gut. Such studies will provide a firmer basis for our understanding of parasite vector cell interactions which are less easily studied than the interactions in mammals but have a greater epidemiological relevance.

ACKNOWLEDGMENT

This investigation received financial support from the UNDP/World Bank/WHO Special Programme for Research and Training in Tropical Diseases

REFERENCES

Bray RS (1983) Leishmania: chemotaxic responses of promastigotes and macrophages in vitro J Protozool 30: 322-329

Brooker BE (1971) Flagellar attachment and detachment of Crithidia fasciculata to the gut wall of Anopheles gambiae Protoplasma 73:191-202

Dwyer DM, Langreth SG, Dwyer NK (1974) Evidence for a polysaccharide surface coat in the developmental stages of Leishmania donovani: a fine structure-cytochemical study Z Parasitenkd 43:227-249

El-On J, Schnur LF, Greenblatt CL (1979) Leishmania donovani: Physicochemical, immunological and biological characterisation of excreted factor from promastigotes Exp Parasitol 47:254-269

Evans DA, Ellis DS (1983) Recent Observaions on the Behaviour of Certain Trypanosomes within their Insect Hosts Adv Parasitol 22:1-42

Farthing MJG (1985) Receptors and recognition systems in intestinal infection Trans R Soc trop Med Hyg 79: 569-576

Gorin PAJ, Previato JO, Mendonca-Previato L, Travassos LT (1979) Structure of the D-Mannan and D-Arabino-D-Galactan in Crithidia fasciculata: changes in proportion with age of culture J Protozool 26:473-478

Gray MA. Cunningham I, Gardiner PR, Taylor AM, Luckins AG (1981) Cultivation of infective forms of Trypanosoma congolense from trypanosomes in the proboscis of Glossina morsitans Parasitology 82:81-95

Hay ED (1981) Extracellular matrix J Cell Biol 91:205s-223s

Heywood P, Molyneux DH (1985) Ultrastructure of the fibrous matrix surrounding cells of Trypanosoma melophagium in the hind gut of the sheep ked, Melophagus ovinus Cytobios 44:183-188

Hendry KAK, Vickerman K (1986) Properties of the attachment site of Trypanosoma congolense epimastigotes J Protozool (in press)

Hirumi H, Nelson RT, Hirumi K (1983) Complete Cyclic Development of Trypanosoma vivax In vitro J Protozool 30:6A

Hommel M, Robertson E (1975) In vitro attachment of trypanosomes to plastic Experientia 32:464-466

Ibrahim EAR, Ingram GA, Molyneux DH (1984) Haemagglutinins and parasite agglutinins in haemolymph and gut of Glossina Tropenmed. Parasit. 35:151-6

Jefferies D, Livesey JL, Molyneux DH (1986) Fluid mechanics of bloodmeal uptake by Leishmania-infected sandflies Acta trop 43:43-53

Killick-Kendrick R (1979) Biology of Leishmania in phlebotomine sandflies, in: Biology of the Kinetoplastida 2:395-460 eds WHR Lumsden and DA Evans, Academic Press, London/New York

Killick-Kendrick R, Molyneux DH, Ashford RW (1974) Leishmania in phlebotomid sandflies. I. Modifications of the flagellum associated with attachment to the midgut and oesophageal valve of the sandfly Proc R Soc Lond B187:409-419

Lainson R, Shaw JJ (1979) The role of animals in the epidemiology of South American leishmaniasis, in: Biology of the Kinetoplastida 2:1-116 eds WHR Lumsden and DA Evans Academic Press, London/New York

Lewis JW, Ball SJ (1979) Attachment of the epimastigotes of Trypanosoma cobitis (Mitrophanow, 1883) to the crop wall of the leech vector Hemiclepsis marginata Z. Parasitenkd 60:29-36

Livesey JL, Molyneux DH, Jenni L (1980) Mechanoreceptor-trypanosome interactions in the labrum of Glossina: fluid mechanics Acta trop 37:151-161

Maraghi S, Mohamed HA, Wallbanks KR, Molyneux DH Scratched plastic as a substrate for trypanosomatid attachment (in preparation)

Mendonca-Previato L, Gorin PAJ, Previato JO (1979) Investigations on polysaccharide components of cells of Herpetomonas samuelpessoai grown on various media Biochemistry 18:149-154

Molyneux DH (1975) Trypanosoma (Megatrypanum) melophagium: modes of attachment of parasites to midgut, hindgut and rectum of the sheep ked, Melophagus ovinus Acta trop 32:65-74

Molyneux DH (1977) Vector-parasite relationships in the Trypanosomatidae Adv Parasit 15:1-82

Molyneux DH (1983) Host-parasite relationships of Trypanosomatidae in vectors, in: Current Topics in Vector Research 1:227-248 ed KF Harris Praegar Publications, New York

Molyneux DH, Takle G, Ibrahim EA, Ingram GA (1986) Insect Immunity to Trypanosomatidae Symp zool Soc Lond No 56 117-144

Pereira MEA, Andrade AFB, Ribeiro JMV (1981) Lectins of distinct specificity in Rhodnius prolixus interact selectively with Trypanosoma cruzi Science 211:597-600

Perrone JB, De Maio J, Spielman A (1986) Regions of mosquito salivary glands distinguished by surface lectin-binding characteristics Insect Biochem 16:313-318

Peters W, Kolb H, Kolb-Bachofen V (1983) Evidence for a sugar receptor (lectin) in the peritrophic membrane of the blowfly larva, Calliphora erythrocephala Mg. (Diptera) J Insect Physiol 29:275-80

Sacks DL, Hieny S, Sher A (1985) Identification of cell surface carbohydrate and antigenic changes between noninfective and infectve developmental stages of Leishmania major promastigotes J Immunol 135:564-569

Suhrbier A, Garrod D (1986) An investigation of the molecular components of desmosomes in epithelial cells of five vertebrates J Cell Sci 81:223-242

Tetley L, Vickerman K (1985) Differentiation in Trypanosoma brucei: host-parasite cell junctions and their persistence during acquisition of the variable antigen coat J Cell Sci 74:1-19

Tieszen KL, Heywood P, Molyneux DH (1983) Ultrastructure and host-parasite association of Blastocrithidia gerridis in the ventriculus of Gerris odontogaster (Hemiptera: Gerridae) Can J Zool 61:1900-9

Tieszen KL, Molyneux DH, Abdel-Hafez SK (1985) Ultrastructure of cyst formation in Blastocrithidia familiaris in Lygaeus pandurus (Hemiptera: Lygaeidae) Z Parasitenkd 71:179-88

Tieszen KL, Molyneux DH, Abdel-Hafez SK (1986) Host-parasite relationships of Blastocrithidia familiaris in Lygaeus pandurus Scop. (Hemiptera:Lygaeidae) Parasitology 92:1-12

Vickerman K, The mode of attachment of Trypanosoma vivax in the proboscis of the tsetse fly Glossina fuscipes: an ultrastructural study of the epimastigote stage of the trypanosome J Protozool 20:394-404

Watt PJ, Ward ME (1980) Adherence of Neisseria gonorrheae and other Neisseria species to mammalian cells, in: Receptors and Recognition Series B6:251-188 Bacterial Adherence ed EH Beachey, Chapman and Hall, London New York

Wallbanks KR, Molyneux DH, The stability of trypanosomatid attachment sites (in preparation)

Whur P, Koppel H, Urquhart CM, Williams DC (1977) Substrate retention of fractured retraction fibres during detachment of trypsinized BHK21 fibroblasts J Cell Sci 24:265-273

Zeledon R, Alvarenga NJ, Schosinsky K (1977) Ecology of Trypanosoma cruzi in the insect vector, in: Pan American Health Organisation Scientific Publication No. 347 pp 59-70

THE MICROECOLOGY OF <u>LEISHMANIA</u> IN THE GUT AND PROBOSCIS OF THE SANDFLY

R. Killick-Kendrick
Medical Research Council External Scientific Staff, Department of Pure and Applied Biology, Imperial College at Silwood Park, Ascot, Berks SL5 7PY, England

INTRODUCTION

In the sandfly, leishmaniae adopt several morphological forms in different parts of the alimentary tract. These changes are presumably in response to different conditions in the various parts of the fly (FIG.1). As with other trypanosomatids, leishmaniae attach to the wall of the alimen-

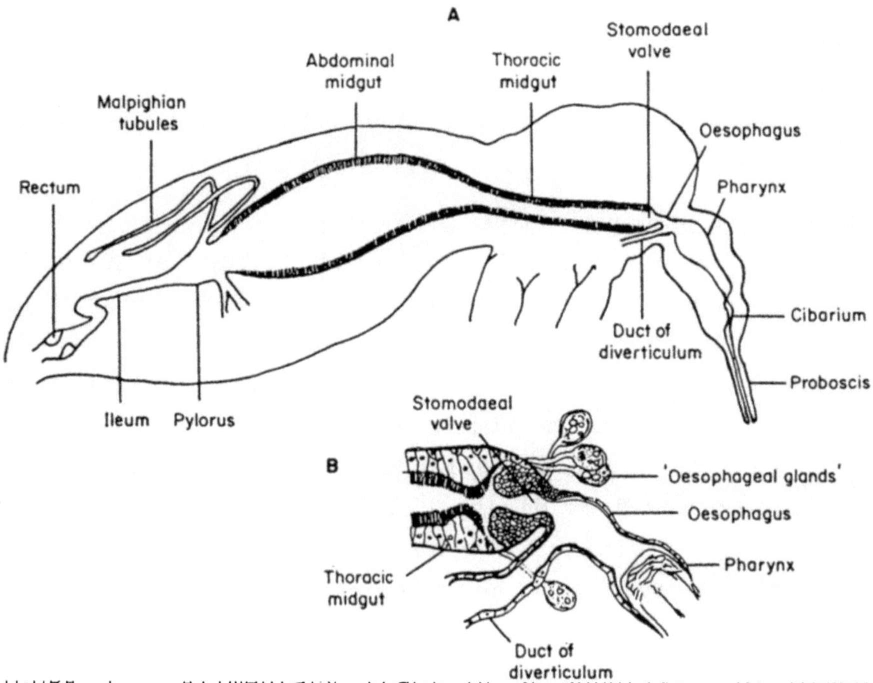

FIGURE 1. Alimentary tract of a sandfly. A. Sites of development of leishmaniae. B. The stomodaeal valve and adjacent parts (from Killick-Kendrick (1979); reproduced by kind permission of Academic Press).

NATO ASI Series, Vol. H11
Host-Parasite Cellular and Molecular Interactions
in Protozoal Infections. Edited by K.-P. Chang and D. Snary

tary tract to maintain position. This intimate contact between parasite and host is poorly understood at the molecular level although it has been postulated that specific receptor molecules between the flagellar membrane and the insect cuticle lining the hindgut and foregut probably play a part (Molyneux & Killick-Kendrick, 1986).

In this paper, a review is given of (a) the forms of leishmaniae in different parts of the alimentary tract, (b) recent observations on metacyclic promastigotes found in the proboscis and elsewhere and (c) new hypotheses on the influence of the digestion of the bloodmeal and of sugar feeding on the development of leishmaniae in the sandfly.

FORMS OF *LEISHMANIA* IN THE SANDFLY

Amastigote

Amastigotes are taken up with an infecting bloodmeal. There is evidence that they divide once or twice before changing into promastigotes (refs in Killick-Kendrick, 1979). Clearly, the stimulus for this change in form in the sandfly is not simply, as *in vitro*, a fall in temperature.

Multiplicative promastigotes

These are the principal forms in all Suprapylarian species. They occur only in the thoracic and abdominal midgut where they attach by inserting their flagella between the microvilli lining the gut. The sequence of division is: (a) production of a daughter flagellum which always remains shorter than that of the parent, (b) replication of kDNA which is not immediately followed by division, (c) division of the nucleus into two, (d) division of the kinetoplast into two, (e) division of the cytoplasm down the midline from the anterior end giving rise to two daughter individuals about half the size of the original promastigote. Division of the kinetoplast before the nucleus, which is commonly seen *in vitro*, appears not to occur in the sandfly.

When promastigotes are in the stomadaeal valve at the anterior end of the thoracic midgut, they attach to the intima by the flagellum, usually at the tip. So-called

hemidesmosomes arise inside the flagellar membrane at the point of attachment. Some species (e.g. L. mexican amazonensis) stop dividing on attachment whereas others (e.g. L. infantum) continue dividing (Killick-Kendrick, et al., 1975). The attachment is preceded by a thickening of the flagellum and, in L. mexicana ssp., a change in the shape and appearance of the parasite from a slender, electron dense promastigote (nectomonad) to a broad, electron lucid promastigote (haptomonad) (Killick-Kendrick et al., 1974).

Paramastigotes

Peripylarian leishmaniae divide as attached paramastigotes in the pylorus (=hind triangle) and ileum (=hindgut) of the sandfly. These forms are round and, unlike the promastigote, have the kinetoplast lying to one side of the nucleus. They attach to the cuticular intima by inserting their flagella into clefts or folds in the gut lining. Again, so-called hemidesmosomes are seen inside the flagellar sheath at points of contact between the tip or the sides of the flagellum and the cuticular lining (Killick-Kendrick et al., 1977).

Paramastigotes are also the usual form of both Suprapylarian and Peripylarian species in the oesophagus, pharynx and cibarium of the sandfly where they appear never to be in division. Characteristically, they are sessile and are attached to the lining of the gut lining by their flagella, again with the production of "hemidesmosomes". A suggestion that these are the sole forms from which proboscis (or metacyclic) forms arise (Killick-Kendrick, 1979) is no longer tenable (Killick-Kendrick, 1986) following the demonstration of metacyclic stages in the midgut of the sandfly (Sacks and Perkins, 1985). It is not clear what part foregut paramastigotes play in the life cycle.

Metacyclic promastigotes

These are stationary phase parasites, never in division, which are adapted to life in the vertebrate. They are the only stages seen in the mouthparts of the sandfly where they neither divide nor attach. Morphologically they are distinguishable from other promastigotes by the small size

of their body and long flagellum (FIG.2). Mean measurements of 50 metacyclic promastigotes of L. major in a stained smear of the proboscis of an experimentally infected specimen of P. papatasi show that this stage of the life cycle is markedly monomorphic (total length (±SD), 30.6 ± 2.75µm; body length (±SD), 10.1 ± 0.89µm; flagellum length (±SD), 20.5 ± 2.16µm). Other characteristics of metacyclic promastigotes are (a) high motility (b) a narrow, pointed posterior end, (c) an absence of inclusions or vacuoles in stained specimens, (d) an oval nucleus lying slightly posterior to the centre of the body and (e) a comparatively deep flagellar pocket (1.5 - 2.5µm) (Killick-Kendrick, 1986). Descriptions of metacyclic promastigotes of "L. tropica" (probably L. major) (Adler and Ber, 1941), L. infantum (Adler and Theodor

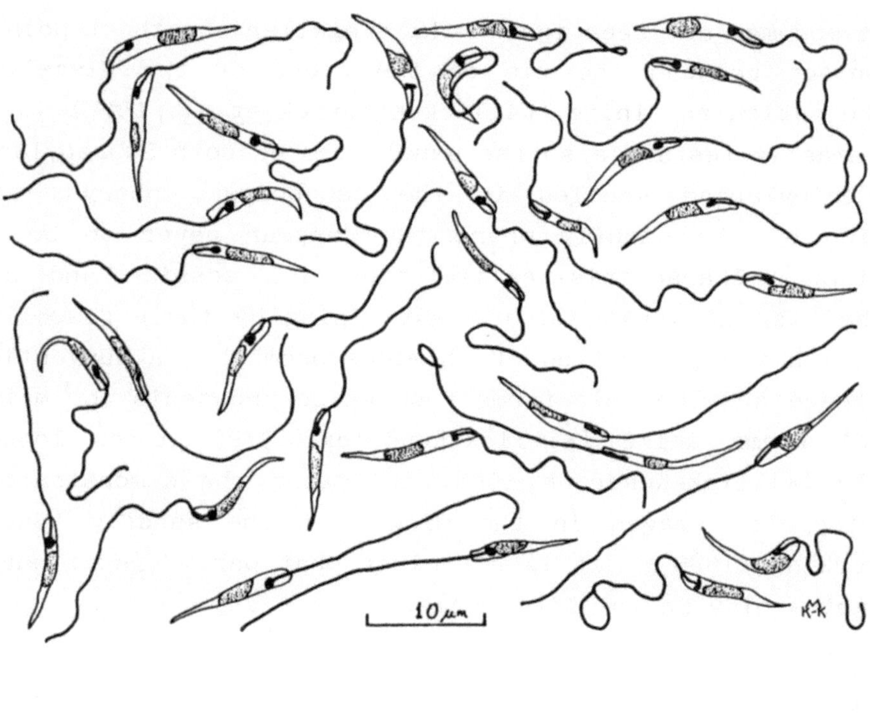

FIGURE 2. Metacyclic promastigotes of Leishmania major in a stained smear of the proboscis of an experimentally infected specimen of Phlebotomus papatasi (from Killick-Kendrick (1986); reproduced by kind permission of the Editor, Journal of the Royal Army Medical Corps).

1931, p.486) and L. donovani (Shortt et al., 1926) suggest that the morphology and behaviour of metacyclic promastigotes of all species will be found to be broadly similar.

Evidence for the adaptation of metacyclic promastigotes for life in the vertebrate host comes largely from the work of Sacks and colleagues. Forms of L. major morphologically indistinguishable from the same parasite in the proboscis of P. papatasi were separated from stationary phase cultures by agglutinating log phase forms with peanut lectin. They were then shown to differ from log phase parasites by (a) a greater or lesser resistance to complement, depending upon strains (Franke et al., 1985), (b) surface antigens and (c) surface carbohydrates (Sacks et al., 1985). Similarly, Sairaiva et al. (1986) showed that the change to infectivity of L. m. amazonensis in vitro is accompanied by changes in surface carbohydrates and, moreover, that many lectin receptors on metacyclic promastigotes are also present on amastigotes of the same species.

Further evidence for the adaptation of metacyclic forms to life in the invertebrate is that, unlike other promastigotes, they do not evoke an oxidative burst when taken up by macrophages (Alexander et al., 1985).

In the sandfly, infective forms of L. mexican amazonensis and L. major were shown to arise first in the gut of sandflies three days after an infecting meal with a peak of infectivity at the time the sandflies were ready for a second bloodmeal. This once again raises the question of whether or not it is essential for the life cycle to proceed to a point where parasites are in the proboscis of a sandfly before transmission can occur (Sacks and Perkins, 1985; Killick-Kendrick, 1986). The question is particularly pertinent in mathematical models of transmission, especially with vectors such as P. papatasi which, because they are gonotrophically discordant, are not inhibited from taking bloodmeals while their eggs are developing. Such species have more frequent contact with their vertebrate hosts than sandflies which are gonotrophically concordant (i.e. which take only one bloodmeal at the beginning of ovarian development and do not engorge again until after the eggs are laid about one week later).

The mechanism which controls the change from a non-infective promastigote to a metacyclic form is not known. Possible stimuli are (a) a shortage of essential nutrients, (b) a threshold level of waste products or (c) a genetically controlled change after a number of divisions. However, both this change and the forward migration to the stomodaeal valve and foregut may be simple general consequences of population pressure in the abdominal midgut as the numbers of parasites increase. The change to a form adapted for life in the vertebrate is not irrevocable. Metacyclic promastigotes from the proboscis of a sandfly or separated from cultures can be cultured in NNN medium as normal multiplicative promastigotes.

THE HABITATS IN THE SANDFLY

Leishmania species have evolved to fit the niches of the alimentary tract of the sandfly. In the Old World species, which are better known than those of the New World, there are many observations showing that the parasites are unlikely to survive and be transmitted by a species of sandfly which is not their natural vector or one of its close relatives (Killick-Kendrick, 1985). Schlein (1986) suggested that one mechanism favouring or hindering the growth of a Leishmania in the gut of a sandfly is the depression of digestive enzymes by a parasite in its natural vector and the stimulation of the enzymes by a parasite in an unnatural vector. The possible way whereby a parasite could influence the production of proteases are unknown. Depression does not, however, appear to interfere with or prolong digestion of the bloodmeal. Nevertheless, Schlein's observations highlight the fact that the habitat in the midgut of the sandfly is not static. It follows that different species of parasites must have evolved to fit the changes in the midguts of their vectors and that this has led to a host restriction. Schlein (1985) also made the interesting speculation that Peripylarian species of Leishmania may have evolved a life cycle with a multiplicative phase in the hindgut to avoid the action of digestive enzymes.

The possible role of enzymes in mediating success or failure in the establishment of infection in the gut was also thought to explain the detrimental effect of turkey blood on the development of L. major in P. papatasi. Schlein et al. (1983) suggested that raised levels of DNAase following ingestion of the avian blood were inimical to the parasite.

Changes in the habitat of the midgut other than those following the ingestion of blood relate to periodical sugar meals taken by both female and male sandflies (Young et al., 1980). Although it has often been suggested that these meals may profoundly affect the development of Leishmania, there are no published observations on the sugars taken by sandflies in nature. Schlein and Warburg (1986) showed, however, that colonized P. papatasi in the laboratory took the sap of plants by piercing stems or leaves. Some plants from the habitat of the fly were prefered to others. They also showed that the same species would take insect honeydews in the laboratory. In similar experiments in the laboratory, Killick-Kendrick and Killick-Kendrick (unpublished observations found that P. ariasi avidly took the honeydew of an oak aphid but did not feed on stems or leaves of many plants of the habitat and only rarely took honeydews of aphids from a bean, hawthorn or Calendula. This suggests a selection by the sandfly.

Plant saps contain a great variety of lectins. Together with observations on the attachment of Trypanosoma cruzi in its vector (Pereira et al., 1981) and lectin mediated attachments between Leishmania promastigotes in vitro (e.g. Schottelius and da Costa, 1982), the findings of Schlein and Warburg (1986) have led to the working hypothesis that attachments in different parts of the alimentary tract of the sandfly are mediated by lectins and sugars taken by the sandfly in sugar meals (Molyneux and Killick-Kendrick, 1986).

CONCLUSIONS

Although the colonization of sandflies in the laboratory remains labour intensive and slow, there is now a range of vector species available in laboratories in many parts of the world. This major advance has led to, firstly, British work on a re-examination of life cycles of leishmaniae, particularly with the electron microscope, secondly, to the Israeli work which has given rise to new ideas on possible ways in which the parasites are restricted to a particular group of vectors and, thirdly, the pioneering American work on the recognition and characterization of metacyclic promastgotes. Progress in understanding more about the development of leishmaniae in the sandfly now depends to a large degree on studies on the physiology of digestion and sugar feeding of sandflies, supplemented by observations on cultures of Leishmania in vitro.

ACKNOWLEDGEMENTS

The financial support of the Medical Research Council, London, and WHO, Geneva, is gratefully acknowledged. I thank Dr J. Alexander, Mrs M. Killick-Kendrick and Mr T. J. Wilkes for the stained preparation of the promastigotes from the proboscis of an infected sandfly.

REFERENCES

Alexander J, Killick-Kendrick R, Wilkes TJ, Bailly M (1985) The interaction of Leishmania major isolated from the invertebrate host, the vertebrate host and cell free culture with mouse macrophages (Abs). International Sympoisium on Macrophages in Infection Immunity, Elsinore, Denmark 17-21 May 1985

Adler S, Ber M (1941) The transmission of Leishmania tropica by the bite of Phlebotomus papatasii. Indian J med Res 16: 803-809

Adler S, Theodor O (1931) Investigations on Mediterranean Kala-azar. IV Infection of sandflies with Leishmania infantum. Proc R Soc Lond B 108:481-493

Franke ED, McGreevy PB, Katz SP, Sacks DL (1985) Growth cycle-dependent generation of complement-resistant Leishmania promastigotes. J Immunol 134:2713-2718

Killick-Kendrick R (1979) The biology of Leishmania in phlebotomine sandflies. In: Lumsden WHR, Evans DA (eds) Biology of Kinetoplastida, vol II. Academic Press, London/New York, pp 395-460

Killick-Kendrick R (1985) Some epidemiological consequences of the evolutionary fit between leishmaniae and their phlebotomine vectors. Bull Soc Path Ex 78:747-755

Killick-Kendrick R (1986) The transmission of leishmaniasis by the bite of the sandfly. J R Army med Corps (in press)

Killick-Kendrick R, Molyneux DH, Ashford RW (1974) Leishmania in phlebotomid sandflies. I. Modifications of the flagellum associated with attachment to the mid-gut and oesophageal valve of the sandfly. Proc R Soc Lond B 187:409-419

Killick-Kendrick R, Molyneux DH, Hommel M, Leaney AJ, Robertson ES (1977) Leishmania in phlebotomid sandflies. V.The nature and significance of infections of the pylorus and ileum of the sandfly by leishmaniae of the braziliensis complex. Proc R Soc B 198:191-199.

Killick-Kendrick R, Molyneux DH, Leaney AJ, Rioux J-A (1977) Aspects of the life-cycle of Leishmania in the sandfly. Proc 2nd europ Multicoll Parasit, Trogir, 1975, pp 89-95

Lainson R, Ward RD, Shaw JJ (1977) Leishmania in phlebotomid sandflies: VI. Importance of hindgut development in distinguishing between parasites of the Leishmania mexicana and L. braziliensis complexes. Proc R Soc Lond B 199:309-320

Molyneux DH, Killick-Kendrick R (1986) Morphology, ultrastructure and life-cycle of Leishmania. In: Peters W, Killick-Kendrick R (eds) The Leishmaniases in Biology and Medicine, Academic Press, London, pp 121-176

Pereira ME, Andrade AFB, Ribeiro JMC (1981) Lectins of distinct specificity in Rhodnius prolixus interact selectively with Trypanosoma cruzi. Science NY 211:597-600

Sacks DL, Hieny S, Sher A (1985) Identification of cell surface carbohydrate and antigenic changes between noninfective and infective developmental stages of Leishmania major promastigotes. J Immunol 135:564-569

Sacks DL, Perkins PV (1985) Development of infective stage Leishmania promastigotes within phlebotomine sandflies. Am J trop Med Hyg 34:456-459

Sairaiva EMB, Andrade AFB, Pereira MEA (1986) Cell surface carbohydrate of Leishmania mexicana amazonensis: differences between infective and noninfective forms. Europ J Cell Biol 40: 219-225

Schlein Y (1986) Sandfly diet and Leishmania. Parasitology Today 2:175-177

Schlein Y, Warburg A (1986 Phytophagy and the feeding cycle of Phlebotomus papatasi (Diptera:Psychodidae) under experimental conditions. J med Ent 23:11-15

Schlein Y, Warburg A, Schnur LF, Shlomai J (1983) Vector compatability of Phlebotomus papatasi dependent on differentially induced digestion. Acta Trop 40:65-70

Schottelius J, da Costa SCG (1982) Studies on the relationship between lectin binding carbohydrates and different strains of Leishmania from the New World. Mem Inst Oswaldo Cruz 77: 19-27

Shortt HE, Barraud PJ, Craighead AC (1926) The life-history and morphology of Herpetomonas donovani in the sandfly Phlebotomus argentipes. Ind J med Res 13:947-959

Young CJ, Turner DP, Killick-Kendrick R, Rioux J-A, Leaney AJ (198) Fructose in wild-caught Phlebotomus ariasi and the possible relevance of sugars taken by sandflies to the transmission of leishmaniasis. Trans R Soc trop Med Hyg 74:363-366

THE CELLULAR AND MOLECULAR INTERACTIONS OF MALARIA SPECIES WITH THEIR MOSQUITO VECTORS

R.E.Sinden

Molecular and Cellular Parasitology Research Group, Department of Pure and Applied Biology, Imperial College, LONDON SW7 2AZ

INTRODUCTION

This paper attempts to review the more recent and limited literature on the interaction of malarial parasites with their mosquito vector Few of these biological studies have offered, or attempted to gather information at the molecular level and this review is therefore deliberately speculative in an attempt to provoke the initiation of such work. Extensive appreciation of the older data is offered in the reviews of Vanderberg and Gwadz (1980), Carter and Gwadz (1980) and Sinden (1984). Within the context of this paper the biology of the transmission blocking determinants on the sexual and post zygotic stages is described.

RESULTS and DISCUSSION

The interaction between *Plasmodium* and Anopheline vector strictly begins when intraerythrocytic gametocytes are ingested as part of the infected bloodmeal by the probing female mosquito. Within the bloodmeal the gametocytes rapidly undergo gametogenesis, this is followed by fertilization and zygote formation. The zygote undergoes meiosis (Sinden and Hartley, 1985) and differentiates into the motile infective ookinete which penetrates the midgut epithelium and transforms into an oocyst 1-2 days after the bloodmeal was taken. Within 6-15 days the vegetative oocyst grows to release up to 10,000 sporozoites into the haemocoele. The sporozoites then accumulate in the salivary glands, where they may persist for many months. The parasite-vector interaction ends only when the sporozoite is released by the infected female into the blood of the next vertebrate host.

This limited 'classical' view of parasite-vector interaction however totally ignores the potential influence of *Plasmodium* on both the vertebrate and invertebrate hosts which might modify the behaviour of either in a manner which enhances transmission.

Modification by parasite of vector and host behaviour

The classic studies of Hawking *et al.* (1971) suggested that the availability of the infective gametocyte to the mosquito is directed by the precise synchronisation of a brief period of gametocyte maturity to the circadian biting activity of the mosquito, this hypothesis has however received only limited experimental support. An alternative hypothesis is based on the control of parasite availability in the

NATO ASI Series, Vol. H11
Host-Parasite Cellular and Molecular Interactions
in Protozoal Infections. Edited by K.-P. Chang and D. Snary

peripheral circulation by a sequestration and release mechanism coordinated from the host pineal gland (Gore and Pitman, 1978;Gore *et al.*,1982). Whilst the latter mechanism was proposed for *Leucocytozoon*, similar 'differential circulation' models have been suggested for *Plasmodium* e.g. Landau *et al.* (1979). Specific support for this second hypothesis is still lacking. Others have argued, particularly with respect to *P. falciparum*, that malaria does not always show a circadian pattern of infectivity to the vector.

It has recently been suggested by Rossignol *et al.* (1985) that the infection of mice by *Plasmodium* renders the host more liable to foster a successful bloodfeed by the female mosquito, due primarily to a reduction in the required probing time by the uninfected mosquito on the infected mouse. Small differences in the efficiency of this event could offer a significant evolutionary (survival) advantage to the parasite.

Development of *Plasmodium* in the Mosquito bloodmeal.

a) Biology. Once within the female mosquito the mature capacitated gametocytes of *Plasmodium* may undergo gametogenesis. The trigger for this *in vivo* is in all probability complex, and *in vitro* various agents have been shown to act as inducers, including :- a drop in temperature (5°C below body temperature of the vertebrate host),a rise in p H. (to an optimum of p H.7.8-8.0 depending on parasite species) mediated by a change in the p CO_2 and hence bicarbonate concentration in the bloodmeal (Nijhout and Carter, 1979). Other necessary mosquito 'factors' were originally recognised by Micks *et al.* (1943), and have been identified as heat stable peptides M.W.<10,000 found in both mosquito head and gut (Nijhout,1979). These molecules could be neurotransmitters controlling bloodmeal digestion to which the parasite has 'tuned-in' The molecular route through which these signals could act may lie in the cyclic AMP cascade (Martin *et al.* 1978) rather than the common alternative - the Ca^{++} mechanism (Knight, unpublished).

In response to the trigger the female gametocyte increases in volume and secretes the contents of the osmiophilic bodies into the surrounding parasitophorous vacuole, the enveloping erythrocyte is weakened and the gametocyte emerges into the bloodmeal as a single macrogamete. Within the same space of time (5-20min. at 22-26°C) the microgametocyte similarly escapes the enveloping erythrocyte, but in addition it undergoes three rounds of mitosis which, contrary to earlier interpretations (Sinden,1983), requires the replication of the entire genome at each round of division (Janse *et al.*,1986). The unusual insensitivity of this process to mitomycin C implies that interesting differences exist between the metabolism of the parasite at this stage of the life cycle from that at other times, and from that of either vertebrate host or mosquito vector (all of which may be susceptible to the drug). Microgamete assembly is dependant upon *de novo* protein synthesis (Toye *et al.*,1977) and upon microtubule assembly and microfilament action (Toye *et al.*, 1977; Sinden *et al.*, 1985b). Not surprisingly microgametogenesis is an incredibly complex process and is therefore readily inhibited by a variety of agents including some produced by the infected-(Petit *et al.*, 1982) or immunised-hosts (Mendis and Targett, 1979)

Once released, the microgamete shows a characteristic oscilliatory locomotion cycle which is dependant upon glucose metabolism. Fertilization is achieved with surprising rapidity (usually within 20 min. of gamete release) but despite this efficiency no tactile response of the male to the female has been noted *in vitro* .

b) Interaction with cells and molecules of the vertebrate host. The intensity of infection of the mosquito is dependant not only upon the number of gametocytes ingested, but also on the availability of nutrients (Gooding,1972) including unspecified molecules from the erythrocyte (Rosenberg *et al*,1984; Rosenberg and Koontz,1984) .

Both macro- and microgamete are extracellular in the mosquito bloodmeal , and are thus susceptible to the activity of host phagocytic cells (Sinden and Smalley,1976), and antibodies ingested with the bloodmeal.

The targets of anti-gamete antibodies have been identified for both male and female in *P. falciparum* (Vermeulen *et al.*1986; Harte *et al*, 1986) and *P. gallinaceum* (Carter *et al.* 1984), and for the male gamete in *P. yoelii* (Harte *et al.*, 1985) and *P.vivax* (Mendis *et al.* 1986). Antibodies against these determinants reduce the number of zygotes formed both *in vivo* and *in vitro*. A point of particular interest to the understanding of the mode of action of these antibodies is that antigamete antibodies to *P. vivax* will at suboptimal antibody levels enhance the efficiency of transmission (Mendis *et al.*, 1986). An additional observation made on *P. falciparum*, is that the transmission blocking determinants seem to be shared by both male and female, it is difficult to reconcile this observation with the commonly held assumption that the transmission blocking determinants are carried on the equivalents of the fertilizin- antifertilizin molecules.

The antigamete antibodies described to date are effective in membrane feeds largely reconstituted in foetal calf serum. Nonetheless the technical difficulties of conducting mosquito infections in the total absence of complement renders the interpretation of the mechanism of antibody blockade difficult in most experiments (see below).

The second group of transmission blocking determinants are those that are expressed (acted upon) post fertilization. These have been described in *P. gallinaceum* (Carter *et al.* 1984), *P.falciparum* (Vermeulen *et al.*,1986), *P.berghei* (Winger *et al.*, 1987) and *P.vivax* (Mendis pers. comm.) All probably are present on molecules of the same 'family;' with M_r's of 21-28 kD. In some species a doublet has been found, of which the smaller carries the transmission blocking determinant. The molecules in some species are described as being both acylated and glycosylated, observations consistent with their location in the parasite glycocalyx. Most of the monoclonal antibodies raised against these determinants recognise conformation dependant epitopes as shown by the antibodies' failure to react with reduced and alkylated molecules

Studies on the 21kD ookinete antigen of Plasmodium berghei. Immunisation of mice with mixed cultures containing ookinetes results in a transmission blocking immunity. Monoclonal antibody analysis shows the predominant transmission blocking antibody response is to a determinant carried on a 21kD. protein. Purification of this molecule by affinity chromatography and subsequent immunisation of mice produces a significant reduction in transmission when the mice are infected and mosquitoes are fed upon them (Winger *et al.* unpublished observations). Biosynthetic and immuno-electron microscopic studies have shown this molecule is only found on the macrogamete and its direct products, and forms part of the glycocalyx of the resulting ookinete . It is first synthesised in the macrogamete 2 hours after the induction of gametogenesis. There is now evidence that expression in the macrogamete is triggered by the onset of gametogenesis and not by fertilization (Dearsly, pers comm.). Synthesis continues throughout the remainder of ookinete differentiation, reaching maximal levels 10-12 hours post induction. The subsequent fall in surface density occurs despite continuing *de novo* synthesis of the antigen, this suggests the molecule is highly labile in the ookinete plasmalemma. This lability is consistent with the apparent shedding of antigen seen in immunogold labelled preparations. The synthesis of the antigen in the mature ookinete is certainly dependant on *de novo* RNA synthesis (Dearsly, pers comm.). The antigen is also found on/in the young oocyst ,where it can be detected for at least 6 days (Sinden *et al.* 1987)

The exact mechanism by which transmission blockade is achieved is unclear. The antibodies examined to date recognise epitopes that are either conformation dependent , or others that are conformation independent. These sites are recognised both *in situ* and *in vitro*. The antibodies although reactive against the ookinete plasmalemma do not prevent the formation of ookinetes *in vitro*, even in the prescence of homologous complement. Oocyst production *in vivo* however is heavily, or totally suppressed (Winger et al. 1987). It has been suggested that these antibodies could interfere with the specific recognition of the midgut wall by the ookinete, however direct evidence for this is still lacking. Possibly we should consider that early post-invasion growth of the oocyst is inhibited?

c) Interactions of the mosquito digestive enzymes on parasite development. Not only are the malarial parasites under potential attack by the immune system of the vertebrate host, but also by the digestive enzymes secreted into the bloodmeal.Amongst these are trypsin-like enzymes which have been shown to have direct deleterious effects on the zygote and ookinete (Gass, 1977; Gass and Yeates, 1979). The synthesis of these enzymes reaches maximal levels approximately 18 hours after bloodmeal ingestion (Briegel and Lea, 1969),just as the ookinete is apporaching maturity. Clearly therefore those parasites maturing more rapidly should be at a biological advantage. Conversely those mosquitoes digesting the bloodmeal more slowly will be at a disadvantage due to the resulting higher parasite burden. This simple arguement however neglects to consider an interesting interplay between the negative pressures exerted by the two hosts.

The interplay concerns the effects of the mosquito proteases with antbodies and complement in the bloodmeal. Evidence for the importance of this interaction on parasite development has recently been offered by Grotendorst *et al.* (1985) who showed that gametocytes and zygotes cultured *in vitro* are infectious to mosquitoes following the first three hours of culture in the presence of heterologous complement but then pass through a period of non-infectiousness. Mature ookinetes grown *in vitro* are fully infective when reconstituted in blood from a susceptible host and fed to the vector (Janse *et al.*,1985; Sinden *et al.*,1985). Grotendorst *et al.* suggest this variability of infectivity is related to parasite sensitivity to components of the alternative complement pathway and that a trypsin sensitive component on the plasmalemma of the gamete/zygote protects the recently emerged parasite from complement activity. Of particular relevance to this type of study is the evidence that within the mosquito bloodmeal complement (specifically C_3 and C_4) are effectively destroyed within the first three hours, and that antibodies (specifically light chains) are detectable for at least 42 hours (Fallon, M. pers. comm.). The correlation of these observations is self evident and suggests that parasite development has successfully adapted to the interplay of the potentially destructive factors derived from each of the two 'hosts'.

Development of *Plasmodium* in the mosquito haemocoele

a) Biology Having developed in the mosquito bloodmeal the ookinete migrates through the peritrophic layer . This layer when 'mature' is regarded by some to be an inpenetrable barrier to the migrating ookinete (Weathersby, 1975). Observations by Freyvogel (1966) indicate however that the ookinete may penetrate this layer under some conditions. This question could be profitably be reexamined.

Following passage through the peritrophoic layer the ookinete encounters the brush border of the midgut epithelium and thence passes through this single cell layer.There is at present no evidence that the ookinete binds to specific receptors on the endothelial cell. The exact route of traverse has been variously described as intra- and intercellular (Garnham *et al.* 1962; Canning and Sinden,1975; Meis pers comm.). Evidence currently suggests that the parasite is not obliged to follow either route exclusively. Comparison of this invasion with the migration of the sporozoite and merozoites would favour an intracellular route, since the intercellular path would necessitate the dissolution of the desmosomal and tight junctions of the midgut epithelium, and therefore require the secretion of radically different parasite molecules to those used by other invasive stages to disrupt the plasmalemma of potential host cells.

The usual though not inevitable site at which the ookinete comes to rest lies between the midgut basal lamina and the midgut epithelial cells. This predictable location suggests the ookinete is unable to disrupt the glycoprotein matrix of the basal lamina. This observation is compatible with current thinking on the mode of action of the rhoptry/ microneme complex in the merozoite and sporozoite stages (Bannister *et al.*,1986; Stewart *et al.*,1985). At this location the developing oocyst will be bathed in the circulating nutrients of the haemolymph, but might be shielded to a degree from the phagocytic haemocytes by the basal lamina.The trigger initiating the transformation of the ookinete into the oocyst is not known, however it clearly cannot be dependent upon reaching a site beneath the basal lamina because gametocytes injected directly into the haemocoele will form normal oocysts (reviewed by Weathersby,1975), but may be associated directly with the molecular environment offered by the haemocoele.

To date the molecular structure and organisation of the oocyst has been studied very little; there is one published work showing, not unexpectedly, that the oocyst expresses antigens that are common to the asexual blood stages (Krotoski *et al.*,1974). Recent data has also shown that the 20-28 kD family of transmission blocking determinants recognised on the ookinete plasmalemma are also present on/in the oocyst (Sinden *et al.*,1987; Ponnudurai, pers. comm.). Despite the extensive efforts of Ball, Chao and collegues to culture *Plasmodium* throughout its sporogonic cycle, the limited successes obtained remain unimproved, and this perhaps more than any other factor limits the understanding of the biochemical organisation of these stages. The few studies made have relied on the use of autoradiography to follow the uptake of radionucleotides, *in situ* histochemistry, and indirect drug studies.

Autoradiographic methods have shown that, like the erythrocytic stages ^{3}H-adenosine, but not ^{3}H-thymidine is incorporated into both DNA and RNA (Omar *et al*,1975 ; Davies and Howells,1973).

It has been shown that the addition of para-amino-benzoic acid to the mosquito diet increases the ooyst rate, and the effect is reversed by sulphadoxine (Peters and Ramkaram,1980).Sporogony is susceptible to inhibitors of dihydrofolate reductase, and gametogenesis is susceptible to 8-aminoquinolines (or most probably their metabolites).The failure to assay for the effects of drug metabolites upon sporogony is easily redressed with the advent of reproducible methods for membrane feeding of mosquitoes (eq.Teklehaimanot *et al.* 1985). It would however seem more profitable to search for compounds that will act on the parasite <u>within</u> the bloodmeal where effective drug concentrations are more readily achieved.

Few enzymes have been sought in the oocyst, those found include acid-phosphatase, beta-glucuronidase, aryl sulphatase and succinic dehdrogenase (Davies, 1974).

b)<u>The interactions of *Plasmodium* with the mosquito haemocoele</u> The interaction of the oocyst with the mosquito has been studied with two major objectives only. First to determine what molecules if any could be shown to change in concentration in the haemolymph in infected, as opposed to uninfected mosquitoes. Recorded differences include a reduction in sugars, and the amino acids valine, histidine, methionine, alanine, aspartic acid and glycine (Gad *et al.*, 1979b; Mack *et al*,1979a, 1979b). It must be remembered that in such studies the haemolymph composition will be affected directly by differences in the composition of the parasitised vs. non-parasitised bloodmeals. Deprivation of host nutrients may result in the recorded reduction in the fecundity (Rossignol *et al.* 1986) and longevity (Gad *et al.*,1979a; Scheifer *et al.*,1977) of the malarial infected mosquito.

The second level of interaction studied between the mosquito host and the developing oocyst, is the activation of the cellular and humoral defence mechanisms by the invader (Weathersby,1975). It has long been known that the developing oocyst can degenerate resulting in the production of Ross' black spores, whether this reflects an innate defect in the parasite or is the result of attack by host molecules is unknown. In a series of ingeneous experiments Weathersby demonstrated the existence of (unidentified) antiblastic factors that could be transfered from resistant to susceptible mosquito by the method of parabiotic twinning. The degenerating parasite is melanised by the mosquitoes humoral defence mechanism. This is first detected at the ultrastructural level as an increase in electron density of the oocyst wall, which then penetrates deeper into the cytoplasm along the internal membrane systems. Melanin is synthesised by the mosquito from tyrosine through the intermediate DOPA (Lipke,1975). No explanation has been offered as to how the observed melanin deposition <u>inside</u> the parasite is acheived.

It has been argued that the mosquito cellular response is not effective against *Plasmodium* . There is no published evidence for attacks on the oocyst, but sporozoites following their ectopic release into the cells of the midgut epithelium are certainly ingested by haemocytes (Sinden,1984).

The large mature malarial oocyst contains many thousand sporozoites and lies under the basal lamina of the midgut epithelium.The lamina has become stretched and raised from the gut wall. When the mature oocyst is examined by scanning electron microscopy the epithelium is seen as a thin and often torn layer, which can offer little if any barrier to the contained parasites. The release of the sporozoites into the haemolymph cannot be shown to be independant of sporozoite penetration activity, but is almost cetainly facilitated by the muscular churning of the midgut contributing to the fragmentation of the basal lamina

The sporozoites now lying free in the haemolymph 'migrate' to the salivary glands. It is seldom appreciated that the motility of malarial parasites (merozoite, sporozoite and ookinete) is expressed only when in contact with a substratum. Thus sporozoites freely suspended in the haemolyph are unable to move by their own activity. How then do the sporozoites come to be found exclusively in the salivary glands? It is most probable that they are carried passively by the circulation of the haemolymph throughout the haemocoelomic cavity of the mosquito (as are the merozoite, and sporozoite in the blood of the vertebrate host). Interestingly salivary glands transplanted into infected mosquitoes become infected, clearly indicating that the sporozoite can recognise the glands irrespective of their site in the host tissue (Betschart and Freyvogel, pers. comm.) . A similar technique has shown that the sporozoites of *P.knowlesi* specifically recognise and invade the transplanted salivary glands of the vector *A.dirus*, but not of the non-vector host of the transplant *A.freeborni* (Rosenberg, 1985). Recent studies show that the different lobes of the salivary glands can be distinguished on their haemocoelomic surfaces by the different affinities to lectins, e.g. the lateral distal lobes, but not the proximal lobes, bind PNA, RCA 120, UEA 1, and WGA, and it is these parts of the lobes that are similarly invaded by the sporozoites (Perrone *et al.*,1986). Thus it might be speculated that the sporozoites are carried to the glands in the haemolymph, and that they specifically recognise and bind to the basal lamina covering the glands, from where their own limited motility is used to invade the tissue. This invasion ,unlike the escape from the oocyst requires the penetration of the continuous barrier of the basal lamina- this is not a unit membrane and will require mechanisms that differ from those needed later to invade the salivary gland cell and the hepatocyte. When the sporozoite subsequently invades the salivary acinar cells it has been described as lying free in the cytoplasm and also in a parasitophorous vacuole. It ultimately crosses the endoplasmic reticlum membrane to lie free, possibly for very long periods, in the secretory droplets prior to their ejection into the salivary duct (Sinden, 1986). In heavy infections cellular damage is readily visible in the electron microscope, this results in increased cell fragility, and may accout for the change in saliva 'quality' reported by Rossignol *et al.* (1984)- although there is no apparent reduction in saliva volume, a factor critcal to the successful delivery of the parasite to the next vertebrate host.An indirect result of infection is that when feeding, the infected mosquito probes more frequently than its uninfected counterpart. This would enhance the probability of sporozoite transmission provided the irritability threshold of the host is not exceeded (Rossignol *et al.*,1986).

It is now clear that the molecular interactions of the malarial parasite in the mosquito are not confined to the numerous well recognised nutritional demands;chemical and spatial recognition of host tissues; evasion of host digestive enzymes; and humoral or cellular responses in the haemolymph; but in addition must include the persisting attack of the antibodies and phagocytes of the vertebrate host within the bloodmeal.

The ability to culture the parasite throughout the entire bloodmeal phase i.e. gametocyte - gamete - zygote - ookinete has allowed penetrating studies on the complex events involved, particularly into the molecular basis of the transmission-blocking determinants. The continued inability to culture the oocyst and to acheive salivary gland invasion *in vitro* continue to frustrate further advances in our understanding of the biology of *Plasmdium* in the mosquito at the molecular level. This deficiency must be overcome to enable and enhance future studies.

The impact of the molecular interactions between parasite and vector upon the behaviour and physiology of the mosquito are only now being recognised in the laboratory. The importance of these findings to our understanding of the transmission and evolution of the parasite may prove to be most significant.

REFERENCES

Bannister,L *et al* (1986) Parasitology 92:291-303
Briegel,J, Lea,J (1969) J.Insect.Physiol. 21:1597-1604
Canning,E, Sinden,R (1975) Parasitology 67:29-40
Carter,R, Gwadz,R (1980) In: Kreier J (ed) Malaria,vol.3.Academic Press, New York. p.263-297
Carter,R *et al* (1984) Phil.Trans.Roy.Soc.Lond.B 307:201-213
Davies,E (1974) Ann.Trop.Med.Parasit. 68:283-290
Davies,E, Howells,R (1973) Trans.Roy.Soc.trop.Med.Hyg. 67:20
Freyvogel,T (1966) Acta Tropica 23:201-222
Gad,A *et al* (1979a) Z.Parasitenkde. 60:249-261
Gad,A *et al* (1979b) Z.Parasitenkde. 60:263-275
Garnham,P *et al* (1962) Trans.Roy.Soc.trop.Med.Hyg. 56:116-120
Gass,R (1977) Acta Tropica 34:127-140
Gass,R, Yeates,R (1979) Acta Tropica 36:243-252
Gooding,R (1972) Quest.Ent. 8:5-60
Gore,T, Pitman,G (1978) Poult.Sci. 57;603-607
Gore,T *et al* (1982) J.Protozool. 29:415-420
Grotendorst,C *et al* (1986) J.Immunol. 136:4270-4274
Harte,P *et al* (1985) Parasite Immunol. 7:607-615
Harte,P *et al* (1986) Proc.6th Int.Cong.Parasit.Brisbane p146
Hawking,F *et al* (1971) Trans.Roy.Soc.trop.Med.Hyg. 65:549-559
Janse,C *et al* (1985) Parasitology 91:219-226
Janse,C *et al* (1986) Trans.Roy.Soc.trop.Med.Hyg. 80:154-157
Krotoski,W *et al* (1974) J.Parasit. 60:344-347
Landau,I *et al* (1979) Ann.Parasitol. 54:145-161
Lipke,H (1975) In: Maramosch,K, Schope,R (eds) Invertebrate Immunity. Academic Press, New York. p327-336
Mack,S *et al* (1979a) J.Parasit. 65:130-136
Mack,S *et al* (1979b) J.Parasit. 65:217-221
Martin,S *et al* (1978) Exp.Parasitol. 44:239-242
Mendis,K, Targett,G (1979) Nature 277:389-391
Mendis,K *et al* (1986) J.Cell Biol. 10A:149

Micks,D *et al* (1943) Am.J.Hyg.48:182-190
Nijhout,M (1979) Exp.Parasit. 48:75-80
Nijhout,M, Carter,R. (1979) Parasitology 76:39-53
Omar,M *et al* (1975) Am.J.Trop.Med.Hyg. 23:334-338
Perrone,J *et al* (1986) Insect.Biochem. 16:313-318
Peters,W, Ramkaram,A (1980) Ann.Trop.Med.Parasitol. 74:276-282
Petit,G *et al* (1982) Tropenmed.Parasit. 26:303-306
Rosenberg.R (1985) Am.J.Trop.Med.Hyg. 34:687-691
Rosenberg,R, Koontz,L (1984) Exp.Parasit 57:234-238
Rosenberg,R *et al* (1984) Exp.Parasit. 57:158-164
Rossignol,P *et al* (1984) Am.J.Trop.Med.Hyg.33:17-20
Rossignol,P *et al* (1985) Proc.Natl.Acad.Sci. 82:7725-7727
Rossignol,P *et al* (1986) Am.J.Trop.Med.Hyg. 35:277-279
Scheifer,B *et al* (1977) Exp.Parasit. 41:397-404
Sinden,R (1983) Adv.Parasitol. 23:154-216
Sinden,R (1984) Experientia 40:1330-1343
Sinden,R (1986) Trans.Roy.Soc.trop.Med.Hyg. 79:598-605
Sinden,R, Hartley,R. (1986) J.Protozool. 32:742-744
Sinden,R,Smalley,M. (1976) Trans.Roy.Soc.trop Med.Hyg. 70:344-345
Sinden,R *et al* (1985a) Parasitology 91:227-244
Sinden,R *et al* (1985b) Int.J.Parasit. 15:211-217
Sinden,R *et al* (1987) Proc.Roy.Soc.Lond.B (In press)
Stewart,S *et al* (1985) J.Protozool. 32:280-283
Teklehaimanot,A *et al* (1985) Am.J.Trop.Med.Hyg. 34:429-434
Toye,P *et al* (1977) Z.Parasitenkde. 53:133-141
Vanderberg,J, Gwadz,R (1980) In: Kreier,J (ed) Malaria,vol.2. Academic Press,New York. p.154-234
Vermeulen,A *et al* (1986) J.Exp.Med. 162:1460-1476
Weathersby,A (1975) In: Maramosch,K, Schope,R (eds) Invertebrate Immunity. Academic Press, New York.p273-288
Winger,L *et al* (1987) Parasite Immunol. (In press)

CONCLUDING REMARKS

W. Trager, E. R. Pfefferkorn and R. T. Schwarz

W. Trager (Parasitology, Rockefeller University)

One unifying thread running through a number of the papers on intracellular protozoa deserves special comment. This is the ever-increasing evidence that all those studied, the amastigotes of *Leishmania*, the erythrocytic stages of *Plasmodium*, and the tachyzoites of *Toxoplasma*, all have special ways to modify their host cells, presumably to make them more suitable for development of the parasite. At least two of the surface proteins of *Leishmania* turn out to be active enzymes of adaptive value, one an acid phosphatase and the other a protease. These may play roles in enabling *Leishmania* to develop within the sandfly gut and a phagolysosome of the mammalian macrophage. For *Plasmodium falciparum* developing in human erythrocytes there is good evidence that it produces a number of specific proteins that are transported through the red cell cytoplasm and come to lie just beneath or within the red cell membrane. The best studied example is the histidine-rich knob protein essential for formation of the knob-like surface structures that function in the sequestration of infected cells in capillaries of deep organs, in this way protecting them from destruction in the spleen. Some malarial proteins are also secreted into the external medium, but their function is not understood. Similarly *Toxoplasma* secretes into its parasitophorous vacuole protein-rich vesicles that form a network between the parasite surface and the boundary membrane of the parasitophorous vacuole. The principal protein of this network is also incorporated into the membrane of the parasitophorous vacuole and may function to protect *Toxoplasma* from intracellular digestion. I would predict that we will be learning some interesting cell biology from further studies of these phenomena.

NATO ASI Series, Vol. H11
Host-Parasite Cellular and Molecular Interactions
in Protozoal Infections. Edited by K.-P. Chang and D. Snary

E. R. Pfefferkorn (Microbiology, Dartmouth Medical School)

Many of us attending this meeting were initially trained in other fields and recently gravitated to parasitology. I grew up intellectually as an animal virologist. Virologists must, perforce, study host cell-parasite interactions. But not always so, parasitologists. The parasitic protozoa encompass a diverse array of organisms interesting in their own right. No doubt, there lurk among them biochemical quirks every bit as unexpected as the remarkable discovery of trypanothione in the trypanosomatids. Yet, the organizers of this workshop were correct in encouraging us to look at interactions of host and parasite at all levels. The result has been an embarrassment of riches.

My overall impression is that molecular interactions between the parasitic protozoa and their host cells are central to our understanding of the entire process of infection, not just the initial steps of attachment and penetration. A theme running through the presentations was that many of these interactions involve complex polysaccharides. These polysaccharide problems will probably not yield readily to the magic bullets of gene cloning, but will require, instead, painstaking analytical chemistry. The answers that emerge are likely to be both interesting and important.

R. T. Schwarz (Lab. Chim. Biol., Univ. Sci. Tech. Lille)

It is clearly noticeable from the Workshop that many investigators have recognized the important roles of glycoconjugates in host-parasite interactions and have therefore started to take a closer look at them. As parasitic protozoa differ from other pathogens (such as viruses) in the molecular strategies for survival and proliferation in the host, it is not surprising that many exciting discoveries have been made. One of them is the finding of a special type of glycolipid anchor for the plasma membrane glycoproteins, e.g. the VSG of the African trypanosomes, gp63 of leishmanias and the major merozoite surface antigen of malarial parasites. There are structural differences in the carbohydrate part of this molecule between different trypanosomes, suggesting that they may vary still further from those found in membrane proteins of some mammalian cells. A detailed analysis and comparison of the lipid anchors of glycoconjugates from parasitic protozoa and vertebrate sources are warranted. Our knowledge of their function, composition, structure and biosynthesis may provide the basis of rational chemotherapy for the diseases caused by these protozoa. Another example relevant to this point is the discovery of L-rhamnose in *Trypanosoma cruzi*. To my knowledge, it does not exist in carbohydrates of vertebrate cells, thereby opening the possibility of chemotherapy using, for instance, specific analogues of sugars.

Apart from these more practical aspects, biological study of the carbohydrates of glycoconjugates reveals that they may serve specific functions in host-parasite recognition and parasite pathogenecity in protozoal infections. Of importance here is also our precise knowledge of the carbohydrate structure and biosynthesis. Although the biosynthesis of protozoal carbohydrates may, to a large extent, follow the model pathway, there may be crucial differences. The best approach to initiate such study is to examine the activated precursors of sugars in the cytosol before studying in detail more complex systems, such as the glycosyltransferases. However, interesting results are produced in a first study of UDP-N-acetylglucosamine: DolP-N-acetyl-glucosaminyl-1-phosphoryltransferases in leishmanias.

I am convinced that our knowledge of the glycoconjugates in general will be benefited by working with parasitic protozoa and vice versa, as has been the case with animal viruses in previous years.

SUBJECT INDEX

Acetylation 69
Acetylcholine receptor 137,142
Acetylcholinesterase 26,315
Acetylglucosaminidase 389
Acid phosphatase 55,175,189,389,412
Acid pyrophosphatase 55
Acinar cell 413
Aconitate 42
Acridine orange 186
Actin 223,318
Actinomycin D 346
Active transport 61,179
Adenine 43
Adenosine 176,412
Adenylate cyclase 55
Adjuvant 322,337
Adriamycin 204
Adsorptive endocytosis 58
Alanine 284,412
Albumin 58,62
Alkaline phosphatase 26
Amantidine 224
Amido black 125
Aminoquinoline 412
Amphotericin B 223
Amylase 389
Ankyrin 318
Antigenic repertoire 9
Antigenic shift 343
Antigenic variation 1,9,19,25
Anti-idiotype 103
Antimonial 223
Antipain 257
Aprotinin 257
Arabinose 83,179
Arginine 323
Arsenical 37
Aryl sulphatase 412
Asialoglycoprotein 322
Asparagine 22,82
Aspartic acid 22,284,412
ATP 34,235,239,362
ATPase 55,179,362
Autoimmunity 98
Azide 184,186
Azocasein 169
B cell 102
Bestatin 171,257
Beta-galactosidase 271
Beta-glucuronidase 412
Beta oxidation 203
Bovine serum albumin 169
Calcium 59,408
Casein 57,125,165,225
Catalase 36,361
Cathepsin D 57,171
cDNA 19,308
Chemotaxis 224,340
Chloramine T 359
Chloropromazine 224
Chloroquine 224,324
Cholesterol 62,210
Chymotrypsin 130,323

Citric acid cycle 42,43
CoA 239,241
Codon 266
Collagen 374
Collagenase 371,389
Complement 19,85,90, 111,170, 213,223, 401,409
Complement receptor 166
Concanavalin A 180,372
Cortisone 242
Cosmid 12
Cyclic AMP 408
Cycloheximide 346
Cyclooxygenase 230
Cysteine 37,269
Cysteine protease 167
Cytochalasin B 180,214, 226,324
Dehydrogenase 41
Deoxycholate 177
Desmosome 411
Dextranase 389
Diacylglycerol 46,177
Dicyclohexylcarbodiimide 184
Difluoromethylornithine 39
Dihydrofolate reductase 412
Diploid 77
Dithiothreitol 31,179,268, 367
DNAase 403
ELISA 99,101,219
Endo-beta-galactosidase 247,251
Endocytosis 48,55,58, 324
Endoglycosidase 124,133, 247,251
Endonuclease 54,130
Endoplasmic reticulum 25,356
Endothelium 90,297,307
Epithelium 3,407,411, 413
Epitope 19,81,119, 130,270, 285,321, 330,336, 339,357, 409
Erythrocyte 90,235,281, 286,291, 300,315, 391
Ethanolamine 19,277
Exocytosis 368
Exoglycosidase 24
Exon 10,309
Exoribonuclease 176
Fab 85
Facilitated diffusion 61
FC receptor 223,226
Ferritin 51
Fetuin 92
Fibrinogen 165
Fibroblast 345,351
Flagellar pocket 55,61,62
Flavoprotein 33
Flow cytometry 119
Fluoro-L-histidine 298
Formamidase 350
Fructose 276
Fucose 82,83,84
Galactose 26,84,179, 197,246, 252,391
Galactosyl transferase 246,252

Gametogenesis	407
Gamma-irradiation	102
Gas-liquid chromatography	20,80,200
Gelatin	190
Gene	1,5,44,63, 73,100
Gene amplification	113
Gene expression	34,41
Gene rearrangement	273,312, 336,342
Gene reassortment	74,76
Genomic fluidity	312
Genomic library	113,154
Glossina morsitans	2,73
Glucosamine	20,82,198, 275
Glucose	84,179,247, 275,298, 391,409
Glucose oxidation	45
Glucosidase	53,389
Glutamate dehydrogenase	315
Glutathione	29,33,37
Glutathione peroxidase	36
Glutathione reductase	29,32,36
Glutathionylspermidine	34
Glycerol	20
Glycerol-3-phosphate oxidase	45
Glycine	412
Glycolysis	41,203
Glycophorin	235,245, 249,324
Glycosidase	62
Glycosome	2,203
Golgi	25
Haemocyte	411,413
Haemolymph	391,411
Hemicellulase	389
Hemidesmosome	3,387,399
Hemoglobin	57,169,194, 236,284,286
Hepatocyte	321,413
Hexokinase	377,379
High-density lipoprotein	59
Histidine	283,297, 307,318,412
Histidine-rich protein	281,300
Histone	54
Homopolymer tailing	339
Honeydew	403
HPLC	20,261
Hyaluronidase	389
Hybrid	73,76,77
Hybrid selection	339
Hydrogen peroxide	29
Hydrophobicity	155
Hydrophobic tail	267
Hypoxanthine	260,298
I-A antigen	231
IgG	120,169
IgM	119
Immunoaffinity	276
Immunoelectron microscopy	46,214,282, 297
Immunoprophylaxis	97,99
Immunosuppression	98
Indoleamine 2,3-dioxygenase	346
Inositol	21,177,199
Interferon-gamma	229,345
Interleukin	229
Intracellular pH	186
Intron	10,68,309

Iodoacetamide 32,132,167
IodoGen 165,247
Ionophore 179
Iron 59,389
Isoenzyme 75,81,377
kDNA 398
Kinetoplast 3,76,398
Kynurenine 346,350
Laminin 102
Lectin 79,91,149, 180,220, 252,391
Leupeptin 132,171, 237,372
Lipoic acid 31
Lipophosphoglycan 197
Lipopolysaccharide 232
Liposome 4
Low-density lipoprotein 59,204
Lymphokine 230,353
Lysine 307,322
Lysosome 51,57,61, 111,175, 197,208, 224,355
Macrophage 25,57,107, 120,131, 154,166, 189,197, 213,223, 321,353, 389,401
Major histocompatibility complex - MHC 229,233
Mannitol 372
Mannose 20,82,111, 150,176, 197,275, 391
Mannose-fucose receptor 149,166,223
Mannose-6-phosphate 22
Mannosidase 52,57,389
Mass spectrometry 21,29,198
Mauer's cleft 287,299,300
Meiosis 77,273,407
Melanin 412
Melarsen oxide 37
Membrane anchor 19,22,25
Mercaptoethanol 247
Mercury 167
Methionine 317,412
Methylamine 224
Methylation 22
Microfilament 67,294,408
Microtubule 67,408
Miniexon 69
Mitochondrion 2,41,184, 356
Mitogen 90
Mitomycin C 408
Mitosis 408
Monoclonal antibodies 2,46,81,99, 111,117, 120,131, 143,170, 199,229, 266,281, 297,321, 329,335, 346,358, 365,409
Monocyte 225
Monokine 229,232
Mosquito 323,329, 408,413
mRNA 19,43,68, 113,131, 339,346

Mutation 312
Myosin 223
Myristic acid 20,121,155, 158,275
N-acetylgalactosamine 252
N-acetylgalactose 246
N-acetylglucosamine 246
N-acetylglucosaminyl-transferase 107,111,115
N-acetylation 22
N-formylkynurenine formamidase 346
Neuraminidase 90,246,389
Neutrophil 175,189,224
N-glycanase 171
N-linked oligosaccharide 171,245
NMR 31,198
Northern analysis 113,336
Nucleotidase 26,53,61, 175,189
Oleic acid 177
Oligomycin 184
Orthovanadate 184
Oubain 55,197
Oxidative burst 224,401
Oxidative phosphorylation 41
Palmitic Acid 158,159
Parasitophorous vacuole 235,282, 289,294, 298,322, 408,413
Peanut lectin 91
Pentamidine 223
Pentose 82
Pepsin 130
Pepstatin 57,132,257
Peptide mapping 129,165,338
Percoll 236,315
Periodate oxidation 21,82
Phagocytosis 167
Phagolysosome 51,123, 137,169, 178,355
Phenanthroline 171
Phenylmethylsulphonyl fluoride 132,171,257
Phosphoenolpyruvate 34
Phorbol ester 241
Phosphatidic acid 21
Phosphatidyl choline 58
Phosphatidyl inositol 20,25,197
Phosphoglucomutase 379
Phospholipase 20,25,46, 57,125,130, 160,176, 198,389
Phosphomonoesterase 175,190
Phosphoprotein 315,316
Phosphoprotein phosphatase 54
Phosvitin 54
Phytohaemagglutinin E 252
Pinocytosis 51
Polyamine 29,34
Polyethylene glycol 91
Polyhistidine 307
Posttranslational modification 275,293,339
Primaquine 324
Proline 45,179,183, 266,317
Pronase 20,389
Prostaglandin 230
Protamine 54

Protease	19,48,92, 107,125, 150,170, 190,210, 241,248, 257,266, 324,388, 402
Protective immunity	86,338
Protein A	140,160, 214,260, 322,365
Proteinase K	130
Protein kinase	25,54,241
Proton pump	180,181, 183,186
Pulse field gel electrophoresis	76,113
Purine	176,362
Pyruvate	235,241
Pyruvate kinase	34
Radioimmunoassay	81,139,271
Receptor	59,111,149, 204,213, 223,245, 251,295, 321,391, 398,411
Receptor-mediated endocytosis	51,62,324
Recombinant plasmid	308
Reduviid bug	79
Restriction mapping	13
Rhamnose	83,84
Rhodamine 123	239
Rhoptry	233,330, 357,365, 411
Ribose	82,178
Ricin	199
RNA	58,176,410
RNAse	57
Sabin-Feldman dye test	359
Salicylhydroxamic acid	45
Salivary gland	1,5,74,391, 407
Sandfly	107,117, 123,137, 197,213, 389,397
Saponin	285,337
Scatchard analysis	323
Serine	82
Serine esterase	167
Serodeme	75
Sexual conjugation	74
Sialic acid	246,249, 252,322
Signal peptidase	261
Signal peptide	10,309,339
Sodium meclofenamate	231
Sorbitol	236
Southern blot	13,113,126, 130,336
Spectrin	235,284,318
Spermidine	29,33
Staphilococcus aureus-V8 protease	125,129,140
Succinic dehydrogenase	412
Succinate oxidase	45
Sulfonic acid	165
Superoxide dismutase	361
Suramin	52,58,63
Symport	180,186
Tandem repeat sequence	284,311
Tannic acid	357
Tartrate	56,190
Taxol	69

T-butylhydroperoxide	36
T cell	100,229, 233,345
T cell hybridoma	103
Thioredoxin	31
Threonine	82
Thymidine	412
Thymocyte	229
Transcription	2,5,44,68
Transferrin	59,62
Transferrin receptor	294
Transformation signal	42
Translation	44,69,339
Trifluoromethane-sulfonic acid	165
Triton X-114	150,157
Trypanothione	30,34
Trypanothione peroxidase	30,37
Trypanothione reductase	32,38
Trypanothione synthetase	35,38
Trypsin	32,46,59, 92,125,245, 285,389,410
Tryptophan	346,349,353
Tryptophan dioxygenase	347
Tsetse fly	1,41,73
Tubulin	67,70
Tunicamycin	107,115
Tyrosine	412
Tyrosinylation	69
Uracil	345
Vaccine	79,102,335
Valine	412
Valinomycin	179
Vanadate	185,186
Variant surface glycoprotein (VSG)	1,6,9,41, 158
Vector	402,407, 410,414
Virulence	107,115
Western blot	103,140, 176,266, 341,367
Wheat germ agglutinin	252
Xenodiagnosis	89
Xylose	82,179,391
Zinc	167
Zymodeme	81,377

NATO ASI Series H

Vol. 1: **Biology and Molecular Biology of Plant-Pathogen Interactions.** Edited by J. A. Bailey. 415 pages. 1986.

Vol. 2: **Glial-Neuronal Communication in Development and Regeneration.** Edited by H. H. Althaus and W. Seifert. 865 pages. 1987.

Vol. 3: **Nicotinic Acetylcholine Receptor: Structure and Function.** Edited by A. Maelicke. 489 pages. 1986.

Vol. 4: **Recognition in Microbe-Plant Symbiotic and Pathogenic Interactions.** Edited by B. Lugtenberg. 449 pages. 1986.

Vol. 5: **Mesenchymal-Epithelial Interactions in Neural Development.** Edited by J. R. Wolff, J. Sievers, and M. Berry. 428 pages. 1987.

Vol. 6: **Molecular Mechanisms of Desensitization to Signal Molecules.** Edited by T. M. Konijn, P. J. M. Van Haastert, H. Van der Starre, H. Van der Wel, and M. D. Houslay. 336 pages. 1987.

Vol. 7: **Gangliosides and Modulation of Neuronal Functions.** Edited by H. Rahmann. 647 pages. 1987.

Vol. 8: **Molecular and Cellular Aspects of Erythropoietin and Erythropoiesis.** Edited by I. N. Rich. 460 pages. 1987.

Vol. 9: **Modification of Cell to Cell Signals During Normal and Pathological Aging.** Edited by S. Govoni and F. Battaini. 297 pages. 1987.

Vol. 10: **Plant Hormone Receptors.** Edited by D. Klämbt. 319 pages. 1987.

Vol. 11: **Host-Parasite Cellular and Molecular Interactions in Protozoal Infections.** Edited by K.-P. Chang and D. Snary. 425 pages. 1987.